Factor Analysis
in International Relations

Factor Analysis
in International Relations

Interpretation, Problem Areas, and an Application

Jack E. Vincent

University of Florida Press / Gainesville / 1971

Contents

Acknowledgments

I would like to thank Gail Glassmeyer, Norm Otto, and Edward Schwerin for their work as coders on the project; Bruce Alper, Carol Jones, and Pauline Kartrude as computer programers; Myrtle Cassel for her typing assistance; the FAU Research Committee for the support of the project out of NSF Institutional Grant Monies; and Rudolph J. Rummel and Michael Leavitt for reading and commenting on portions of the manuscript.

An abbreviated version of this monograph was first given as a paper under the title "Factor Analysis as a Research Tool: Some Problem Areas, Some Suggestions, and an Application," presented at the 1969 meeting of the American Political Science Association in New York.

Factor Analysis
in International Relations

Introduction

The purposes of this monograph are (1) to explain certain aspects of factor analysis regarding its use as a research tool, (2) to suggest "solutions" to some of the problems that occur in connection with its applications, (3) to critique the bulk of the factor analytic studies that have been done in international relations, and (4) to generate, through an application, a number of indices that can be used in empirical research.

Factor analysis is presently employed in a number of disciplines. With the advent of high-speed computers, barriers to its use by the ordinary researcher, in connection with the laborious computations involved, have been removed, and a number of recent studies have been based on the results of some factor analytic technique.[1] In spite of these important contributions, however, this researcher has felt a considerable dissatisfaction. In conjunction with a strong empirical orientation, he had hoped that indices would be forthcoming from such studies that could be employed in various kinds of cross-national research. In contrast, these studies, for the most part, have focused upon the questions of "factor loadings" and, thus, they have neglected the computation of factor scores which could be used in this regard. In those studies where factor scores have been computed, a number of limitations can be noted.

1. In International Relations, for example, such studies include Adelman and Morris' *Society, Politics, and Economic Development*; Jack Sawyer's "Dimensions of Nations: Size, Wealth, and Politics"; Alker and Russett's *World Politics in the General Assembly*; Rummel's "The Relationship Between National Attributes and Foreign Conflict Behavior"; Banks and Gregg's "Grouping Political Systems: Q-Factor Analysis of *A Cross Polity Survey*"; Rummel's "Dimensions of Conflict Behavior Within and Between Nations"; Alker's "Dimensions of Conflict in the General Assembly"; Gregg and Banks' "Dimensions of Political Systems: Factor Analysis of *A Cross Polity Survey*"; and Russett's "Discovering Voting Groups in the United Nations."

These are (1) the N is usually quite small, that is, the study usually embraces fewer than the "universe" of states, a fact which inhibits usage; (2) the study may focus on a narrow category of variables; (3) "incomplete methods" of calculation may be used so that the scores do not possess the characteristics that they should have in terms of the factor model; and (4) the "factor scores" may relate only to a small fraction of the total variance in the original variables.[2]

One of the purposes of this monograph is to supply factor scores which do not have these "deficiencies"; that is, the purpose is to create indices based on a diversity of variables, cutting across the major social, political, and economic features of nations, indices based on the largest possible N, indices calculated from a formula which yields scores consistent with the factor model employed, and indices which relate to the bulk of the original variance of the variables treated.

An Overview and Problem Areas

To introduce factor analysis and set the stage for the interpretation of the results, however, it will be useful to discuss (1) the purposes of factor analysis, (2) the problem of entries in the principal diagonal, (3) the problem of rotation, (4) the problem of factor score calculation, (5) the problem of variable selection, (6) the problem of data transformation, (7) the problem of missing data, (8) the problem of the number of factors, and (9) the problem of factor interpretation.

THE PURPOSES OF FACTOR ANALYSIS

To understand the value of and the confusions concerning factor analysis, it should be remembered that the methods that fall under the name of factor analysis originally emerged from attempts by some psychologists to account for human abilities as evidenced by test performance. If tests are given, performance usually varies. The primary question that these psychologists have directed themselves to is,

2. See, for example, the factor scores generated by Alker 1964, Rummel 1963a, Sawyer 1967, and Alker and Russett 1965. All projects, other than Sawyer's, tend to focus on a fairly narrow category of variables, and the N is fairly small. In the case of Sawyer's project, scores of certain heaviest loading variables are used to "represent" factor scores. If the scores were actually calculated on the dimensions presented by Sawyer they would relate to approximately 40 per cent of the variance of the original variables.

"What stands behind or accounts for observed differences in test performance?"

Initially, the search concentrated on "g," a general intelligence factor posited by Spearman in 1904. Spearman also recognized that beyond this general ability there seemed to be "specifics," i.e., particular abilities, also related to the tests. This gave rise to Spearman's famous two-factor theory with one factor being "g" and the other "factor" referring to the specifics that were not "g." Performance, then, on any single test could be viewed as resulting from a combination of the general factor "g" and a specific or specifics.

Somewhat later the concept of "group factors" developed. These were considered abilities relevant to more than one test but less than all tests. In general, however, the adherents of the Spearman school tried to explain as much variance as possible by "g" and as little as possible in terms of group factors.

Sophisticated methods, consistent with these conceptualizations, were developed. For example, there is Holzinger's Bifactor method, which identifies the degree to which each test in a battery is saturated by "g" and then, after the explanatory power of "g" is calculated, the extent to which "test clusters" are saturated by mutually exclusive "group factors."

Attention to the problem of many factors (as opposed to two factors) gave rise to the famous "centroid method" of factor analysis, primarily developed by Thurstone. This method accounts for the relationships between tests without the conceptual restriction of a general factor. In Thurstone's words:

> When multiple-factor analysis was first developed, it consisted essentially in a shift of emphasis. The dominant question in previous factorial studies had been as to whether the correlations could be accounted for by a single general factor and whether the general-factor residuals were large enough to justify the recognition of "disturbers," which were called "group factors." These were sometimes frankly admitted at the start of a study, but always with the reservation that they should be regarded as secondary to the general factor, which was invariably postulated. With multiple-factor analysis the fundamental question was formulated without this restriction, namely, to determine how many factors were indicated by the experimentally given correlations. When that question was answered, the next question was naturally to inquire about the nature of all the common

factors that determined the correlations. This extension of the factor problem to any number of dimensions made it a question of fact as to whether one or more of the common factors were general in the sense of participating in the variance of every test in the battery.[3]

Spearman criticized these initial efforts, however, on the grounds of "indeterminancy," that is, the methods failed to solve the problem of locating a "reference frame" in the test configuration. This problem stemmed from the fact that there are an "infinite" number of solutions to the basic factor analytic problem in the sense that an infinite number of reference frames can account for the original correlation matrix. Thurstone's "answer" was to locate the first reference frame in such a way as to maximize the sum of the squares of the factor loadings. In everyday language, this means that the first factor would account for as much of the test performance as possible. The second factor was located in exactly the same manner, explaining test performance not accounted for by the first factor. This process is continued, that is, factors are generated, until all or most of the variation in test performance is accounted for.

Thurstone referred to this solution as the "principal axis solution" and demonstrated that it was unique for any correlation matrix. That is, two independent workers using the same matrix, following the methods, would not get different results. Thurstone viewed this solution, however, only as a first step to the final identification of the "underlying abilities" causing the test performance. Thus, Thurstone tended to view the principal axis solution as the most parsimonious description of the correlation matrix but argued that rotation of the reference vectors to new positions is necessary to identify psychologically meaningful factors. In this connection, Thurstone developed his famous "simple structure" criterion to guide his rotations of reference frame.

Basically, the idea behind simple structure is that variables should either load very high or very low on any single factor but should not load moderately on several factors. In this connection the final solution might either be orthogonal (uncorrelated) or oblique (correlated) factors, the final pattern being determined by the rules of simple structure.

Thurstone's original method for finding the principal axes was to form an arbitrary factor matrix and then rotate to the principal axes.

3. Thurstone 1947, p. 473.

Hotelling (1933) showed how the principal axes could be calculated directly from the correlation matrix. This approach generally replaced the centroid method with the introduction of high-speed computers. This was because the centroid method, although easier to calculate by hand, is only an "approximation" of the principal axes in the sense that each successive factor in the centroid method will account for most but not all of the variance in the residual matrices.

Although many factor analysts agreed with Thurstone that some rotation criterion must be established, if factors are to be rotated, "simple structure" as developed by Thurstone incorporates subjective judgments which seem to open the door to experimenter bias. Since then, numerous efforts have been made to establish an objective basis for simple structure. One such method referred to as the "quartimax criterion" was developed by Carroll in 1953. Basically, the application of the method gives a simplification of rows in the factor matrix. In Kaiser's words, "In the light of later developments, Carroll's criterion should probably be relegated to the limbo of 'near misses'; however, this does not detract from the fact that it was the first attempt to break away from an inflexible devotion to Thurstone's ambiguous, arbitrary, and mathematically unmanageable qualitative rules for his intuitively compelling notion of simple structure."[4]

Kaiser's proposed substitute, the varimax criterion, has achieved considerable popular acceptance and has the advantage of simplifying columns for factors. In short, then, the application of the method tends to maximize and minimize loadings in the sense that if a variable loads heavily on one factor, it is not likely to load heavily on another (at least in the orthogonal case). The most important characteristic of the varimax solution, however, is that it maximizes factorial invariance. Simply stated, this means that if a number of tests are factor analyzed, the loadings of the tests on the rotated factors will vary but very slightly if additional tests are added to the battery. In his words:

> Although one often gets the impression that simple structure is the ultimate criterion of a rotational procedure, it is suggested here that the ultimate criterion is factorial invariance. The normal varimax solution was originally devised solely for the purpose of satisfying the simple structure criteria. But the fact that it shows mathematically this sort of invariance suggests that Thurstone's reasoning was basically directed toward factorial invariance. The

4. Kaiser 1958, p. 188.

principle of simple structure may probably be considered inciden-
tal to the more fundamental concept of factorial invariance. This
viewpoint renders meaningless the arguments concerning "psychological reality" of general factors, bipolar factors, simple
structure factors, etc.[5]

Although the varimax criterion is usually applied to create rotated
factors that are orthogonal, Kaiser has shown that an alternative form
can be used to obtain oblique factors (additional techniques, in the oblique case, will be taken up under "The Problem of Factor Rotation").

The purpose of the above discussion is to do more than simply give
a brief, though admittedly inadequate, history of the development of
factor analysis. Its primary purpose is to indicate the degree to which
the development of factor analytic techniques has been conditioned—
perhaps a better word in such cases is "burdened"—by the needs of
some psychologists. That is, some psychologists, in applying factor
analytic techniques, attempt to identify fundamental human abilities
which cause test performance. The tendency to think of these abilities
as causes is perhaps carried farthest in the work of Cattell where he
argues, "The present writer's theoretical position, in contrast to that
of many factor analysts . . . has always been that a factor . . . is a
cause."[6] Basically, and this will be further explained in subsequent discussion, this school sees factor analysis as a device which reveals that
which identifies the causes of observed common variance. Factors, then,
are the causes of the variance which is shared between the variables
under consideration.

The development of factor analytic techniques, however, has yielded
powerful mathematical tools from a purely descriptive and/or data
reduction point of view. This usage allows one to transform a large set
of variables, many of which are intercorrelated, into a smaller set of
variables which are uncorrelated. Because the latter variables, called
factor scores, are usually far fewer in number than the original variables,
this second use of factor analysis may be referred to as "Data Reduction." (The advantages of using uncorrelated factor scores in place
of the correlated original variable scores for subsequent correlational
analysis will be taken up later.) The legitimacy of this second use of
factor analysis is frequently clouded by the concerns of those who
endeavor to use factor analysis in the investigation of causes.

5. Ibid., p. 195.
6. Cattell 1962, p. 684.

As will be seen, one set of techniques seems best suited for the identi-fication of causes and another set of techniques for the purposes of descriptive and/or data reduction. Unfortunately, and this will be illustrated later, sometimes we find investigators employing techniques best suited for the identification of causes when they seem to be pri-marily interested in descriptive and/or data reduction, and investiga-tors using techniques best suited for descriptive and/or data reduction when they talk as if they were identifying causes.[7]

It should be clear from the outset that, regardless of method, I see the primary value of factor analysis as falling in the realm of descriptive and/or data reduction, and I am particularly skeptical of studies in the international relations area which attempt to identify "causes" (or at least seem to). I see a clear distinction between the identification of fundamental abilities which *cause* test performance and the *description* of variables in terms of factors. That is, when a psychologist is able to predict the performance of an individual on a number of eye tests in terms of a factor that he might call "visual acuity," I may be prepared to accept the idea that visual acuity is the cause of the performance on the tests; but when I call a factor dimension "Economic Development" because per capita GNP loads on it, Newspaper Circulation loads on it, and the Number of Pupils in Higher Institutions loads on it, I am hesitant to view Economic Development as the *cause* of high newspaper circulation, high per capita GNP, etc., in the same sense that I might view visual acuity as the cause of eye test performance. That is, I would prefer to view the loadings of such variables as simply descriptive, not causal (i.e., one way of describing "Economic Development" is in terms of high newspaper circulation, high per capita GNP, a large number of pupils in secondary school institutions, etc.).

I feel that the failure to appreciate some of these distinctions may explain some of the confusions that seem to surround the present use of factor analysis as an analytical technique in international relations.

THE PROBLEM OF ENTRIES IN THE PRINCIPAL DIAGONAL

One of the first and perhaps most important decisions facing a factor analyst is the one relating to the principal diagonal of the correlation

---

7. The concept of "best suited" needs to be emphasized here. That is, it is possible that authors may use methods "best suited" for the identification of causes but treat the results from a descriptive point of view and vice versa. When this is done, however, they should be aware of certain conceptual difficul-ties that arise in this regard. This will be more fully developed in the next section.

matrix. Generally speaking, the option is viewed as twofold, either to place estimates of communalities[8] in the diagonal or to place unities. If estimates of communalities are entered, and are less than 1, only a fraction of the variance of each test will be accounted for by the factor analysis regardless of how many factors are computed. The variance of the tests, in this case, is viewed as being divided into common variance and specific variance, and the factor matrix will only account for the common variance. The distinction between common variance and specific variance can be understood from definition of common variance, for a particular test, as the variance the test shares with other tests. The remaining nonshared variance, then, is specific.[9]

There are various methods of estimating communalities, one estimate being the multiple correlation squared of the test in question with all other tests in the battery. An important consideration when using the entry of estimates of communalities concerns the fact that, conceptually, the total number of factors exceeds the total number of tests, except in the case when all estimates of communalities are equal to 1, which would hardly ever be the case.[10] Thus, in all other cases, some of the variance of each test, with a communality less than 1, is relevant to the common factors and the rest of it relevant to a specific factor. Theoretically, then, if a factor analysis is performed on 25 tests, each with a

8. It is important to realize that the terms "estimates of communalities" and "communalities" can have two distinct meanings. Typically "estimate of communality" refers to the estimate of the variance a variable shares with other variables in the analysis, while "communality," for a variable, refers to the amount of variance accounted for by the factors computed, taken collectively. That is, each factor accounts for some of the variance and the sum of the variance explained for all of the factors computed equals the "communality" (technically the sum of squares of rows in a n x m matrix where n equals variables and m equals factors). It should be apparent, from subsequent discussion, that the term "communality," in this second usage, can refer to variance *that is "unique"* in the sense that *factors can account for nonshared variance* when 1's are placed in the principal diagonal. I believe this latter point is frequently misunderstood by those who proceed by putting 1's in the principal diagonal.

9. Sometimes the concepts of "unique," "specific," and "error" variance are developed. In such cases the variance left over, that is, not explained by the factors, is viewed as unique and subdivided into specific and error factors. The above discussion, then, ignores the problem of error factors and treats all of the unique variance as specific for the purposes of simplification. See Cattell 1966a, pp. 177, 200–211.

10. In the discussion that follows it is assumed that the estimates of communalities will not be equal to 1. If they are, of course, the comments concerning causes, factor scores, and loadings do not apply.

communality less than 1, there are 25 specific factors and a lesser number of common factors.

An important consequence of this approach is that factor scores (explained in detail later) can never be directly calculated, only estimated. Godfrey Thomson has pointed out:

> As soon, however, as any fractions, minimum or not, are placed in the diagonal cells, we have thereby decided to use, in describing our tests, more orthogonal axes than there are tests; for each test has then a specific factor, and there are in addition the common factors. This means in terms of our spatial model that none of the axes, neither the common factors nor the specific factors, are in the test space at all (except at the origin where they all cross). It is only about the test space, of dimensions equal to the number of tests, that we have any information from our battery. These axes are away in outer darkness and we cannot know them, but only their projections or shadows on the test space. Psychologists invariably confine their attention, after making an analysis using communalities, to the "common factor space," of a comparatively small number of dimensions, without, I think, being usually aware that this space is not in the test space at all. . . . The effect of all this is that the factors arrived at by an analysis which has begun by placing fractions in the diagonal cells can never be measured in any man, but only vaguely estimated, and with maximum vagueness if minimum communalities are used.[11]

According to Cattell, those concerned with causes behind test performance should use estimates of communalities in the principal diagonal.[12] The purpose is to examine and compare the common factors which generate the intercorrelations in tests. Such common factors, even if considerably smaller than the number of tests, have the property that they can reproduce the intercorrelation matrix with a small margin of error. They cannot predict the original test scores as accurately, however, because the specific variance in each test is not accounted for by the factors. The linkage between putting estimates of communalities in the principal diagonal and the "search for causes" can be understood if it is realized *that factors can account for nonshared variance* (as well as shared variance), if unities are placed in the principal diagonal.[13] In such a case, if the factors are viewed as causes, variables, in a sense,

11. Thomson 1951, pp. 336–37.
12. See note 16.
13. See note 10.

would be (partially) "causing" themselves. In short, in using unities, the factor analysis *does not* deal with that which is *common*. Most, I feel, would want to restrict the concept of "cause" as referring to "that which explains that which is common."

Such features and considerations stand in contrast to an analysis where unities are placed in the principal diagonal. Here the factor matrix can account for, if all factors are computed, the complete variance of each test. That is, the number of factors will not exceed the number of tests and all of the test variance can be accounted for by such factors because they will be equal to or less than the number of tests. Generally speaking, a factor analyst using unities in the principal diagonal will not calculate all of the principal components and therefore some of the test variance will be unaccounted for. If he did calculate all of the principal components, however, as stated above, every bit of the variance in the tests could be accounted for by the factors. Placing unities in the principal diagonal, then, is particularly compatible with the descriptive and/or data reduction objective, in contrast with the search for causes, in that the original data can be described completely by computing all of the principal components, or almost completely by computing somewhat fewer principal components, and does not deal solely with variance that is common, which is the case when estimates of communalities are placed in the principal diagonal.

The principal advantage of "unity approach" is that factor scores can be calculated exactly and there is no need for factor score estimation. Also, original scores can be predicted with a small degree of error, although, if fewer components than tests are extracted and the potential number of components is equal to the number of tests, the intercorrelation matrix cannot be reproduced as exactly as in the case of the common factor analysis.

To summarize, if the object of the factor analysis is primarily descriptive and/or data reduction, the most appropriate entries in the principal diagonal are unities. That is, to use estimates of communalities (less than 1) would leave some of the information out of the analysis, although it would be possible, of course, to view the results as "descriptive of that which is common," remembering, however, the effects on factor scores. On the other hand, if the principal objective is to identify causes, standing behind similarities in test performance, the most appropriate entries are estimates of communalities, unless the term "cause" is used to refer to "that which, in part, is unique."

THE PROBLEM OF FACTOR ROTATION

Another problem confronting a factor analyst concerns factor rotation. If a researcher has used the principal components solution, he will have an unrotated factor loading matrix where the first factor accounts for the maximum variance that any one factor can account for, the second factor the maximum of the remaining variance, and so forth. If the objective of the analysis is to identify the factor with the greatest predictive power back to the original variables, the factor with the second greatest predictive power, etc., this solution can be considered final. That is, the first two or three factors emerging from a battery of 15 or 20 tests could possibly account for something in the order of 70 per cent to 80 per cent of the original variance. The factor analyst, at this point, could compute factor scores and these scores have the property, if unities were placed in the diagonal, that they will predict (explain the variance of) the original variables exactly to the extent of their loadings. For example, if the first variable loaded on the first factor .90, the first factor, that is, factor scores, can account for $.90^2$ or 81 per cent of the variance of the first variable. In the unrotated matrix, however, loadings tend to be moderately high on the first principal component, less high on the second principal component, and so forth, until, depending on how far out the principal components are computed, the loadings may become very small. That is, the predictive power of the last principal component in respect to the heaviest loading variable might be something on the order of 5 per cent of the variance of the variables, or less. Because of this, many researchers feel that it is desirable to rotate the factors in such a way that each has important predictive power for at least one or more of the variables.

Rotation, then, may be thought of as an effort to re-align the reference axes (geometric representatives of the factors) through closely related clusters of variables. At first, rotation was accomplished through analytical and graphic techniques that incorporated subjective evaluations on the part of the investigator. The first completely objective technique was developed by Carroll in 1953 and is referred to as the quartimax method. Basically the quartimax method may be viewed as a technique which minimizes the summed product (across columns) of the squared factor loadings, when this calculation is carried out on all possible pairs of factors. It can be seen that this has the effect of making loadings either large or small. Consider, for a moment, the pairs of squared loadings equal to .9, .1 and .5, .5. Both pairs represent an equal amount of variance explained. However, the multiplication of .9 by .1 only yields .09, whereas the multiplication of .5 by .5 yields .25. It should be evident, then, that the quartimax criterion tends to maximize *or* minimize loadings compared to a situation of having moderate loadings. It should be noted that in satisfying the quartimax criterion there is no longer any need that the first factor explain most of the variance, the second factor the second most variance, and so forth. Typically, then, in a rotated factor matrix the variance explained by the factors varies according to the solution of minimizing the magnitude of cross products.

One of the main limitations of the quartimax technique is its tendency to develop a general factor and to simplify rows rather than columns. As noted earlier, dissatisfactions with this solution led Kaiser to develop the varimax technique[14] which is based on an effort to maximize the sum of the squares of the columns. The original solution is called the raw varimax criterion and does not normalize loadings by dividing them by communalities. In contrast, the normal varimax criterion does normalize loadings, through division by appropriate communalities, and has the effect of weighting each variable equally for the purposes of rotation. Both the raw and normal variance technique can be applied in the orthogonal case of rotation.

Oblique solutions which have been developed include Oblimax, Quartimin, Covarimin, Biquartimin, and Direct Oblimin. Each criterion yields somewhat different results. For example, in comparing Quartimin, Biquartimin, and Covarimin, the Quartimin solution tends to produce the most oblique factors, Biquartimin somewhat less oblique factors, and Covarimin the least oblique factors, when used on the same data.[15]

As noted earlier, the selection of one of these various techniques depends in part upon the research needs of the investigator. Depending on the data input, all techniques may lead to fairly similar or divergent results. For example, rotation to an oblique solution may still leave the factors basically uncorrelated. The primary value of oblique rotation is that it *may* reduce the number of moderate loadings more than orthogonal rotation. In short, oblique rotation offers the possibility of even simpler structure than it is possible to achieve in the case of orthogonal rotation. The primary difficulty of oblique rotation concerns the problem of using correlated factor scores in further analysis.

14. Kaiser 1958.
15. Harman 1967, pp. 314–41.

It might be recalled that one of the principal (possible) values of factor analysis is to reduce correlated indices into a set of uncorrelated indices or factor scores. Uncorrelated indices are highly desirable in two ways. First, if such indices are correlated with some third variable, the variance explained in each case is known to be unique. Say, for example, a factor analysis yields two uncorrelated factor dimensions which we label "Development" and "Democracy." If factor scores are calculated and the two sets of factor scores are run against some other variable, say, "Number of Riots," we know that any variance accounted for by "Democracy" cannot be accounted for by "Development" and vice versa. This will not be the case if the predictors are correlated, i.e., are redundant. In this case the predictors may be "explaining" the same variance in the dependent variable. This point will be expanded further in the illustrative example of factor analysis.

Second, uncorrelated scores facilitate the use of techniques such as multiple regression analysis and canonical correlation. If correlated variables are inserted, in either case, interpretation becomes difficult. This point will also be discussed in detail in the illustrative example which includes a demonstration of the use of factor scores.

All of this is to say that orthogonal rotation has many advantages over oblique rotation when the purpose is to use factor scores as variables in further analysis.

Why, then, do many factor analysts choose to rotate to an oblique solution? Generally speaking, they are either attempting to compare the results of an oblique solution to the orthogonal solution and draw some sort of conclusions from this, or they are primarily interested in the identification and interpretation of the simplest possible structure. This approach has frequently been associated with those analysts who concern themselves with the identification of causes. Thus, it is not unusual to see a researcher who is primarily interested in descriptive and/or data reduction placing 1's in the diagonal and rotating to an orthogonal solution, and a researcher primarily interested in "causes" placing communalities in the diagonal and rotating to an oblique solution. Both Cattell[16] and Thurstone with their interest in causes tend to prefer the

16. Cattell, for example, argues, "The components model [putting unities in the principal diagonal] must be rejected for general scientific investigation, because it would be most unlikely that *n* variables would contain within themselves *all* the causes for accounting for their own variances. To do this, they would have to lie in a completely self-explanatory subuniverse, self-sufficient as a system entirely isolated from the rest of the universe" (Cattell 1966a, p. 177). Concerning rotation, he has argued, "It needs to be said clearly at this point—despite

latter procedures.[17] In the interest of simplicity of data reduction, however, I prefer orthogonal rotation, particularly if the objective is to create a set of factor scores for additional predictive purposes.

THE PROBLEM OF FACTOR SCORE CALCULATION

Subjects can be located on factor dimensions in a manner conceptually similar to the way that they are located on original variables. Considerable confusion exists on this point, however, because before the advent of large-scale computers, shortcut hand methods were frequently employed which resulted in factor scores with properties different from those of the true factor scores. Many of these methods are still used, and an investigator may be in doubt as to the proper procedures even when adequate computer facilities are available. In addition, factor analysis programs frequently do not include a routine for the calculation of factor scores.

To illustrate some of the problems connected with factor scores, the procedures suggested and used by certain factor analysts, Harman and Rummel, will be examined in detail. In Harman's case, I include the example because I feel there are certain errors in his work that need to be corrected. In Rummel's case, my major objective is to use his work to clarify a few points about factor scores that are sometimes misunderstood. In this connection, the discussion does not mean to imply that Rummel, himself, misunderstands the distinctions that are made.

Starting with Harman, he correctly defines factor scores (components model) in the unrotated case as $F = (A'A)^{-1} A'Z$, where F is an m x N matrix of factor scores, A is an n x m matrix of unrotated loadings, and Z is an m x N matrix of the standard scores of the subjects on the original variables.[18] He also gives the proper algebraic equivalent as:

$$F_p = \sum_{j=1}^{N} \frac{a_{jp}}{\lambda_p} z_j$$

17. Thurstone develops arguments similar to Cattell's given in the preceding note (Thurstone 1947, pp. 176–93, 473–510).

18. Harman 1967, p. 348.

(p $= 1,2,...,$m), where $F_p =$ factor scores, $a =$ the unrotated loadings, $\lambda =$ the eigenvalue, and z $=$ the standard scores of the subjects.[19]

However, he incorrectly states, "The necessary eigenvalues are the square roots of the variances in the second to last row of that table. Then dividing the numbers in the columns by the respective eigenvalue produces the required coefficients. For example, the resulting equation in the case of the second component (P2) is: $F_2 = 1.0809_{z1} - .7302_{z2} + .9731_{z3} - .1398_{z4} - .7482_{z5}$."[20]

In the table referred to by Harman the eigenvalues are equivalent to the variance explained by the various factors. *It is quite improper to take the square root of these eigenvalues.* That is, the eigenvalues themselves must be divided into the loadings. *Further, even if one does take the square root of the eigenvalues, make the necessary divisions, and sum, one will not arrive at the answers Harman provides.* This, of course, is very confusing to the uninitiated. In brief, Harman is simply in error (in his verbal description and example) although his matrix algebra expressions are correct.

Even more serious, perhaps, for the uninitiated, is the possible impression that Harman gives that the same procedures are applicable in the rotated case as in the unrotated case. It is true that his example, of factor score calculation, shows values other than zero off the principal diagonal, created by B′B (B is the rotated matrix), but the reader is given no real warning that, in the rotated case, it is no longer proper to simply divide the loading by the "variance explained" by a particular factor (before multiplying the loadings by the standard scores and summing). In short, unless one follows the mathematical example very closely, one might be led to believe that the proper algebraic equivalent in the rotated case is

$$F = \sum_{j=1}^{N} \frac{bjp}{\lambda p} zj$$

where $F =$ the factor scores, b $=$ the rotated loadings, $\lambda =$ the variance explained by a factor, and z $=$ the standard scores of the subjects. As will be shown in the illustrative example, such a formula cannot be applied in the rotated case because values other than zero occur off the principal diagonal when the transpose of the rotated matrix is multiplied times the matrix.

Rummel, in explaining factor scores in connection with an orthogonal

19. Ibid.
20. Ibid.

rotation, suggests, "To determine the score for a case on a pattern, then, the case's data on each variable is multiplied by the pattern weight for that variable. The sum of these weights-times-data products for all the variables yields the factor score."[21] Rummel's objective is to give a "feeling" for factor scores and, if computed in this way, they will "approximate" factor scores with the proper values. However, it should be stressed that such scores are not "true" factor scores in the sense that, if correlated with the original variables, they will reproduce the factor structure. Rummel proceeds along similar lines to calculate his "factor scores" in his "Dimensions of Conflict Behavior Within and Between Nations," where he add up z scores of variables with the heaviest weights on the factors.[22] When scores are calculated in this way, they tend to become correlated and, in addition, will not reproduce factor structure. For example, Turmoil correlates .37 with Revolutionary and .34 with Belligerent, War correlates .40 with Belligerent, and so forth. Also, the degree to which such scores actually predict the original variables becomes an empirical question. As for this particular application, Rummel admits that such factor scores are not true scores: "The factor scores were calculated by adding the standard scores of each nation on those measures high on the particular dimension and low on the others. The result is probably a very close approximation to the actual factor scores a country would have if one were to use more precise techniques involving the data, correlation, and factor matrices. The additional accuracy one would thus gain, however, did not seem worth the expenditure in time and resources."[23] Although the "factor scores" calculated were not highly correlated *in this particular case*, the degree to which they become correlated becomes an empirical question and should probably be reported when incomplete methods are used. Otherwise, the user of such scores might incorrectly assume that they have no redundancy. Also, the degree to which such scores actually predict the original variables should probably also be reported. Finally, *the degree of correlation among the factor scores can be much higher*.[24]

It is the opinion of this researcher that factor scores are not only an important ingredient of the basic analysis but shed a great deal of light on the nature and value of factor analysis itself, and for this reason confusions of the kind indicated above should be cleared up.

21. Rummel 1967a, p. 469.
22. Rummel 1963a, p. 15.
23. Ibid., p. 21.
24. See Glass and Maguire 1966.

If communalities (less than 1) are placed in the principal diagonal, factor scores can only be estimated, that is, not precisely calculated as explained above. One estimation formula is $F = ZR^{-1}A$, where N are subjects, n original variables, m factors, and F is an N x m matrix of factor scores, Z is an N x n matrix of scores on the original variables in standard score form, R^{-1} is an n x n inverse of a matrix of inter-correlations among variables, and A is an n x m factor structure matrix.[25] This formula yields scores for both the orthogonal and oblique cases. In this connection it is necessary to distinguish between factor pattern and factor structure. Factor structure refers to the correlations of the original variables with the factors, while factor pattern refers to the coefficients assigned to the factors in predicting the original scores. It should be noted that it is the factor structure, not the factor pattern, which is employed in this formula, when the factors are oblique, although it does not matter which is employed in the orthogonal case because in such a case factor pattern and factor structure coincide. If the original scores are correlated with the factor scores, using this formula, they usually come fairly close but do not perfectly coincide with the factor structure. That is, such factor scores predict the original variables almost to the extent they should, as given by the factor structure. How far off they are becomes an empirical question and depends upon the magnitude of the communalities. As the communalities approach unity, such correlations come closer and closer to the factor structure.

If unities are placed in the principal diagonal, factor scores may be calculated precisely (orthogonal case) and have the quality of predicting the original variables exactly to the extent they load on the factors. A proper formula is $F = ZA (A'A)^{-1}$ where F is an $N_{.}$ x m matrix of factor scores, Z is an N x n matrix of scores on the original variables in standard score form, A is an n x m matrix of factor coefficients (loadings), and N equals subjects, n equals variables, and m equals factors.[26]

This formula applies to both the unrotated and rotated cases. In the unrotated case, A refers to the unrotated matrix of loadings, and in the rotated case to the rotated matrix of loadings. In the unrotated case, the mathematical operation for one case is equivalent to multiplying each z score for a subject by the appropriate variable loading, given in the factor matrix for a particular factor, summing, and then dividing

25. See Horn 1965a.
26. See Glass and Maguire 1966; Kaiser 1962.

the obtained quantity by the variance explained (sum of column of the entries squared) by the factor. In the rotated case the procedure is more complicated because A'A does not, as in the unrotated case, produce a diagonal matrix. That is, values other than 0 occur as entries off the principal diagonal and these affect the magnitude of the factor scores as well as the variance explained, given in the principal diagonal. In both cases, however, the scores have exactly the same properties after calculation, i.e., they are standardized, they predict the original variables exactly to the extent that the variables load on the factors, and they are uncorrelated. In contrast, incomplete methods (such as $F = ZA$ in the rotated case) usually yield correlated scores. Further, correlations of such scores with the original variables will not reproduce the factor loading matrix, although such scores may still have other uses. The value of uncorrelated scores has been suggested above and is covered in the illustrative example while the descriptive value of scores which predict the original variables in a known fashion should be evident. That is, on the larger factors, a great deal of information may be summarized by correctly calculated factor scores.

THE PROBLEM OF VARIABLE SELECTION AND THE "IMPORTANCE" OF FACTORS

The justification for the inclusion of a certain number of variables in a factor analysis, out of the potential universe of available variables, usually varies somewhat from one investigator to another. For example, Sawyer justifies the variables he selected in his "Dimensions of Nations: Size, Wealth, and Politics" in terms of "conceptual relevance," "prior use," "representativeness," "availability," and "objectivity."[27] Adelman and Morris in their *Society, Politics, and Economic Development* justify their variables in terms of "inclusiveness" and "parsimony."[28] Russett, discussing this problem in his "Delineating International Regions," argues, "They [the variables] were selected to provide information on a wide variety of social, cultural, political, and economic conditions. Though it is easy to think of other variables that might have been included, many would correlate quite highly with some already in the *Handbook*, and I believe we were reasonably Catholic in our tastes."[29]

It should be apparent that variable selection really has to do with what the investigator has in mind. If his focus of attention is quite

27. Sawyer 1967, pp. 149–50.
28. Adelman and Morris 1967, p. 15.
29. Russett 1968, p. 322.

narrow, say, concerns only economic growth, then he would naturally attempt to accumulate a number of indices of an "economic growth nature" for his study. On the other hand, an investigator might limit himself to just "political indices," etc. Whatever the investigator chooses to study, then, guides his selection of variables in the broad sense, although he may attempt to narrow his choice in terms of the quality of the data, degree of redundancy, etc. However, the really important point remains that the degree to which indices are in fact correlated affects directly the interpretation of the results.

When an investigator is primarily interested in causes, he tends to focus his attention on the "big factors," that is, those factors which account for a large portion of the original variance, whether unrotated or rotated, oblique or orthogonal (although generally rotated and oblique factors). This is generally in line with a scientific interest in parsimony. For this reason, small factors are usually relegated into the realm of "unimportant," "uninterpretable," and so forth.

It should be understood, in this connection, that the size of a factor, in terms of variance explained, is a direct function of the degree to which the original variables correlate with one another. For example, say three factor dimensions emerge from the study of a number of variables, i.e., the dimensions of "Development," "Authoritarianism," and "U.S. Relations." If most of the variables loaded heavily (average .90) on "Development," it would explain most of the original variance. If the second largest number of variables loaded heavily (same average) on "Authoritarianism," it would explain the second most variance, and so forth. It should be apparent that either of the second two dimensions could be made "most important" simply by variable manipulation. For example, Newspaper Circulation, Radios per 100,000, Number of Inhabitants per Physician, Per Cent Literate in Population, Per Capita Gross National Product, and similar variables tend to load very heavily on a "Development Dimension," which emerges from a number of studies. If all of the heavy loading variables except one, say, Newspaper Circulation, were dropped from a "modified study," then "Development" would emerge as a very weak factor. "Authoritarianism" would become the most important factor, "U.S. Relations" the second most important, and, of course, "Development" the third most important. Similarly, if a number of variables loading heavily on "Authoritarianism," such as Competitive Electoral System, Representativeness of the Regime, Freedom of Group Opposition, etc., were dropped, leaving only one or two political indices, Authoritarianism would drop greatly in importance, and U.S. Relations would now emerge as the "most important" factor.

From this discussion it should be clear that the importance of certain dimensions is a direct function of original variable selection. In the present study, for example, "Communist China Economic Relations" emerges as a separate dimension, with Exports to China and Exports from China loading heavily on it. This could have been made a more important factor if other variables could have been found that correlate highly with these two variables. Put in these terms, then, even the smallest factors can have considerable "importance" in the sense of understanding the original correlation matrix. This is because the factor analysis is telling us, in the case of the smaller factors, that the variables that load on them do not correlate well with any of the other variables in the analysis, although those on the same dimension correlate fairly well with each other if their loadings are quite high. Further, if only one variable loads heavily on a dimension, the analysis tells us that this single variable correlates poorly with all of the other variables considered. *As predictors, such "small" factors may be extremely important.* This is why, in the present study, factor scores are computed for all the rotated dimensions. Although Development as a dimension has emerged as the most important predictor of the variables contained within the study, it is entirely possible that a predictor such as Diplomatic Activity, which accounts for only 2.3 per cent of the variance in the original variables, may prove to be an important predictor of variables outside of this study. Thus, unless one assumes that a certain factor analysis included a set of variables that are truly representative of all the variables that can be generated, it is somewhat premature to try to identify certain dimensions which tend to emerge repeatedly in various cross-nation studies as *the* important dimensions of nations. In short, the sets of factor scores generated by a study, viewed as potential predictors of variables *outside of the study*, can all be treated as having roughly equal importance. That is to say, until we run such predictors against variables not included in the study, there is no way of knowing what the magnitude or the frequency of the associations with other variables is likely to be.

In closing on this point, it should be evident that factor analysis is capable of describing any collection of variables no matter how diverse or redundant. Thus, it is perfectly possible to describe the relationships of a correlation matrix made up of Moon Phases, Suicides, Sun Spots, Erosion Statistics, Bird Migration Statistics, Human Birth Rate Statistics, etc. Factor analysis is capable of describing any set of correlations and "rational" connections need not be assumed. Thus, if Births and Sun Spots just "happen" to be highly related, but unrelated to any of the other indices included within a study, they would emerge together loading heavily on a factor and a set of factor scores could be calculated with a specified degree of predictive power in respect to these variables.

THE PROBLEM OF DATA TRANSFORMATION

Most investigators use Pearson's r to describe the relationships in their data at the beginning of a factor analysis and, therefore, they may legitimately become concerned if their data do not fit the Pearson r model. Pearson's r only measures linear relationships and, therefore, it may underestimate other kinds of nonlinear relationships, such as curvilinear ones which may be present in the data. It is usually too laborious to determine whether the relationships are basically linear, curvilinear, etc., when the number of variables is very large. Further, even if the data are checked, the results may create many difficult decisions for the investigator. For example, a certain variable may be basically linear in its relationship to some other variable, curvilinear to still another, and have an even more complex relationship with still another. Should such a variable be retained or cast out? Perhaps the best way to proceed is to simply take the position from the outset that *Pearson's* r *measures only the linear relationship between the two variables.* This is to recognize that other kinds of relationships may be present but "these are not the focus of attention at this time in this particular study." In effect, then, this is the position that most investigators actually take when they ignore the problem of checking all of their variables against one another for linearity. To me, with the above limitation made clear, this is a reasonable position.

Most investigators realize that Pearson's r describes data in terms of deviations from the mean. Thus, a Pearson r description is most interpretable when distributions are normally distributed and less so when distributions are not normal. Although, by formal definition, r^2 equals the "proportion of variance explained," in both normal and nonnormal distributions, it should be clear that r^2 can have a different meaning in different cases. That is, a single deviant case, or cases, can make a tremendous contribution to r^2 when distributions are badly skewed.

Consider Diagram 1. The line (m) in this diagram can be imagined as cutting through the mean, and cases to the right are positive deviations and those to the left negative deviations. If a single case with enough magnitude is added to this distribution, it might be re-represented as

Diagram 2. Note that in this distribution all but one case become negative, and the single deviant case is positive. In what sense, then, can we view this distribution as being described in terms of deviations from the mean when the mean is understood to describe the central tendency? If two such distributions were correlated and deviant cases matched, they could make a tremendous contribution to r^2. If the deviant cases were removed and the remaining distributions re-correlated, the correlations could change considerably. In short, deviant cases can markedly affect Pearson's r. Realizing this, many investigators attempt to transform data toward normality. Sawyer, for example, uses various log and

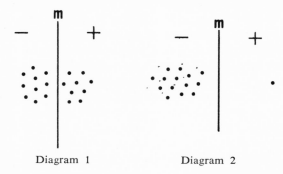

Diagram 1   Diagram 2

square root transformations in an effort to normalize his data.[30] One problem with this approach is that each case tends to be different and may require different techniques. Furthermore, such transformations are usually quite laborious and the resulting distributions may still not be normal.

One fairly easy solution to the problem of distribution is to reduce data to rank data. This is the solution I prefer when distributions are greatly skewed (or assume other nonnormal forms) and the character of the skew differs from variable to variable. When data are ranked, of course, the numbers representing the data run from 1 to N. This places the mean or median in the exact center of each distribution and the mean or the median is exactly the same in each case. Also, in each case, there are exactly as many deviations on one side of the mean or median as there are on the other. Perhaps the greatest advantage of this transformation technique is that one is applying exactly the same kind of transformation to each variable. That is, one does not have to check to see whether a log transformation was made in this case, a square root

30. Sawyer 1967.

transformation in that case, and so forth. Further, researchers frequently wish to employ ordinal and interval data. It usually does not seem reasonable to promote ordinal data to interval data. In such cases, by demoting interval data to ordinal data, all data become measured on the same scale.

It should be apparent that rank distributions, in the absence of ties, cannot be normal, that is, they are flat.[31] However, if normal distributions are ranked, the resulting distributions will come within .02 of the correlation generated between the normally distributed data. That is, if two normal distributions are correlated and yield a correlation of .80 and these distributions are then ranked, the resulting correlation will fall between .78 and .82. Also, as soon as the N of a distribution exceeds 30, the t test, to test significance, may be calculated from the same formula in the case of Spearman's *rho* as Pearson's *r*. The only major objection that can be raised to this manipulation of the data is that ranking "throws away data," that is, original distances as measured by interval measures become ordinal directions as measured by rank numbers. This seems to be a small price to pay to obtain distributions which have the sterling quality possessed by rank numbers.

Still another possible transformation is to convert all scores either to 1 or 0. This can be accomplished by dichotomizing a distribution by placing an observation in the 1 or 0 category, because it is either above or below the mean, median, or mode of the distribution, or, by arbitrary assignment, in the case of an already dichotomous classification (such as male-female). In effect, then, such transformation creates a whole number two-step scale which can be viewed conceptually in the same terms as the N scale created by rank numbers. By this I mean that the scale can be viewed as moving toward maleness in the case of a male-female dichotomy or linguistic homogeneity in the case of the linguistic-homogeneity–linguistic-heterogeneity dichotomy, etc. Obviously, this technique, applied to interval data, throws away considerably more information than the ranking techniques but, by careful assignment, where distributions are "divided" by the investigator, the number of + and − cases, as deviations from the mean, can be made to come into a rough balance.

It might be wise to remind the reader at this point that Pearson's

31. In a flat, continuous distribution one standard deviation on either side of the mean will account for approximately 60 per cent of the cases and 1.732 standard deviations, on either side of the mean, for all of the cases. Thus, the z scores of a flat, continuous distribution will always be smaller than ± 1.732.

r formula applied to rank numbers yields Spearman's *rho* and, applied to 1 or 0 distributions, yields the *phi* coefficient.[32] Thus, when such conversions are made, it means that the factor analysis takes place on Spearman *rho* coefficients or *phi* coefficients instead of Pearson *r* coefficients. In practical terms, this simply means that the correlation of the variables with the factors will be one kind of numbers on one side (i.e., the factor scores) and another kind on the other (i.e., the variables). That is, the factor scores may not assume a rank or a dichotomous form.

In the first case, the conversion of the factor scores to rank numbers, of course, will then yield legitimate Spearman *rho* correlations between the rank factor scores and the original variable scores. Similarly, in the second case, the factor scores can be converted into a dichotomous form and correlated with the original variables to yield *phi*.[33]

To summarize, there is a variety of ways to manipulate data to meet the model of a correlation coefficient, so that the *r* description of the relationship between the variables "makes sense." Because Pearson's *r* describes data in terms of deviations from the mean, it is probably wise to transform data before the factor analysis, if markedly deviant cases are known to exist. Transformation such as square root, log, etc., are possible, but the transformation usually must vary from variable to variable. Further, such transformations seldom yield perfectly normal distributions. Rank transformation is relatively easy, the treatment does not vary from variable to variable, and ranking seems to eliminate a *minimum* of information (from interval data) compared to a transformation yielding *phi*. An important consideration in this regard is that demotion of interval data to ordinal data seems more "conservative"

32. See Baggaley 1964, Ch. 3. Baggaley demonstrates that the application of Pearson's *r* formula to rank numbers will yield the same coefficient as the application of the ordinary Spearman *rho* formula to such numbers, and that its application to 1/0 distributions will yield the same coefficient as the application of the ordinary *phi* formula.

33. If factor scores are converted, as suggested above—that is, either converted to rank numbers or 1/0 distributions—such distributions, if correlated with the original variables, will no longer produce the factor structure as the raw factor scores would. The relationship of such converted scores, then, to the original variables becomes an empirical question and cannot be deduced from the factor structure. Exploring this question empirically I have found that if the "large" dimensions (many variables load heavily) are ranked, such rank numbers will correlate very highly with the original raw factor scores and, therefore, they will come very close to producing the factor structure for such factors. On the smaller factors greater divergency between such correlations and the factor structure is usually evident, and considerable divergency may exist when the conversion of the factor scores is to dichotomous distributions.

than promotion of ordinal data to interval data, if mixing of both kinds of data occurs in the original variables. In short, the rank transformation is preferred by this investigator when dealing with the kinds of variables (mixed and nonnormal) explored in this study.

THE PROBLEM OF MISSING DATA

It is not uncommon for a factor analyst to have incomplete information on his subjects. That is, some of the subjects that he wishes to include in the analysis may be missing numerical values on some of the variables. To overcome this problem at least three possibilities seem evident: (1) the missing values may be estimated by regression techniques, (2) the missing values may be estimated by assigning mean values, or (3) the subjects with missing data may be excluded from the correlation matrix on those variables where information is missing. Each solution will be discussed in turn.

If one variable is known to have a high degree of relationship with another variable and a subject has a score on such a variable, it is evident that we can predict his missing score through simple regression. It would be, of course, possible to view two or more variables as predictors and thus move into the realm of multiple regression to estimate the missing score. One problem connected with such an approach, however, is that each case probably has to be treated by itself. Subject X might have values missing on variables 4, 16, 37, and 55, while subject Y might have values missing on variables 1, 3, 16, 72, and 84. One subject may have missing values on some of the best predictors of a particular variable and another subject may not. Although regression techniques are probably "theoretically best" when information is missing, the application of such techniques can be extremely laborious. Furthermore, the "power" of the estimation will typically vary from case to case. Thus, if multiple regression techniques are employed, the multiple correlation in the case of one variable might be .83, in the case of another variable .63, and so forth. That is, basis for the prediction may be firmer in one case than in another.

Another possibility is to assign mean values in the absence of information.[34] This manipulation has the advantage of applying the same

34. See Guertin 1968 for an argument for using mean estimates instead of multiple regression estimates. Guertin maintains, "At least in this study, the extra labors of obtaining multiple regression estimates were not worthwhile even for a variable with 40 per cent missing scores and samples of S's of around 50" (p. 896).

standard for each variable, in contrast to the use of multiple correlations, and the manipulation is quite simple. For example, if data are standardized for the known subjects, the unknown subjects may be assigned a mean value by simply assigning them 0. If such subjects are included in the correlation matrix to be factor analyzed, then the correlations in each case are based on distributions running from 1 to N, although the effect will be to reduce the magnitude of the correlation, over the alternative of leaving such subjects out of the correlation matrix. That is, such subjects cannot contribute to accumulated + or − deviations from the mean of a variable, because they have no deviation from the mean, but they reduce the magnitude of the correlation because they are part of N. Mean values, then, are treated as real scores, but we know such an assignment reduces the magnitude of the correlations.

Still another possibility is to drop no-information subjects from the correlation matrix. Each correlation, then, will be based on a different N, but the correlation represents only "good," not estimated, information. Many researchers have adopted this practice in computing their correlation matrix. It should be understood, however, that the interpretation of communalities and eigenvalues is affected by this manipulation. Thus, a communality of .80 no longer refers to variance explained relative to the total N, but only variance explained relative to a reduced and not clearly identifiable N. Thus, on the first factor, a loading may have pertinence to one set of subjects less than N, on the second factor another set of subjects less than N, on the third factor still another set of subjects less than N, and so forth. The eigenvalue representing the variance explained in the variables by the factors is to be similarly interpreted. Thus, on the first factor, working down the column, correlations of the variables with the factors may all represent different N's less than the total N.[35]

In spite of these difficulties, there is a great deal to be said for this approach. Most important, the correlation matrix represents only the ascertainable universe. Generally speaking, if more information were known, it is likely that the correlations would not change greatly, whereas to assign mean values as if they were true scores, as treated above, has the known effect of driving down the correlations.

At this point, one might question why subjects that have no information on some of the variables are included in the analysis. The answer

35. The general effect of dropping no-information subjects from the correlation matrix is to inflate both the communalities and the variance explained.

is usually fairly simple. In most studies of cross-national character, almost every subject has a value missing here and there on some of the variables. If we set the requirement that every subject has to have a true value on every variable, we would eliminate all of our subjects. The "dropping" technique has the advantage of wringing out the maximum amount of information from the data, without making estimations. If factor scores are of interest, the mean estimate can then be made for the variables where information is missing. This will have the effect of driving subjects with a large number of missing values to the center of the factor score distributions. If such factor scores are then correlated with the original variables, there will be a discrepancy between the correlations and the loadings, which would have defined the actual predictive power of the factors, relative to the variables, if no information were missing. Generally speaking, when the N is large and the amount of missing data small, the predictive loss is fairly inconsequential.

THE PROBLEM OF THE NUMBER OF FACTORS

Like any set of observations, the factor matrix in a factor analysis may be viewed as affected by sampling variability. In such a case, statistical tests may be employed to determine the number of factors to be retained. That is, factors not retained are viewed as probably arising primarily from random variation. Bartlett, Hotelling, and Rippe,[36] for example, provide tests which make such a discrimination. Frequently investigators use rule of thumb cutoff points to determine the number of factors to be retained, such as the factors should account for at least 75 per cent of the variance (collectively) and no one factor should account for less than 5 per cent of the variance. A frequently used rule, generated by Kaiser, is that the number of factors retained should be equal to the number of eigenvalues greater than 1 in a correlation matrix. This means that the smallest factor will account for at least as much variance as is contained in one variable.

Concern with the problem of random variation is of particular importance to those who are looking for causes. They would like to feel that repeated samples will yield similar results and they don't want to deal with factors generated primarily by random fluctuation.[37] In this connection, the distinction is frequently made between statistical significance and practical significance. For many investigators, even though

36. Harman 1967, p. 197.
37. See Cattell 1966a, pp. 200–211, for perhaps the best discussion of the problem of interpreting sampling effects.

a factor is statistically significant, it still may be viewed as uninterpretable, nonsense, etc. Thus, as noted earlier, frequently researchers just deal with the largest factors, i.e., the ones most likely to emerge again with new samples.

If the factor matrix is viewed as describing a universe (ignoring the sampling problem), then the number of factors retained depends upon the investigator's purposes. The factor analyst could ask for all of the principal components and thus account for all of the variance in each of the tests, or he could content himself with fewer components. Thus, he might ask the solution to be computed out to an eigenvalue of 5 or larger, 3 or larger, 1 or larger (Kaiser's suggestion), .5 or larger, or 0 or larger. It should be understood in this connection that, although the unrotated factor matrix (those factors computed) is unaffected by this choice, the rotated factor matrix can be greatly affected. That is, the first, second, third, etc., unrotated factors have exactly the same loadings no matter how far out the factors are computed. Rotating from the first three factors, however, may produce a completely different rotated structure than rotating from the first four, five, six, etc., factors. This will be treated in detail in the illustrative example.

If a researcher desires that a great deal of the variance of his variables should be accounted for by his factors, then he may compute out to whatever percentage level he wishes to accept. If less variance is acceptable, he may calculate fewer factors. It should be understood, however, that if he chooses to calculate fewer factors, some variables may be hardly accounted for by the factors that he has chosen to retain. Remembering that there are as many principal components as there are variables (assuming the rank of the correlation matrix to be equal to the number of variables), Kaiser's rule for cutoff point seems to be a good one, even from the purely descriptive and/or data reduction point of view. In its application, then, typically, the analysis may be expected to account for 75 per cent to 90 per cent of the variance and produce one-third to one-sixth of the number of factors as there are variables. To put it another way, in the typical case, to describe the remaining 25 per cent to 10 per cent of the variance would require three to six times as many factors as have already been calculated. The inefficiency, implied by these facts, leads most researchers to cut off at some point less than all of the principal components. This is not to say that such a procedure would be meaningless if the objective of the investigator is to completely reduce the original variables into orthogonal predictors which can account for all of the variance.

The Problem of Factor Interpretation

The discussion here must be related to our previous distinction between use of factor analysis as a method to identify causes and its use as a descriptive and/or data reduction technique. If a researcher is interested in causes, then the research will require very careful identification of factors which may be arrived at only after inspection of repeated factor analyses with numerous samples. In such cases, although the loadings of the tests may suggest the names to be given to the factors, the factors themselves are assumed to have independent existence beyond the tests. Certain psychologists, for example, have developed such ideas as "visual acuity," "mechanical aptitude," "fluency," as "standing behind" psychological test performance. I have already indicated that I am suspicious of efforts to proceed with such assumptions in treating cross-national data. I wish those well who would attempt to use factor analysis for this purpose, but the complexities of such a search should be recognized and, as Cattell has pointed out, it's not simply a matter of "seeing" similarity in various studies.

In the search for factors (viewed as causes) where oblique rotation is used, Cattell maintains that it is necessary to evaluate at least the following matrices: reference vector structure, reference vector pattern, factor pattern, factor structure, factor estimation weights, and reference vector estimation weights.[38] He criticized most factor analysts who search for causes for usually dealing with only one of these matrices: "Probably four-fifths of all published factor analyses talk about factor loading patterns when strictly their tables present only reference vector correlations. But, in any case, there are at least six alternative possibilities of expressing the profile of variables upon a dimension, and it is time that they become clearly recognized. For they could have different degrees of usefulness for *identifying*, i.e., matching, discovered factors, and they certainly have different meanings for purposes of *interpreting* the influence revealed by a factor."[39]

Further, Cattell does not view any of the oblique solutions (oblimax, etc.) as final if the investigator's quest is for simple structure. "All *present* analytical solutions fail because they are not working with a criterion rightly demanded by the model."[40] The search for simple structure, as developed by Cattell, is an extremely complicated matter and may prove to be very time consuming. In his words:

38. Cattell 1962, pp. 673–76.
39. Ibid., pp. 671–72.
40. Cattell 1966a, p. 187.

Formerly, a large factor analysis might take a month's calculation for two clerks, and the rotation to simple structure four or five months. Nowadays the former time can be reduced to minutes, and the latter to perhaps a month, but not, as some imagine, to a moment. And it remains true that three-fourths of published factor analyses are demonstratably nowhere near simple structures (editors having no standards or resources to check this) and constitute no contribution (except confusion) to the field that they are intended to clarify. What these casual and unworkmanlike studies have actually done in the last decade is to create an atmosphere of pointlessness and disillusionment by cumulating the junk heap in which factors can rarely be matched from any one research to another.[41]

To me his case is reasonable and *the burden is on those to use such procedures if they view factor analysis as a search for causes.* If we accept his arguments, the following kinds of practices prevalent in certain factor analytic studies are relatively meaningless: (1) "seeing" similarity in an orthogonal and oblique solution and then concluding that the factors are basically uncorrelated; (2) "seeing" similarity in the factors generated from different methods that emerge across several studies and concluding the same fundamental causes are operative; and (3) rotating to one of the oblique solutions and then concluding that the result is simple structure. Fortunately, those that follow the descriptive school of factor analysis escape the procedural burdens, imposed by Cattell, for the reasons developed in the preceding pages.

When factor analysis is used to describe and/or reduce data, it matters little what names are given to factors as long as the loadings are given. In the descriptive case, I think of factors as being nothing more than sets of scores that have certain properties. That is, if properly computed, they are standardized, they are uncorrelated (assuming orthogonal rotation), and they predict the variables in the study to the extent that the variables load on the factor. Any naming of these sets of scores, then, is simply a device to simplify presentation. Thus, when I call a factor "Authoritarianism," I simply mean that the scores of this particular dimension are defined primarily in terms of certain variables. In the present study, then, I choose to call as "Authoritarian" states which tend to have ineffective constitutional limitations, noncompetitive electoral systems, nonrepresentative regimes, ineffective legislatures,

41. Ibid., p. 186.

and so forth. I do not mean to imply that "Authoritarianism" causes ineffective legislatures, noncompetitive electoral systems, etc.

If the factor scores generated here are used beyond the study, that is, used to predict variables that were not included in the study, then the usage might be conceptualized as follows: "Those with high scores tend to have the following characteristics, i.e., have ineffective legislatures, etc. Those with low scores tend to have the opposite characteristics. I note a relationship between such factor score values and the values I obtained when I scored the states on such-and-such variable." Further, given the orthogonal nature of the scores, an investigator could go on to say: "I know that the variance accounted for by what I call 'Authoritarianism' (defined primarily in terms of the following variables) cannot be accounted for by the sets of scores I call 'Militarism' (defined in terms of the following variables) nor 'Economic Development' (defined primarily in terms of the following variables), etc."

Thus, factor scores in such usage are treated just like original variable scores (they describe something about the subjects) but they have the important advantages of being uncorrelated and few in number. *Factors, then, in descriptive and/or data reduction usage are the factor scores.* They have certain predictive power relative to the original variables, and they may have predictive power to other variables, but the latter is an additional empirical question.

An Illustrative Example

To illustrate the above observations and to facilitate an understanding of the crucial aspects of factor analysis and factor scores, an analysis of 39 subjects (states) and 21 variables will be presented. In this connection, *the example and discussion will be limited to the problem of unities in the principal diagonal and the use of the varimax criterion to illustrate rotation.*[42]

SOURCES OF DATA

The data employed here come primarily from the *Cross Polity Sur-*

42. I limit the discussion, as suggested above, for reasons of space because of the complexity of common factor analysis and oblique solutions and because placing "unities" in the principal diagonal and using the varimax criterion for rotation is by far the most popular technique, at least in IR studies. See Table 28.

vey and the *World Handbook of Political and Social Indicators.* Sources and their origins are listed in Appendix A.

DATA MANIPULATIONS

Because of the skewed nature of many of the distributions, all data were ranked as the first step in the analysis with average ranks assigned in cases of ties. Because factor scores require standardized scores, the rank numbers were standardized; Appendix B gives final scale values for the 39 subjects on the 21 variables.

THE ANALYSIS

The first step in a factor analysis is the computation of a correlation matrix. From one point of view, the purpose of the factor analysis is to re-represent the correlation matrix in another form. Basically, the object is to eliminate redundancy. For this reason, an examination of the simple correlation matrix frequently gives clues as to the way variables will load on the factor dimensions. Our illustrative data generated the correlation matrix shown in Table 1.* It can be seen that certain variables correlate very highly with certain other variables but that some of the variables do not associate very well with any of the other variables. For example, "Many Inhabitants per Physician" correlates − .69 with "Big Gross National Product." We can probably expect these two variables to load together (but with opposite signs) on the same factor dimension. (An examination of Table 2 will show that this is indeed the case in the actual analysis.) Also, "Small Per Capita Gross National Product" correlates strongly with "Many Inhabitants per Physician." Again, we would suspect that the two variables will likely load heavily together on the same factor dimension. (Table 2 shows that they indeed do.) "Large Percentage Export of Total to U.S.," on the other hand, correlates strongly with "Large Distance from USSR" but both correlate weakly with "Many Inhabitants per Physician." We would predict that it is likely they will load heavily on the same dimension but not heavily on the same dimension as "Inhabitants per Physician." (Table 2 again confirms our predictions.)

In short, an examination of the correlation matrix frequently gives many clues as to which variables are likely to load together on the factor dimensions, and this is ascertained by examining which variables have strong correlations with one another and which do not.

More technically, the factor analysis starts with a solution to a char-

* Tables begin on p. 63.

acteristic value problem,[43] involving the correlation matrix. The roots
of the characteristic equation, denoted by λi where $i = 1, \ldots, M$, are
called the eigenvalues, and the vectors, not equal to 0, which satisfy the
equation are called the eigenvectors. In a sense, the eigenvectors re-
represent the correlation matrix with the difference that the columns are
orthogonal. The unrotated factor loadings matrix is the eigenvector
matrix scaled by the eigenvalues. This is accomplished when each eigenvector entry is multiplied by the square root of the appropriate eigenvalue. When properly computed, the loadings become correlations of
the original variables with the factor scores. The actual resulting loadings matrix, however, depends upon the magnitude of the last eigenvalue retained in solving the equation. For example, employing our
sample data we get different matrices computing out to different size
eigenvalues, say, 3.0, 1.0, or .00, as shown in Table 2.

When we compute only out to an eigenvalue with a value 3.0 or larger,
we are asking, in effect, that a particular factor dimension account for
at least as much variance as is contained in three variables. The amount
of variance that a factor dimension explains is given by the sum of the
column entries squared. Thus, the first factor dimension in the above
example accounts for 7.23 units of variance of a total of 21 units of
variance, or approximately 33 per cent of all of the variance considered
in the 21 variables. The second dimension accounts for 4.47 units of
variance, and when we only compute out to an eigenvalue with a value
of 3.0 or larger, we can get no additional factor dimensions because no
additional factor dimension is capable of accounting for 3.0 or more
units of variance.

When we compute out to an eigenvalue of 1.0 or larger, on the other
hand, we get four factor dimensions, that is, two in addition to the two
already computed, with one accounting for 2.68 units of variance and
the other 1.52 units of variance.

If we wish to account for all of the variance in the variables, then we
compute to an eigenvalue of .00 or larger.

The sum of squares *by rows* gives us the total amount of variance
accounted for by the dimensions taken collectively. Thus, in the case
of computing out to an eigenvalue of 3.0 or larger, we can account for
70 per cent of the variance of the first variable, 87 per cent of the variance of the second variable, etc. By computing out to an eigenvalue of
1.0 or larger, however, we can account for 80 per cent of the variance of

43. See Hadley 1964, Ch. 7.

the first variable, 93 per cent of the variance of the second variable, etc. Naturally, as already indicated, the sum of squares by rows equals 1 when we compute out to an eigenvalue of .00 or larger. (In other words, we account for all of the variance of each variable.)

The really crucial aspect of this analysis is to note that early factor dimensions, that is, "big ones," are in no way affected by the calculation of later factor dimensions, that is, "small ones." The biggest factor has exactly the same loadings whether we compute out to an eigenvalue of 3.0 or larger, 1.0 or larger, or .00 or larger. It should be apparent that, in this particular analysis, no factor dimensions would have emerged if we had asked that all eigenvalues be at least 8.0 or larger.

The situation in respect to the *unrotated* matrices stands in considerable contrast to that of the *rotated* matrices when we compute out to different size eigenvalues. Thus, rotating, according to Kaiser's varimax criterion, with each of the above examples, we get the matrices shown in Table 3.

First, we should note that the rotation does not affect the amount of variance explained by the factors taken collectively or the sum of squares of the rows. Thus, all these values remain exactly the same in spite of rotation. For example, the value for "Many Inhabitants per Physician" in the 3.0 eigenvalue case is .70 in both the unrotated and rotated matrices.

The sum of the column entries squared, however, is affected. For example, in the 3.0 eigenvalue example, the "first dimension" accounts for 7.23 units of variance in the unrotated case, whereas in the rotated case the "first dimension" accounts for 6.50 units of variance. Naturally, of course, the loadings also change and the degree of change is a function of the application of the varimax criterion.

This leads us to a very important observation about factor analysis. In the unrotated case, early dimensions—that is, those "big" dimensions—initially computed can in no way be affected by the computation of later dimensions; therefore, the decision in respect to the eigenvalue can in no way affect the magnitude of the loadings on such dimensions. For example, the first dimension loadings stay the same regardless of whether we compute out to an eigenvalue of 3.0, 1.0, or .00. In the rotated case, on the other hand, the loadings are considerably affected by the eigenvalue decision. Thus, the "first dimension" in the rotated case, in computing out an eigenvalue of 3.0 or larger, *is not the same dimension* as the "first dimension" in the case of computing out to an eigenvalue of 1.0 or larger.

This brings us to an extremely important point concerning factorial invariance. It has been pointed out by Kaiser that adding tests to a battery of tests that have been factor analyzed will not greatly affect loadings if we then re-factor analyze and apply Kaiser's varimax criterion.[44] It should be apparent, however, *that factorial invariance is hinged to decision concerning the eigenvalue.* That is, unlike the unrotated case, manipulations in respect to the eigenvalue will have direct and immediate effects on loadings. Factorial invariance, then, is hinged to holding the eigenvalue computed out to constant.

COMPUTING THE FACTOR SCORES

As indicated above, there is some confusion as to the proper formula to apply to generate factor scores so they are consistent with the factor model. "Shorthand methods," typically, have the following flaws: (1) scores representing presumably orthogonal dimensions are actually correlated, and (2) the correlation of the factor scores with the original scores will not reproduce the factor loadings matrix.

These observations will be illustrated by applying one of the incomplete formulas $F = ZA$ (ZA defined above) to the sample data.[45] When this is done, it yields the results given in Table 4. One glance at these results shows immediately that the scores lack one of the qualities of factor scores computed by the complete formula, that is, the scores are not standardized. However, for reasons previously explained pertaining to the differences between the unrotated and rotated cases, the scores are valid scores in the unrotated case, even though they are not standardized. That is, there are zero order correlations between such "factor scores" and they will predict the original variables to the extent such variables load on the dimensions. The rotated scores, on the other hand, are invalid in terms of the factor model. For example, these particular scores correlate .40 and, as will be seen shortly, they do not predict the original variables to the extent of their loadings. If we standardize these scores we get the results shown in Table 5. These observations as to the correctness of the unrotated scores can now be checked against the scores computed (shown later) using the complete formula. As will be seen, the application of the $F = ZA$ formula and standardizing yields exactly the same scores as the application of the

44. Kaiser 1958.
45. For the next several examples, data will be used generated from the 3.0 eigenvalue example to save space.

$F = ZA (A'A)^{-1}$ formula.[46] Standardization, of course, can in no way affect the distributional properties of scores and, therefore, the standardized rotated scores *remain invalid* in terms of the factor model.

If we correlate the above scores with the original variable scores, we get the results shown in Table 6. It can be seen, in the *unrotated case*, by reference to Table 2, that the correlations of the factor scores with the original variable scores are exactly equal to the loadings of the variables in every case. In contrast, in the *rotated case* (see Table 3), the correlations are the same in some cases but different in others. For example, "Many Inhabitants per Physician" loads .83 on the first factor dimension, and the first factor dimension scores, computed by the shorthand method, correlate with the original variable scores .83. The correlation and the loading are also the same in the case of "Big Gross National Product." In the case of the third variable, however, "Large Percentage Increase in Population," the loading is .55 but the correlation is .49. The fourth variable loads $-.42$ but correlates $-.52$, etc. Thus, the predictive power of the factor scores to the original variables cannot be ascertained from the loadings, an error some researchers fall into when they use this incomplete formula or one of its variations in the rotated case.

One of the most popular variations, in applying shorthand methods, is to ignore loadings below some value, such as $\pm .50$, in the calculation of the scores using $F = ZA$. It should be evident that in this later application only the heaviest loading variables are used to weight the z scores and the lesser loading variables are ignored. Such an application has the effect of increasing the correlations of the heaviest loading variables with the factor scores but usually makes the correlations between the factor scores even more pronounced.

Still another variation is to ignore all loadings below some arbitrary figure and set all other loadings to ± 1.0. This application further enhances the two tendencies noted above of maximizing the correlations of the original variables with the factor scores, but usually further increases the correlations between the factor scores themselves.

From this discussion it should not be concluded that shorthand methods are without any value. It is possible that a researcher may wish

46. It should be noted, in this connection, that $F = ZR^{-1}A$ will yield the same results as $F = ZA (A'A)^{-1}$ if both are applied to data where unities have been placed in the principal diagonal. As explained above, however, $F = ZR^{-1}A$ is the proper formula to apply when estimates of communalities have been placed in the principal diagonal.

to generate scores which have the very qualities such scores possess and thus are presented in the application that follows. However, a frequent "error" in this regard is to assume that the scores possess the qualities they would have had if the complete formula had been applied, i.e., they are uncorrelated and predict the original variables to the extent of their loadings.

As indicated above, the application of the $F = ZA \, (A'A)^{-1}$ formula produces scores consistent with the factor model. It should be noted that the first part of the formula, ZA, is actually the same as the first incomplete method discussed above. Obviously, then, what happens in the application of the complete formula is an alteration of the ZA (product) matrix to get a set of scores which have proper values. The matrix which results from $(A'A)^{-1}$, then, is the matrix of primary interest. In the unrotated case, using our test data $(A'A)^{-1}$ yields the results given in Table 7. It can be seen that the matrix which results from $(A'A)$ is nothing more than a diagonal matrix with the sum of squares of the columns of the A matrix along its principal diagonal. Thus, if we check Table 2 we see that the first entry in the diagonal of Table 7 (7.23) is the sum of squares of the first unrotated factor dimension. Similarly, the second diagonal entry, 4.47, is the sum of squares of the second unrotated dimension. When this matrix is inverted, it yields the reciprocals of the original entries. Thus, .14 is another way of expressing $1 / 7.23$. Multiplying by a reciprocal, of course, is the same thing as dividing by the original value. Each entry in the product of ZA, then, is simply scaled, in the application of the complete formula, by the appropriate column sum of squares entry. To put it another way, all the application of $(A'A)^{-1}$ amounts to, in the unrotated case, is the division of each entry in a column of the ZA (product) matrix by the appropriate column sum of squares. The effect is to standardize.

In the rotated case, the matter is more complicated because $A'A$ does not yield the diagonal matrix. Proceeding as before, we get, in the rotated case, the results shown in Table 8. It can be seen that values other than 0 occur off the principal diagonal after the multiplication, although the entries in the principal diagonal are still the sum of squares of the columns (see Table 3). Thus, the diagonal entries are the mathematical equivalents of the unrotated case, but instead of 0's occurring in all other elements except the diagonal, other values are evident. When we invert $(A'A)$ the entries in the principal diagonal of the new matrix become the reciprocals of the diagonal entries in the principal diagonal of the previous matrix. Again, however, values other than 0 occur off of the principal diagonal. Thus, when we post-multiply the ZA (product) matrix by $(A'A)^{-1}$ we are doing more than simply dividing each set of column entries by a constant for the purposes of standardization. The additional effect insures orthogonality. The multiplication of ZA by $(A'A)^{-1}$ in the unrotated case, then, only has the effect of standardization, but in the rotated case has the additional effect of insuring orthogonality.

In applying the complete formula to our test data, we get the results given in Table 9. We can see that our previous observations were correct. The standardized scores of the incomplete formula (Table 5) are equal to the scores computed by the complete formula. Thus, Albania has a score of 1.57 in both cases, and so forth. In contrast, there are considerable differences in the rotated case. Albania has a score of 1.23 in Table 5, but 1.00 in Table 9 on the first dimension. Argentina has a score of $-.76$ in Table 5, but $-.60$ in Table 9 on the first dimension. In fact, not a single score in Table 5 is correct, using Table 9 as the standard. In contrast with the rotated scores of Table 5, the scores of Table 9 will correlate with the original scores exactly to the extent of their loadings as given in Table 3. These correlations are shown in Table 10. Simple inspection shows a perfect correspondence in every case.

The Value of Orthogonal Factor Scores

As suggested above, it is the opinion of this researcher that one of the most valuable aspects of factor analysis, beyond simplification—that is, beyond reducing a large number of variables to a lesser number with little loss of information—is related to the fact that the resulting factor scores, using the proper techniques, are uncorrelated. With the application of such techniques, then, any set of partially redundant data can be made nonredundant, while most of the original information is still retained. The resulting factor scores lend themselves to an ease of interpretation, that the original scores lack, in using such scores in additional correlational analysis.

For example, consider the alternative of using factor scores, representing original data, or the original data itself, in a simple correlation with some third variable. In the case of correlated predictors, if two of the variables have a good relationship with a dependent variable, this may be because they are strongly intercorrelated with one another (are basically the same measure) or it may be because one is related to one part of the variance of the dependent variable while the other is related to another part. It is hard to untangle relationships of this type as long as the predictors are intercorrelated. In contrast, if orthogonal factor scores, representing the original variables, are correlated with a dependent variable, because the scores are uncorrelated, each has unique explanatory power. For example, using the factor scores of our test data (1.0 eigenvalue) and correlating them with the variable of "Radios per 1,000 Population" yields the results given in Table 11.

The first set of factor scores correlates with the dependent variable $-.88$. If we square this, we get the variance explained, or .77. The second set of factor scores correlates with the dependent variable $-.17$. Again, if we square this, we get the variance explained for this variable, or .03. If we want to know how much variance these two variables together account for, we simply add together the individual variance explained figures and get .80. The total of all of the variance explained by all of the predictors will equal R^2 (that is, the multiple correlation squared). In this case it equals .82. In fact, the above procedures, of reducing dependent variables by first factor analyzing them and then running the resulting factor scores against a dependent variable, actually make multiple regression analysis unnecessary. For example, putting our test data in a multiple regression analysis, we get the results given in Table 12. We can see R^2 is exactly equal to the sum of the simple correlation coefficients squared. It can also be seen that when predictors are orthogonal, simple correlations are exactly equal to the "weights" of the multiple regression analysis. In short, we can get just as much information from a simple correlation analysis as from a multiple correlation analysis when we reduce data to orthogonal dimensions before applying such correlational procedures. Also, if we do not factor analyze, that is, if we use original predictors that are correlated, we cannot easily assess the importance of the predictors from the beta weights because the multiple regression analysis is concerned with unique explanatory power of the variables in assigning the beta weights. Thus, we might have two very good predictors (individually considered) of the dependent variable, but if one's unique contribution (in a group of variables) is considerably better than the other, it will receive a heavy weight and the other little weight. In other words, intercorrelations among the variables will affect the interpretation of the "importance" of the variables.

To illustrate this (Table 13) we will take ten variables out of our test data and run them against our dependent variable. It can be seen that Variable 2 and Variable 10 are both good predictors of the dependent variable, in terms of the simple correlations, but Variable 2 is a better unique predictor (in association with these variables) than Variable 10; hence, Variable 10 receives considerably less weight in the regression analysis. In other words, when variables are correlated in a multiple regression scheme, what we get is a good picture of the overall relationship between the predictors and the dependent variable, but we may have trouble assessing the predictive importance of variables by inspecting weights. That is, the "importance" indicated by the simple correlations may diverge markedly from the "importance" as indicated by the beta weights. On the other hand, if the predictors are orthogonal, then multiple regression analysis is not really necessary because the sum of squares of the correlations will equal R^2 and the simple correlations will equal the beta weights (if computed).

Similar observations pertain in respect to the use of canonical correlation.[47] A canonical correlation analysis assigns weights to two sets of variables so that the scores generated from the two sets are maximally correlated. The canonical variate scores themselves, for one set, can be computed by the formula $C = SW$, where C is an N x 1 vector of canonical variate scores, S is an N x n matrix of subjects' standard scores on the original variables, and W is an n x 1 vector of canonical weights, where N = subjects and n = variables.

This formula is given here to make clear the meaning of canonical variate scores. The actual calculation of the weights, of course, is the solution to the basic canonical problem, which is to maximally correlate two such sets of scores, one generated from one set of variables and the other generated from the other set of variables. When weights are calculated where this is the case, the canonical variate scores predict the original variables to the extent of their loadings, if the original variables are mutually orthogonal in each set. Like factor scores, a great deal of information can be summarized by properly computed canonical variate scores.

If correlated data are used, then the individual importance of the predictors is not clear by examination of canonical weights, for the same reasons given in the application of multiple regression. That is, the weights are assigned so that only unique variance is considered. If the variables are uncorrelated, however, weights will be equal to the correlations of the variables with the canonical variate scores.

47. Hotelling 1936; Hotelling 1935; Anderson 1958, Ch. 12; Bartlett 1941; Horst 1961a; Horst 1961b; Kendall 1957, Ch. 5; Thompson 1947; Clyde, Cramer, and Sharin 1966, pp. 4–8.

To illustrate these observations, we will take ten variables from our original data set and ten other variables. (See Appendix C.) Applying a canonical correlation analysis we get the results shown in Table 14. From the simple correlation matrix we can see that Variable 12 and Variable 17 are highly correlated. However, Variable 12 (Variable 2 in the second set) is given heavy weight in the analysis, while Variable 17 (Variable 7 in the second set) is given little weight. Thus, as in the case of the multiple regression analysis, intercorrelations may obscure the individual importance of the predictors (relative to the canonical variate scores), although the overall importance is given by the canonical correlation. If we correlate the canonical variate scores with the original variables, we get the results given in Table 15.

It can be seen by inspection that the weights and the correlations are not the same. However, if we use scores in the canonical correlation without redundancy, in this case using factor scores calculated from our original 21 variables along with those calculated from our 10 additional variables, we get the results in Table 16.

If we run the canonical variate scores against the original variable scores, we get the results in Table 17. Here we see an exact correspondence between the weights and the correlations and thus the importance of the predictors viewed individually is made clear through a simple examination of weights, as in the case of multiple regression (when it is used with uncorrelated predictors).

From the above it should be clear that a first step in many kinds of multivariate analysis could be factor analysis. In short, redundancy, one of the primary problems of multivariate analysis, can be eliminated through such techniques. In this connection, as explained above, the "price" to achieve orthogonality is usually minimal and typically involves only 10 per cent to 15 per cent of the aggregate variance, if we use Kaiser's recommendation in respect to the eigenvalue decision. It should be clear in this respect that one need pay no price whatsoever if one wants to merely change the data from a correlated to uncorrelated form by computing out to an eigenvalue of .00 or larger. In such a case, for the reasons explained above, all of the original variance will be retained. In the case where we apply Kaiser's rule, we eliminate some of the variance but greatly gain in terms of simplification by having fewer, usually many fewer, than the number of original variables to work with. Finally, once the data have been rearranged in the form suggested here, they lend themselves to an ease of interpretation which the original correlated data do not, as illustrated in the above examples.

An Application

The present application grew out of an earlier one.[48] The major problem of the previous project was to relate a number of independent variables concerning United Nations delegates' home states to the delegates' attitudes toward the primary organs of the United Nations. Because of confusing interrelationships among the independent variables, they were factor analyzed to render them down to a number of orthogonal predictors. The indices that were developed were given names, such as "Economic Development," "Authoritarianism," "U.S. Relations," "Size," etc. However, the 68 states treated in the project fell short of the universe of states. It became evident to this researcher that if he or others were to gather new samples, it would be necessary to re-do the factor analysis (unless exactly the same states were drawn) which would entail considerable effort. Thus, to minimize labor for future projects, it was decided to factor analyze such indices for the universe of states, so that when future samples were drawn, there would be a ready-made set of predictors. Also, it was decided to expand the number of variables. (The previous study dealt mostly with *Cross Polity Survey* variables, whereas the present project almost doubles the number of variables considered and expands resources.)

From the introductory statements about factor analysis, the major concerns and methods of this project should be clear. That is, the primary purpose is to reduce the 91 variables under consideration to a lesser number of orthogonal predictors. For this reason no "casting out" of information occurs as is the case when communalities are inserted in the principal diagonals. In other words, this researcher wanted the principal components to describe as much of the variance in the variables as possible, ignoring the distinction between common and specific variance.

The question of rotation was again basically dictated by the primary purpose. Oblique rotation, which may lead to correlated factors, was viewed as undesirable. Hence, it was decided to create orthogonal rotated factors and to apply the varimax criterion for column simplification.

Because unities were placed in the principal diagonal, factor scores were calculated directly, not estimated. Hence, the formula, $F = ZA(A'A)^{-1}$ was used.

48. Vincent 1968.

Variables were selected with a view that they should cut across a number of economic, social, geographic, and political areas.[49] In this sense, the basis for selection was similar to the approach used in other studies, i.e., Rummel, Russett, and Adelman and Morris. Particularly important in this regard was a concern for "accessibility." That is, the variables should be of a kind that other researchers can validate. The safeguards to insure the "quality" of the data base are fully discussed in the *CPS* and the *World Handbook*, two of the principal resources, and will not be repeated here. The sources of all variables are given in Appendix D, and, for the most part, in the opinion of this researcher, these sources are "reliable" publications.

Data transformation was handled in the way suggested above, that is, all data were transformed to rank numbers. This was done because this researcher saw little justification for elevating the *CPS* variables to the class of interval data, and yet he wanted to include such data in the study.

Missing data were treated by dropping no-information subjects in the correlation matrix and using mean estimates for the calculation of factor scores. The resulting scores, then, came close to, but fell somewhat short of, predicting the original variables to the extent indicated in the factor loading matrix.

This study had approximately 15 per cent missing data. Appendix E gives the number of zero values (mean estimates) both for subjects and for variables. States such as India, Poland, Belgium, and Finland, then, were the best subjects, having fewest estimates, and states such as Western Samoa, Maldive Islands, and Rwanda were the worst subjects. On some of these variables a sizeable number of subjects fell on the true mean, when the numbers were standardized, and thus the number of zeros do not accurately reflect the number of estimates that were made in such cases. That is, fewer mean estimates had to be made than indicated by the actual number of zeros. This only occurred to an appreciable extent on Variables 53 and 73. Ambiguous codings in the

49. It should be apparent, by examination of Appendix D, that some of the original variables are factor scores generated from other studies. Such scores are treated as ordinary variable scores in this study. The predictive relationship of such scores to "original variables" may be ascertained by reference to the appropriate study. Thus, by reference to *World Politics in the General Assembly*, it will be seen that Alker and Russett's "East-West" factor scores have a specified degree of predictive power relative to roll-call votes in the General Assembly. In studies where more than one factor analysis was done over time on the same kind of data, the most recent scores were taken, i.e., Alker and Russett's 1961 factor scores.

Cross Polity Survey were given zero codings, after standardization, as in the case of no-information codings.

The following states were coded by research assistants: Botswana, Gambia, Kuwait, Lesotho, Malawi, Malta, Tanzania, Zambia, Kenya, West Samos, Singapore, Guyana, Malaysia, Maldive Islands, Barbados, and Taiwan. Each coder worked independently of the others, and unless all three agreed upon the code, using the *Cross Polity Survey* coding scheme, the country was given an ambiguous coding and thus a mean value (after standardization of the rank scores for those with non-ambiguous codings). It should be apparent, then, that reliability of the coders is 1.0 on the variables not assigned mean values.

Original interval numbers for non–*Cross Polity Survey* variables for such states were taken from the resources which are listed in Appendix D.

Although it was tempting to try to "update" the *CPS* codings, it was decided to let them stand as is in all cases. This decision was made for two reasons: (1) the labor involved exceeded resources presently available to this research (*CPS*, of course, is a massive undertaking); and (2) any "changing" would immediately raise additional validity and reliability problems. Fortunately, the *CPS* categories are "broad" enough so that dramatic shifts in categories are unlikely for most states for several years. There are some obvious exceptions where *coups d'état* have occurred since the *CPS* codings were done but the "stability" of the changes in many cases, i.e., Greece, is far from clear and the best course seemed to be to stick with the *CPS* codings. Thus, except for the 16 states coded by research assistants, the bulk of the "base" of the factor analysis can easily be inspected by interested researchers by reference to the appropriate resources (see Appendix D).

Principal components were computed out to an eigenvalue of 1 or larger, which yielded 19 factors, explaining 89.2 per cent of the original variance. It should be evident, then, that a complete solution would have required (assuming the rank of the matrix is not smaller than the number of variables) 72 additional factor dimensions which would have accounted for only 10.8 per cent of the variance.

Factor interpretation will be handled as suggested above. Labels are given in terms of the heaviest loadings, but with the understanding that this refers to a set of scores with certain predictive qualities relative to the original scores. That is, as explained above, the factors are not treated as "explanations" in the sense of identifying causes, but as convenient orthogonal predictors of the original variables.

Results

If the steps outlined in the preceding pages are followed, both rotated and unrotated findings will result. The unrotated findings will be discussed first.

UNROTATED DIMENSIONS

Table 18 gives the first six principal components of the unrotated factor matrix. Variables may be identified by their numbers, given in Appendix D. It can readily be seen that the first few factors have the heaviest loadings, and loadings fall off rather rapidly as the dimensions account for less and less variance. The table can be "interpreted" by indicating the attributes of those with high factor scores and by concentrating on loadings of $\pm$ .50 or above. The results are shown in Table 19.

This analysis, then, yields some dimensions with certain useful properties. On the first factor dimension a state with a high factor score tends to have the characteristics indicated. Because of the mixing of economic and political variables on this dimension, it might be called an "Economic-Political" dimension. The meaning of dimensions, of course, as already indicated, is quite arbitrary and every variable loads to some extent on every factor. The important point is that the factor scores can be used to predict the original variable scores to the degree of their loadings. Thus, Upper Volta, with the highest score on this dimension, tends to have most of the attributes indicated in the table, while the United Kingdom, with the lowest score on this dimension, tends to have the opposite characteristics.

In terms of their loadings, the remaining dimensions might be named "Politics," "U.S. Relations," "Belligerency," "Dispersion," and "Population Pressure."

The predictive value of the factor scores relative to the original variables becomes quite weak as soon as we move beyond the third factor, and the scores can only account for approximately 35 per cent or less of the variance of even the heaviest loading variables. The interested researcher, then, would probably only want to utilize the factor scores on the first two or three dimensions, although all six are given in Table 20. The unrotated analysis, then, has produced a small number of scales of considerable predictive importance relative to the original variables. Collectively they account for 71.8 per cent of the original variance.

ROTATED DIMENSIONS

When the dimensions are rotated, a factor matrix is generated which has considerably more value, from the viewpoint of the factor scores predicting the original variables, than in the case of the unrotated matrix. The rotated matrix (all 19 dimensions) is given in Table 21, and, as before, the variables can be identified by the variable numbers given in Appendix D. It can be seen that, generally speaking, in contrast to the unrotated matrix, loadings tend to be either high or low. For example, even on the nineteenth dimension, which explains only 2.3 per cent of the variance in the variables, at least one variable, 33, loads − .80. In fact, on every dimension, at least one or more variables have substantial loadings. As before, the matrix can be "interpreted" by indicating the propensities of subjects with high factor scores concentrating on loadings ± .50 or above (or the largest loading for a variable, if it does not reach ± .50). The results are given in Table 22.

In this table the naming of factors becomes considerably easier because the clusters of the heaviest loading variables in many cases seem to be categorically similar. For example, on the first factor dimension, labeled "Underdeveloped," many of the variables seem to have an economic-technological character, and political variables seem to have little weight on this dimension.[50] Similarly, on the second dimension, labeled "Democracy," many of the heaviest loading variables may be categorized as "political"; on the third factor, many of the variables seem to measure something concerning the United States; on the fourth variable, something connected with magnitude, and so forth. Such naming, however, should not confuse the really crucial point of the analysis. On pain of being repetitious, factor scores on these dimensions predict original variables basically to the extent of their loadings. Every factor, then, has some predictive power in respect to every variable (except in the case of a .00 loading), but each factor predicts certain variables much better than others. Subjects may be located on these dimensions in the same way as in the unrotated case. Table 23 contains the results.

These scores, then, locate the states on the dimensions. Considering the first dimension, states such as Nigeria, Afghanistan, Sierra Leone, Cambodia, and the Malagasy Republic, with high factor scores, tend to have Low Newspaper Circulation per 1,000 Population, Few Radios

50. The naming of variables was done with a view of making the name reflect the characteristics of those with high factor scores.

per 1,000 Population, Large Number of Inhabitants per Physician, and so forth. States on the opposite ends of the dimension, such as the United States, Australia, Czechoslovakia, Hungary, East Germany, and New Zealand, tend to have the opposite characteristics. Each of the other scales can be similarly interpreted, remembering that states which have a number of mean estimate predictions on the original variables tend to move to the middle of the factor score distributions.

·The question now arises as to whether these scores possess the qualities indicated earlier as being desirable, that is, are orthogonal, and predict the original variables basically to the extent the variables load on the factors.

To answer the first question, we simply run a correlation of each set of factor scores with every other set of factor scores. To the extent that non-zero correlations occur, there will be a degree of redundancy in the scores. The results are given in Table 24. It can be seen that the largest correlation is only −.12. We can conclude from this that our mean estimates did not destroy, except slightly, the quality of orthogonality, and that when such scores are used as predictors in respect to variables not included in the study, the variance explained by any one set of scores can be viewed as basically unique relative to the rest of the predictors in the study.

In order to answer the second question, we correlate the factor scores with the original variable scores. The results are shown in Table 25. It can be seen that these scores predict the variables almost to the extent of their loadings. The discrepancy between the loadings and the actual correlations arises, of course, because of the method of assigning mean values, in cases of no information, to calculate the factor scores and treating now for correlational purposes mean estimates as real values on the original variables (i.e., subjects are *not* dropped in this analysis as they were in the correlation matrix upon which the factor analysis was performed). It should be clear that if only the ascertainable universe were considered, the loadings would exactly equal the correlations. In this connection, we see that Military Personnel as a Percentage of Total Population loaded −.45 on the first factor dimension. When we run Military Personnel as a Percentage of Total Population against the set of factor scores that include calculations from mean values, in the absence of information, we find the correlation is −.45. That is, there is perfect agreement here (to two decimal places) between what we observe and what we would expect if only the ascertainable universe had been considered. On most of the remaining variables we see that the correlations are either very close to or equal to the loadings. For example, with respect to Variable 2, Population per 1,000 Hectares of Agricultural Land, the loading was $-.16$ and the correlation $-.14$. In the case of Inhabitants per Physician, the loading was .91 and the correlation .89, and so forth. In general, then, inclusion of mean scores for subjects on the variables where information was not obtainable does not seem to greatly reduce the predictive power of the factor scores relative to the original variables.[51]

It can be seen by reference to Table 21 that some variables are much better accounted for by the factors on which they load heavily than other variables. For example, the variance of Newspaper Circulation is almost completely accounted for by Factor I. On the other hand, UN Supranationalism only loads .60 on its heaviest factor, and does not load, except weakly, on other factors. The communality, of course, indicates the degree to which the variables have been accounted for by the factors computed. In Table 26 the variables are listed in rank order in terms of their communality. Variables 1, 11, 12, and 13, then, are best accounted for by the factors computed, and Variables 24, 34, 37, and 88 are least accounted for by the factors computed. It will be recalled that it is basically uneconomical to attempt to account for all the variance in all the variables because 72 additional factors would have to be computed to account for the remaining 10.8 per cent of the variance (unexplained by the factors). At this point the descriptive approach to factor analysis runs into a difficult problem. If we are economical, the factors may not relate well to some of the variables. If we are uneconomical, we will wind up with about as many factors as we have variables, which is hardly consistent with our objective of data reduction. Those variables with small communalities, of course, do not relate well to the other variables in the analysis. That is, they tend to be orthogonal to all other variables considered. Knowing this, a researcher might want to use in such cases the original variables as predictors instead of the factor scores, although he will sacrifice the advantages

51. Where predictive power from the factor scores to the original variables is fairly poor, it is possible to use an incomplete method of factor score calculation, such as $F = ZA$, to enhance the relationship of the heaviest loading variables to the factor scores. Such scores, of course, will no longer be uncorrelated with the factor scores calculated with the complete method, nor will a correlation of such scores with the original variables reproduce the factor structure. The effect is *similar* to rotating to an oblique solution for these variables. The enhancing effect can be made most pronounced when low loading variables, i.e., those below some arbitrary point, such as $\pm$ .50, are set to zero. (See Appendix H for details and uses.)

explained, concerning completely unrelated predictors. (For example, Rummel's Diplomatic Factor Scores, loading on Factor 19, are virtually unrelated to any other variables within the study. That is, the highest correlation is only −.33 with Political Modernization and they correlate up to the .25 level only with a few of the remaining variables in the study.)

In the above discussion concerning the relationship of the factor scores to the original variables, *raw* factor scores were treated as meaningful predictors. Such scores do not have the property of rank scores, of course, and, therefore, Table 25 cannot be interpreted as a Spearman's *rho* correlation matrix. Hence, it is of interest to determine the predictive power of *rank* factor scores to the original variables, in that some researchers may wish to make only rank assumptions concerning the location of states on the factors. Table 27 contains the results. It can be seen that the predictive power of the factor scores relative to the original variables remains basically the same (compare Tables 25 and 27), regardless of whether the raw or the rank factor scores are used to predict the original variables. In this connection, each rank factor score distribution correlates .95 or above with its corresponding raw factor score distribution. Appendix F gives the rank position of every state on the 19 factors for the interested researcher.

Uses

Some may agree, at this point, that the procedures followed here "make sense" in terms of the previously developed arguments, and yet may still question the value of the end product. In short, what is the value of the factor scores that have been generated here?

It has already been pointed out that an earlier analysis of 68 states and 48 variables proved to be of considerable value for the purposes of ascertaining the relationship of national attributes to United Nations delegate attitudes. It should be apparent that any similar cross-national project might profitably employ the scores generated in this study as relevant predictors. The scores, then, lend themselves to various kinds of hypothesis testing as new data are collected.

In this connection the scores provide a quick test of "representativeness" if less than the universe is being considered. Researchers frequently find it difficult to obtain a random sample and may wish to know if a sample is basically "representative." That is, in respect to these factor scores, does the sample deviate markedly from the universe concerning such categories as "Underdeveloped," "Democracy," etc? To ascertain this, the states in the sample can be scored from the information in Table 23 and the scores correlated. If the correlations deviate considerably from zero, the sample may be viewed as "unrepresentative" from a descriptive point of view. For example, if 30 states are selected from the universe of states and their factor scores on the Underdeveloped and Democracy dimensions correlate at the .50 level, it can be concluded from this that the sample does not "represent" the universe with respect to these characteristics. That is, if the correlation is positive, the sample overrepresents the developed, authoritarian states and underdeveloped, nonauthoritarian states; but it underrepresents the developed, nonauthoritarian states and underdeveloped, authoritarian states, compared to the "balance" of these characteristics in the universe.

Finally, the scales summarize a great deal of information from a descriptive point of view. That is, even if scales are not used as predictors with respect to other kinds of data, they can be used to describe the location for classification purposes of the majority of important political units in the world today. In this connection, the scales readily lend themselves to broader classifications than given by the raw scores themselves. For example, considering the Underdeveloped dimension, cutoff points with respect to the raw scores could be established to classify states as highly developed, moderately developed, etc. In this connection it should be remembered that "developed" is being defined here by a particular set of variables, given certain kinds of weightings, and that information on some subjects is considerably better than on others. If one wanted to proceed very conservatively, making such classifications, one could set aside in a separate "ambiguous" category states that have zero values (see Appendix E) above some arbitrary cutoff point (i.e., more than 20 zero values).

Relationship to Other Projects

In the beginning of this study it was argued that, in part, this research could be justified because the problem of factor score computation has been neglected in cross-national studies, at least in those analyses cutting across a broad spectrum of variables, and because the N of most studies falls considerably short of the universe of states. Table 28 shows the relationship of the present project to the bulk of the projects done in the IR area, using factor analytic techniques.

EXPLANATION OF TABLE 28

Concerning references.—I have included, I hope, the bulk of the "important" studies done in the international relations area. One important exception is Brian Berry's "An Inductive Approach to the Regionalization of Economic Development" which, because of its application of a "direct factor analysis,"[52] was difficult to compare with the other studies.

In the case of the studies that were included, a very brief description, such as "voting data," "rotated," "unrotated," etc., is given to help characterize the study under consideration. Complete references on all of these studies are given in the Bibliography. In most cases where various analyses are done, these are presented as separate studies, such as Tanter's separate analyses of "Foreign" and "Domestic" conflict variables. In some cases it was not possible to do this for reasons of space limitation, as in the case of Adelman and Morris' study where 13 different factor analyses are done, in addition to their "long-run analysis" given here.

Concerning the number of subjects.—This refers to the maximum number considered in the study. This number falls off, of course, for individual correlations in the correlation matrix in those studies which drop no-information subjects. The subjects are, for the most part, independent states, although, in some studies, units other than independent states are treated. For example, Alker and Russett (Studies 4–11) include Byelorussia and the Ukraine in their analysis. (Subjects in all cases refer to such units, although in the case of a Q-analysis, the subjects and variables, naturally, are reversed in the actual analysis—see "factor technique" below.)

Concerning factor method.—"Prin. Comp." stands for "Principal Component" and indicates the initial solution.

Concerning entries in the principal diagonal.—"Comm." stands for "Communalities" and "Un." for "Unities." In some cases where the entry is not made clear, this is shown by a question mark.

Concerning factor technique.—R refers to an analysis where the variables load on the factor dimensions, and Q refers to an analysis where the subjects load on the factor dimensions. The Q-analysis, then,

52. Berry 1960, p. 82. Russett also has some direct factor analyses, not compared here, in his *International Regions*.

reverses the subject-variable role through a matrix transpose. The value of such an analysis is fully explained in the studies where it is used.

Concerning factor interpretation.—The classification of studies on this point proved to be the most difficult aspect of table construction. In most cases, the results of the factor analysis could be viewed in purely descriptive terms. Nevertheless, most authors seem to imply that something more is going on than just descriptive and/or data reduction, i.e., they seem to be searching for "underlying dimensions," "causes," "reasons," etc. Very common in this regard is the tendency to compare the "factors" that emerge from one study with those that emerge from other studies and then conclude that some sort of "basic pattern" or "dimension" has been uncovered. In addition, factors are frequently "tossed out" on the grounds that they are "uninterpretable," *something that cannot be the case when factor analysis is used descriptively.* That is, the relationship of the factor scores to the original variables is just as "interpretable" on small factors as it is on large factors, even though most loadings are smaller.

In this connection, it should be clear that I have used the term "causes" to describe studies which *seem* to be looking for "something" "standing behind" observed relationships, and the term "descriptive" to indicate that the purpose is descriptive and/or data reduction. A question mark indicates that the author *seems* to be going both ways at once or is unclear. In this connection, if an author is *not* interested in description and/or data reduction and yet is *not* searching for causes, then to me the purposes of his analysis seem obscure. Also, efforts at "classification," I assume, fall into the "descriptive and/or data reduction" category.

To justify these classifications, I will treat each briefly in turn.

Concerning Study 1.—I label it as a search for causes, because the authors argue, "Furthermore, it seems unlikely that the patterns of relationships observed in these three chapters could arise from the existence of a common cause (or causes) that is neither explicitly included in the analysis nor so all-inclusive as to be devoid of operational meaning. After all, the list of phenomena represented by the included variables is so extensive that, in general, only extremely broad attitudinal characteristics of nations (such as national spirit or ethos) are not reflected explicitly to some degree in the indicators. We shall, therefore, use a causal interpretation whenever it appears justified."[53] This is

53. Adelman and Morris 1967, p. 175.

stated in spite of the fact that the authors say earlier, "It should be emphasized that, like all multivariate analysis, factor analysis is a study of mutual association rather than a study of causality."[54] Because the authors do in fact seem to make some "causal interpretations," I believe "causes" to be a more appropriate classification than a question mark.

Concerning Studies 2–3.—Alker states, "Factor analysis will be used in an attempt to uncover the principal dimensions of conflict underlying votes at the Sixteenth General Assembly and the location of states on these main issues."[55] From this, I am not sure of the proper classification, so I have assigned a question mark.

Concerning Studies 4–11.—Alker and Russett seem to begin their study with a descriptive emphasis. "The inductive problem is to go back from the correlation among roll calls to parsimonious and comprehensive descriptions of these common elements, which will be interchangeably described as underlying voting 'factors,' 'components,' 'dimensions' or 'super-issues.' In the language of the philosophy of science, this is the task of reducing a large number of operational indicators (e.g., roll calls) to a smaller number of basic conceptual variables."[56] However, their search for factor similarity across various years *suggests* more than a data reduction objective and, therefore, I have assigned a question mark.

In the case of Study 12.—Banks and Gregg argue, "The factors reported on in this paper are nonetheless intuitively suggested and would appear to delineate significant components of difference among major political groupings of the contemporary world."[57] By this I assume that their objective may be more than data reduction although their results can be interpreted in such terms, i.e., as a classification scheme through the application of the Q-technique.

Concerning Studies 13–15.—Cattell seems to operate in a way consistent with his previously given arguments in favor of a "causal" interpretation: "Descriptions of the factors in national syntality will be undertaken more confidently when we see how the factor patterns 'stand up' in the two further analyses now being made."[58]

Concerning Study 16.—The question mark is assigned because Gregg and Banks seem to move both ways: "One important considera-

54. Ibid., p. 148.
55. Alker 1964, p. 642.
56. Alker and Russett 1965, p. 33.
57. Banks and Gregg 1965, p. 6.
58. Cattell 1959, p. 458.

tion is whether to adopt the *descriptive* or the *inferential* approach in interpreting the meaning of the factors. The former considers the factor to be a set of coefficients which conveys only descriptive information about the clustering of variables on a factor. The inferential school takes one step further and assumes that the factor indicates the presence of a basic dimension of 'latent variable' which 'causes' the array of variables along the factor. The composition of the variables loaded by the factor and the percent of total variance which the factor explains determine the appropriateness and strength of the inferential interpretation. . . . In light of these criteria we first employed the descriptive and then when appropriate, the inferential approach for interpreting the Survey factors."[59]

Concerning Studies 17–21.—Rummel also seems to move in both directions. In one study he argues:

> There are possibly two approaches to naming a factor or dimension. One may try to label a factor with what one considers to be the underlying source of the clustering of measures (or variables). Hence, I might call the first rotated factor a "rapid industrial growth factor" implying that in those countries in which rapid economic growth is taking place, one finds many riots, demonstrations, etc. This I call source labeling, and it is with respect to source labeling that the term "factor" appears to me most appropriate. On the other hand, I might label the factor in terms of the measures entering into the cluster and what they may represent in terms of typology. This kind of labeling I call type labeling and it is the approach used here. I prefer to leave source labeling to the future, when systematic analysis of the relationships between the dimensions of this study in such things as rapid industrial growth, technological change, political system, and so forth, will indicate some of the possible sources of such a cluster of conflict behavior as indicated by the turmoil dimension.[60]

In Study 17, he argues, "The delineation of these three dimensions does not suffice. They must be compared with the results of other analyses— linked with similar studies so that our convergence upon reliable dimensions of domestic conflict behavior may be assessed."[61] In short, in his numerous studies concerning domestic and foreign conflict vari-

59. Gregg and Banks 1965, p. 606.
60. Rummel 1963a, p. 11*n*.
61. Rummel 1966b, p. 68.

ables, Rummel seems to be pursuing something more than data reduction as evidenced by his statement, "One can conclude that the same *turmoil* dimension has been extracted from the three sets of data."[62] However, in his "Indicators of Cross-National and International Patterns" embracing 236 variables, he argues, "The names given to the seven patterns are based on those variables most highly involved in each. They are meant to be descriptive and mnemonic labels only and are not to be interpreted as presumed causes underlying patterns."[63] Rummel then proceeds, however, to compare his most important factors with those emerging from other studies which to me indicates a search for "underlying dimensions." For example, he argues, "The DON patterns as a whole have also been delineated in other empirical studies."[64] (In this connection, if we factor analyze virtually the same variables in different studies, we, of course, expect highly similar patterns of loadings. The important point at issue here is whether Rummel and others are attempting to identify some fundamental "factors" "behind" the observed correlations or whether they are attempting to produce nonredundant scales, derived from a large number of redundant scales.)

Concerning Studies 22–23.—Russett seems to frequently take a very strong descriptive orientation: "How many groups ('regions') are necessary for an adequate summary description of similarities and differences among types of national and social systems?"[65] However, the search for fundamental dimensions also seems to be evident. "Furthermore they correspond quite closely to the factors that have emerged from a variety of other studies. . . . Although we might well have added other variables, the dimensions we have are neither fortuitous, unique to this study, nor theoretically irrelevant."[66]

Concerning Study 24.—Russett's descriptive emphasis is strong enough here so that I believe "descriptive," in this case, is the proper classification: "Factor analysis is then a data reduction technique . . . the procedure is inductive in that it involves no prior specification of the groups to be looked for nor is even the number of such groups specified in advance."[67]

Concerning Study 25.—Sawyer also seems to be going two ways. In describing factor analysis, he argues, "Factor analysis is specifically designed to solve problems just like the present one. It identifies the minimum number of dimensions that summarize, to any given extent,

62. Ibid., p. 70.
63. Rummel 1969, p. 132.
64. Ibid., p. 134.

65. Russett 1965, p. 320.
66. Ibid., p. 327.
67. Russett 1966, pp. 328–29.

the variation of a larger number of original variables."[68] Later, however, in his discussion of his "size" factor he maintains, "The conceptual distinction among these variables emphasizes that their empirical relations are more than definition or artifact but represent, instead, part of the determinant system constituting the structure of societies."[69] Still later he compares his factors with those found in a number of other studies and concludes, "Thus, size, wealth, and politics account for substantial portions of the variance, not only in the present analysis of 236 variables but also in studies with fewer, and less broadly representative variables."[70]

Concerning Studies 26–28.—Tanter sets about to replicate the work of Rummel in respect to measures of domestic and foreign conflict, using data from a later time base. The distinction between the use of factor analysis as a search for causes and as a technique of data reduction is not discussed by Tanter nor does he discuss the "meaning" that he gives to his factor labels. In the absence of such a discussion, I have put a question mark, although the "cause orientation" seems to be considerably stronger than the "descriptive orientation."

Concerning rotation.—"Orth." equals Orthogonal, "Obliq." equals Oblique.

Concerning matrix presentation.—I have left a blank where appropriate when the unrotated and rotated matrices are treated as separate studies.

Concerning missing data.—I have put a question mark if I suspect missing data, but the presentation is not completely clear on this point.

Concerning the percentage of missing data.—I have put a question mark if I suspect missing data or if missing data are indicated but the per cent is not given.

Concerning how missing data handled.—I have put a question mark if there is the possibility that there are missing data but this point is not discussed. "Est." means that the missing data were estimated and the techniques are given as "Mean," "Regression," or "Other." "Drop" means that the no-information subjects were dropped from the correlation matrix.

Concerning data transformation.—"Rank" refers to some ranking scheme other than rank numbers. "Rank No." means the conversion was to rank numbers. "No" means that no conversion was attempted.

68. Sawyer 1967, p. 147.
69. Ibid., p. 152.
70. Ibid., p. 156.

"Logs & Other" indicates the kind of conversion that was attempted. A question mark indicates that I am not sure if the data were transformed. "Yes?" indicates the data were transformed but the technique was not made clear.

Concerning the smallest N in the correlation matrix.—A question mark indicates either that the N of some of the correlations may be smaller than the largest N, or that it is definitely smaller than N but the figure is not given.

Concerning the inflation of the communalities and variance explained.—This refers only to those studies where subjects are dropped in the correlation matrix, so that all other cases are classified as not applicable (NA).

Concerning the calculation of factor scores.—"Comp." means a complete method was used. "Incomp." means an incomplete method was used. If $F = ZA (A'A)^{-1}$ was used but certain loadings were set to zero, I treated this as an incomplete method because the correlation of the "factor scores" with the original variables will not reproduce the factor structure. This only occurs in the case of the rotated factor scores computed by Alker and Russett in their voting studies (see Studies 5, 7, 9, and 11). "No" means factor scores were not computed. "Yes?" means factor scores were computed but the methods were not made clear.

In the case of the Sawyer and Tanter studies, where they have allowed the heaviest loading variables to represent the dimensions, I have viewed this as an incomplete method of factor score calculation.

Concerning how far out factors are computed.—A question mark indicates that this information was not given.

Question marks, not applicable (NA), Yes, and No may be interpreted as before in the next seven categories of analysis.

In the last two categories of analysis.—Concerning the homo-heterogeneity of variables, I have attempted to indicate whether the author seems to be working basically in one conceptual class, such as "conflict variables," or whether he views himself as working with variables that cut across classes, such as "social," "economic," "political," etc.

DISCUSSION

Table 28[71] lends itself to numerous ways of classifying existing studies. For example, there are studies where the N is relatively small, such as Studies 4, 5, 14, and 15. Obviously for such studies a large

71. Appendix G gives the factor names generated in each study.

number of states will not have "scale values" even if factor scores are computed, and this, of course, limits usage if factor scores are to be used as predictors.

Only two studies exhibit the scope of the present study where the N is quite large (over 100) and a large number of heterogeneous variables are considered. These are Studies 12 and 16. Neither, however, computes factor scores. In this sense, then, this study is "unique" compared to all other studies considered here.

The table also highlights some of the problems concerning the uses of factor analysis. If we accept Cattell's and Thurstone's arguments that communalities must be inserted in the principal diagonal, if we are searching for "causes," the following studies—2–12, 16–23, and 26–28—seem to use methods most appropriate to data reduction, but either imply they are engaged in a search for causes or are unclear in their presentation on this point.

It is apparent that the vast majority of studies enter unities in the principal diagonal and rotate to an orthogonal solution, according to Kaiser's varimax criterion. The resulting matrix is frequently either assumed to be "simple structure" or it is compared with an oblique solution, and one or the other is concluded to be "simple structure" in spite of Cattell's protestations to the contrary. Surprisingly, Cattell doesn't even follow his own (admittedly later) advice.[72] That is, it will be recalled that he maintains that at least six matrices need to be interpreted in the identification of causes, and yet in Studies 13, 14, and 15 he presents only one oblique matrix for interpretation. The same is true (only one rotated matrix presented where rotation occurs) in all the remaining studies. Also, in all other studies, just one of the oblique rotating techniques is adopted (where oblique rotation occurs) and the month or two search for "simple structure," which Cattell insists is necessary, is not followed. For example, Tanter employs the biquartimin rotation technique. Because the resulting factors he generates using this technique are "basically" orthogonal, he concludes that the factors are not related. Yet, as he must obviously recognize from this passage, "The *biquartimin* solution is selected over the *quartimin* or *covarimin* because the quartimin solution is generally biased toward factor axes which are too highly correlated, while the covarimin is almost invariably biased toward factor axes which are too orthogonal,"[73] the degree of orthogonality is a function, in part, of the

72. See notes 16, 38, and 39.
73. Tanter 1966, p. 52.

technique employed. In other words, one can apply here the same kind of criticism that is sometimes leveled against those who use Kaiser's varimax criterion and conclude that the "underlying" factors are orthogonal. As I have indicated, Cattell maintains that none of the present solutions is adequate, either in the sense of allowing a proper interpretation of factors or in providing "simple structure" as he defines it. Thus, the conclusions drawn by Tanter and others, using such "incomplete" procedures, seem to be unwarranted *if we accept Cattell's arguments.*

All of this *is not to say* that I consider the bulk of the above studies to be meaningless. Quite the contrary: I view the propensity toward quantification as one of the most healthy developments in the discipline of international relations. Rather, it is simply to point out that there seems to be some confusion over the use of the factor analytic technique as a search for "causes" as opposed to its use as a technique of descriptive and/or data reduction. In developing these distinctions I have *allowed the "chief spokesman" for that group of factor analysts, who view factor analysis as a useful tool in the search for causes, to establish the guidelines in respect to proper procedures.* My observation is, simply, that if the Cattell-Thurstone school is correct, then many of these studies may follow inappropriate procedures and possibly arrive at erroneous conclusions. If, as a line of defense, some authors claim that in fact they are only engaged in the business of description and/or data reduction, then a number of studies are "incomplete" in the sense of not generating factor scores, the final stage of data reduction, or are misleading because they do not use complete methods of factor score computation.

In conclusion, I have attempted to distinguish between the use of factor analysis as a search for causes and as a technique for descriptive and/or data reduction. One set of techniques and procedures seems most appropriate in the former case and another set in the latter. This study has clearly opted to follow the descriptive and/or data reduction school and in doing so has developed scale values for 129 political units on 19 factor dimensions. The relationship of these scale values to the original 91 variables has been made explicit and the use of the scales as predictive indices, as well as devices for classifying the units, has been indicated. To justify this expenditure of effort and to set this study apart from other apparently similar studies, some of the problems concerning the use of factor analysis in international relations have been identified and suggestions have been made concerning both procedures and interpretation.

Tables

TABLE 1
Correlation Matrix

#	Variable											
01	MANY INHABITANTS PER PHYSICIAN	1.00 / 0.54	-0.69 / 0.19	-0.49 / -0.11	-0.49 / -0.22	-0.36 / -0.06	0.48 / -0.39	0.09 / -0.14	0.46 / 0.0	-0.54 / 0.32	-0.41 / -0.34	-0.46 / 0.38
02	BIG GROSS NATIONAL PRODUCT	0.54 / 0.04	0.19 / 0.19	-0.18 / 0.51	-0.22 / -0.42	-0.40 / -0.29	0.53 / -0.54	0.58 / 0.14	0.30 / -0.36	0.58 / -0.57	-0.64 / 0.21	0.58 / -0.39
03	LARGE PERCENTAGE INCREASE IN POPULATION	-0.49 / 0.77	-0.22 / -0.06	-0.39 / 0.58	-0.53 / 0.60	-0.64 / -0.13	0.25 / 0.32	0.47 / 0.69	0.22 / 0.57	-0.13 / 0.10		
04	LARGE PERCENTAGE CHRISTIANS OF TOTAL POPULATION	-0.49 / -0.09	-0.28 / 0.42	-0.13 / -0.06	0.17 / 1.00	0.52 / -0.09	0.10 / -0.07	0.47 / -0.61	0.22 / -0.42	0.25 / -0.63		
05	BIG POPULATION PER SQUARE KILOMETER	-0.36 / -0.14	-0.06 / 0.02	1.00 / -0.16	-0.03 / 0.10	-0.64 / -0.66	0.14 / 0.10	-0.25 / -0.10	-0.34 / -0.05	0.05 / 0.12		
06	BIG AREA	0.05 / 0.05	0.34 / 0.42	-0.07 / 0.34	0.23 / 0.07	0.14 / -0.06	0.26 / -0.13	0.31 / -0.19				
07	LARGE PERCENTAGE EXPORT OF TOTAL TO US	0.47 / 0.69	0.25 / 0.59	0.39 / 0.60	0.22 / 0.57	-0.25 / 0.01	1.00 / -0.01	0.26 / 0.21	0.14 / -0.06	0.82 / 0.08	0.17 / -0.33	
08	LARGE PERCENTAGE IMPORT OF TOTAL TO US	0.22 / 0.57	-0.06 / -0.13	0.36 / -0.22	-0.34 / -0.07	0.26 / 0.21	1.00 / -0.22	0.82 / 0.10	0.57 / -0.28	0.10 / -0.19		
09	LARGE NUMBER OF PERSONS IN UN MISSIONS	-0.54 / 0.56	-0.01 / 0.05	0.25 / 0.32	0.05 / -0.45	0.31 / -0.37	0.17 / -0.41	0.10 / -0.28	1.00 / -0.40			
10	RECENTLY ENTERED UN	-0.41 / 0.09	-0.54 / -0.45	-0.41 / -0.57	0.07 / 0.26	-0.17 / -0.42	-0.56 / 0.54	-0.49 / 0.37	-0.42 / 0.39			
11	ALLIED WITH USSR	0.04 / 0.51	-0.14 / -0.18	-0.34 / -0.64	0.23 / 0.07	-0.20 / 0.21	-0.60 / 0.24	-0.55 / 0.04	-0.22 / 0.33			
12	LARGE DOLLAR IMPORT FROM US	0.19 / -0.22	1.00 / -0.40	-0.13 / 0.53	0.42 / 0.94	-0.06 / -0.33	0.29 / -0.31	0.59 / -0.40	0.60 / -0.34	0.36 / -0.31		
13	LARGE DOLLAR EXPORT TO US	0.30 / 0.58	-0.64 / 0.34	-0.15 / 0.94	0.52 / 1.00	-0.16 / -0.31	0.34 / -0.36	0.69 / -0.50	0.57 / -0.28	0.32 / -0.33		
14	SMALL POPULATION	0.21 / 0.36	-0.64 / 0.07	0.09 / -0.33	0.10 / -0.31	-0.03 / 1.00	-0.66 / 0.10	0.01 / 0.21	-0.07 / 0.26	-0.45 / 0.18		
15	SMALL PER CAPITA GROSS NATIONAL PRODUCT	0.24 / -0.74	0.26 / 0.07	-0.33 / 0.28	-0.79 / 0.55	-0.31 / 0.10	-0.66 / 0.10	-0.10 / 0.87	0.21 / 0.68	-0.37 / 0.36		
16	NON-WESTERN	0.42 / 0.21	0.21 / -0.22	-0.28 / 0.54	-0.61 / -0.36	0.10 / 0.10	-0.10 / 1.00	-0.01 / 0.87	0.21 / 0.60	-0.41 / 0.38		
17	POLITICALLY NOT MODERN	0.54 / -0.73	0.24 / -0.12	-0.39 / -0.37	-0.61 / -0.50	-0.10 / 0.21	-0.06 / 0.87	-0.22 / 1.00	0.10 / 0.60	-0.28 / 0.33		
18	LARGE DISTANCE FROM US	-0.56 / -0.60	0.28 / -0.12	0.61 / -0.61	-0.42 / -0.28	-0.63 / -0.33	-0.13 / 0.68	-0.19 / 0.36	-0.33 / 0.36	0.08 / 0.60	0.12 / 1.00	-0.40 / 1.00
19	LARGE DISTANCE FROM USSR	0.51 / 0.45	-0.39 / 0.31	-0.19 / 0.38	0.28 / 0.21	0.12 / 0.18	-0.02 / 0.21	0.51 / 0.28	0.54 / 0.31	-0.23 / -0.08		
20	LARGE DISTANCE FROM CHINA	0.14 / -0.05	0.69 / 1.00	-0.18 / -0.40	0.69 / 0.36	-0.38 / 0.28	0.00 / -0.22	0.45 / -0.12	0.44 / -0.19	-0.01 / 0.47		
21	MANY MEN UNDER ARMS	-0.49 / -0.55	-0.45 / -0.13	0.77 / -0.29	-0.06 / 0.32	0.17 / -0.79	0.42 / -0.28	-0.07 / -0.39	-0.06 / -0.37	0.56 / -0.21		

TABLE 2
Unrotated Loadings, Computed Out to 3, 1, and 0 Eigenvalues

Unrotated factor loadings matrix/ 1.0

						ROW SQ SUM
01	MANY INHABITANTS PER PHYSICIAN	0.76	-0.36	-0.31	0.08	0.80
02	BIG GROSS NATIONAL PRODUCT	-0.84	0.40	-0.24	-0.08	0.93
03	LARGE PERCENTAGE INCREASE IN POPULATION	0.34	-0.51	-0.37	0.05	0.52
04	LARGE PERCENTAGE CHRISTIANS OF TOTAL POPULATION	-0.66	-0.27	0.58	-0.04	0.85
05	BIG POPULATION PER SQUARE KILOMETER	0.14	0.44	0.13	0.78	0.84
06	BIG AREA	-0.39	-0.04	-0.64	-0.62	0.95
07	LARGE PERCENTAGE EXPORT OF TOTAL TO US	-0.47	-0.74	-0.08	0.24	0.84
08	LARGE PERCENTAGE INPORT OF TOTAL TO US	-0.31	-0.80	-0.31	0.15	0.86
09	LARGE NUMBER OF PERSONS IN UN MISSIONS	-0.60	0.20	-0.23	0.05	0.46
10	RECENTLY ENTERED UN	0.74	0.15	0.08	-0.27	0.65
11	ALLIED WITH USSR	0.54	0.55	0.10	-0.16	0.62
12	LARGE DOLLAR INPORT FROM US	-0.74	-0.33	-0.21	0.27	0.78
13	LARGE DOLLAR EXPORT TO US	-0.79	-0.36	-0.14	0.19	0.81
14	SMALL POPULATION	0.47	-0.35	0.69	0.09	0.83
15	SMALL PER CAPITA GROSS NATIONAL PRODUCT	0.72	-0.32	-0.46	0.19	0.88
16	NON-WESTERN	0.80	-0.31	-0.35	-0.09	0.87
17	POLITICALLY NAT. MODERN	0.61	-0.35	-0.23	0.12	0.57
18	LARGE DISTANCE FROM US	0.59	0.15	-0.18	0.17	0.43
19	LARGE DISTANCE FROM USSR	0.05	-0.86	0.14	-0.16	0.78
20	LARGE DISTANCE FROM CHINA	-0.33	-0.64	0.45	-0.28	0.81
21	MANY MEN UNDER ARMS	-0.57	0.49	-0.51	0.10	0.83

COLUMN SQ SUM
7.23 4.47 2.68 1.52

Unrotated factor loadings matrix/ 3.0

				ROW SQ SUM
01	MANY INHABITANTS PER PHYSICIAN	0.76	-0.36	0.70
02	BIG GROSS NATIONAL PRODUCT	-0.84	0.40	0.87
03	LARGE PERCENTAGE INCREASE IN POPULATION	0.34	-0.51	0.38
04	LARGE PERCENTAGE CHRISTIANS OF TOTAL POPULATION	-0.66	-0.27	0.50
05	BIG POPULATION PER SQUARE KILOMETER	0.14	0.44	0.21
06	BIG AREA	-0.39	-0.04	0.16
07	LARGE PERCENTAGE EXPORT OF TOTAL TO US	-0.47	-0.74	0.77
08	LARGE PERCENTAGE INPORT OF TOTAL TO US	-0.31	-0.80	0.74
09	LARGE NUMBER OF PERSONS IN UN MISSIONS	-0.60	0.20	0.40
10	RECENTLY ENTERED UN	0.74	0.15	0.57
11	ALLIED WITH USSR	0.54	0.55	0.59
12	LARGE DOLLAR INPORT FROM US	-0.74	-0.33	0.66
13	LARGE DOLLAR EXPORT TO US	-0.79	-0.36	0.76
14	SMALL POPULATION	0.47	-0.35	0.34
15	SMALL PER CAPITA GROSS NATIONAL PRODUCT	0.72	-0.32	0.63
16	NON-WESTERN	0.80	-0.31	0.74
17	POLITICALLY NAT. MODERN	0.61	-0.35	0.50
18	LARGE DISTANCE FROM US	0.59	0.15	0.37
19	LARGE DISTANCE FROM USSR	0.05	-0.86	0.73
20	LARGE DISTANCE FROM CHINA	-0.33	-0.64	0.52
21	MANY MEN UNDER ARMS	-0.57	0.49	0.56

COLUMN SQ SUM
7.23 4.47

Unrotated factor loadings matrix/ 0.0

																						ROW SQ SUM
01	MANY INHABITANTS PER PHYSICIAN	0.76	-0.36	-0.31	0.08	-0.02	-0.27	0.00	-0.13	0.02	-0.18	-0.05	-0.19	0.13	0.05	0.11	-0.06	-0.03	0.04	-0.04	0.01	-0.01	1.00
02	BIG GROSS NATIONAL PRODUCT	-0.84	0.40	-0.24	-0.08	0.04	-0.11	-0.05	0.06	0.01	0.10	-0.11	-0.03	0.07	-0.01	0.08	0.09	0.04	-0.02	0.00	-0.01	-0.08	1.00
03	LARGE PERCENTAGE INCREASE IN POPULATION	0.34	-0.51	-0.37	0.05	-0.36	0.26	-0.44	-0.12	0.10	0.03	-0.24	0.06	0.01	0.04	0.03	0.03	0.03	-0.03	-0.00	-0.02	1.00	
04	LARGE PERCENTAGE CHRISTIANS OF TOTAL POPULATION	-0.66	-0.27	0.58	-0.04	-0.16	-0.15	-0.05	-0.08	0.05	0.12	-0.09	-0.03	-0.09	0.21	-0.07	-0.06	0.03	0.01	0.08	0.05	0.05	0.00	0.02	1.00
05	BIG POPULATION PER SQUARE KILOMETER	0.14	0.44	0.13	0.78	-0.24	-0.24	-0.09	0.09	-0.02	0.09	-0.09	-0.02	-0.05	-0.01	0.02	-0.00	0.00	0.13	0.06	0.03	-0.01	0.03	1.00
06	BIG AREA	-0.39	-0.04	-0.64	-0.62	0.10	-0.03	-0.04	-0.05	0.03	-0.04	0.05	-0.01	0.02	-0.00	-0.09	0.04	-0.08	0.04	0.04	-0.03	0.01	1.00
07	LARGE PERCENTAGE EXPORT OF TOTAL TO US	-0.47	-0.74	-0.08	0.24	0.02	0.10	-0.12	-0.19	0.07	0.14	-0.19	-0.05	0.12	-0.04	-0.02	-0.06	0.04	0.02	0.00	0.03	-0.04	1.00
08	LARGE PERCENTAGE INPORT OF TOTAL TO US	-0.31	-0.80	-0.31	0.15	0.02	0.07	-0.07	-0.03	0.24	0.05	0.03	0.15	-0.02	-0.14	-0.14	-0.02	-0.06	0.04	0.02	0.00	-0.03	-0.04	1.00
09	LARGE NUMBER OF PERSONS IN UN MISSIONS	-0.60	0.20	-0.23	0.05	-0.42	0.12	0.26	0.14	0.46	-0.12	-0.15	-0.05	0.02	-0.01	-0.00	0.03	-0.01	-0.00	0.01	0.01	-0.02	1.00
10	RECENTLY ENTERED UN	0.74	0.15	0.08	-0.27	0.04	-0.44	-0.00	0.23	0.20	-0.01	-0.17	0.05	-0.01	-0.12	-0.12	-0.05	-0.02	0.01	0.01	-0.01	0.01	1.00
11	ALLIED WITH USSR	0.54	0.55	0.10	-0.16	-0.08	-0.11	-0.04	-0.42	0.25	0.30	0.04	0.14	0.02	-0.01	0.09	-0.01	-0.01	-0.00	-0.00	-0.01	0.01	1.00
12	LARGE DOLLAR INPORT FROM US	-0.74	-0.33	-0.21	0.27	0.24	-0.27	-0.11	0.02	0.11	-0.18	0.04	0.13	-0.07	-0.01	0.05	-0.03	0.00	-0.01	-0.04	-0.06	0.02	1.00
13	LARGE DOLLAR EXPORT TO US	-0.79	-0.36	-0.14	0.19	0.30	-0.23	0.01	0.01	0.08	-0.01	-0.07	0.13	0.06	0.08	0.02	0.02	-0.02	-0.04	0.03	0.07	0.02	1.00
14	SMALL POPULATION	0.47	-0.35	0.69	0.09	0.10	0.22	0.05	-0.00	0.10	-0.20	0.02	0.16	0.14	-0.00	0.01	0.00	0.06	0.01	-0.00	-0.02	1.00	
15	SMALL PER CAPITA GROSS NATIONAL PRODUCT	0.72	-0.32	-0.46	0.19	-0.16	-0.15	0.05	-0.11	-0.07	0.05	0.05	0.05	0.06	0.03	-0.13	0.09	0.09	-0.06	-0.05	-0.00	0.00	1.00
16	NON-WESTERN	0.80	-0.31	-0.35	-0.09	-0.04	-0.12	-0.04	-0.01	-0.03	-0.10	0.21	0.09	-0.08	0.13	-0.00	0.07	-0.07	0.05	0.07	-0.01	-0.02	1.00
17	POLITICALLY NOT MODERN	0.61	-0.35	-0.23	0.12	0.06	0.03	0.59	0.15	-0.06	0.15	-0.08	0.05	-0.04	0.02	0.08	-0.07	0.02	-0.04	0.02	-0.01	-0.01	1.00
18	LARGE DISTANCE FROM US	0.59	0.15	-0.18	0.17	0.59	0.17	-0.16	0.13	0.24	0.18	0.17	-0.13	0.01	0.08	-0.05	-0.02	0.02	-0.01	0.00	-0.01	-0.01	1.00
19	LARGE DISTANCE FROM USSR	0.05	-0.86	0.14	-0.16	-0.05	-0.04	-0.08	0.32	0.00	0.29	0.01	0.07	0.03	-0.01	-0.04	0.08	0.08	-0.01	0.07	0.05	0.01	0.02	1.00
20	LARGE DISTANCE FROM CHINA	-0.33	-0.64	0.45	-0.28	-0.14	-0.23	-0.07	0.03	-0.03	0.05	0.27	-0.10	0.11	-0.03	0.01	-0.04	0.03	-0.08	0.05	-0.02	0.01	1.00
21	MANY MEN UNDER ARMS	-0.57	0.49	-0.51	0.10	-0.13	0.01	-0.02	0.12	-0.18	0.12	0.00	0.20	0.13	0.14	0.02	-0.06	-0.12	-0.02	0.05	-0.01	0.00	-0.01	1.00

COLUMN SQ SUM
7.23 4.47 2.68 1.52 0.99 0.76 0.70 0.57 0.47 0.43 0.41 0.20 0.14 0.12 0.10 0.07 0.06 0.04 0.02 0.01 0.01

TABLE 3
Rotated Loadings, Computed Out to 3, 1, and 0 Eigenvalues

Rotated factor loadings matrix/1.0

	F1	F2	F3	F4	ROW SQ SUM
01 MANY INHABITANTS PER PHYSICIAN	0.85	0.10	0.25	-0.01	0.80
02 BIG GROSS NATIONAL PRODUCT	-0.60	-0.20	-0.72	-0.07	0.93
03 LARGE PERCENTAGE INCREASE IN POPULATION	0.65	-0.25	0.10	-0.15	0.51
04 LARGE PERCENTAGE CHRISTIANS OF TOTAL POPULATION	-0.72	-0.47	0.32	-0.11	0.85
05 BIG POPULATION PER SQUARE KILOMETER	-0.02	0.13	-0.07	0.91	0.84
06 BIG AREA	0.02	-0.12	-0.63	-0.74	0.95
07 LARGE PERCENTAGE EXPORT OF TOTAL TO US	0.03	-0.90	0.10	-0.09	0.84
08 LARGE PERCENTAGE INPORT OF TOTAL TO US	0.28	-0.86	0.02	-0.22	0.86
09 LARGE NUMBER OF PERSONS IN UN MISSIONS	-0.35	-0.25	-0.53	-0.01	0.46
10 RECENTLY ENTERED UN	0.38	0.64	0.30	-0.10	0.65
11 ALLIED WITH USSR	0.10	0.77	0.03	0.12	0.62
12 LARGE DOLLAR INPORT FROM US	-0.24	-0.79	-0.31	0.02	0.78
13 LARGE DOLLAR EXPORT TO US	-0.31	-0.80	-0.26	-0.06	0.81
14 SMALL POPULATION	-0.06	0.09	0.69	0.14	0.83
15 SMALL PER CAPITA GROSS NATIONAL PRODUCT	0.93	0.05	0.09	-0.07	0.88
16 NON-WESTERN	0.87	0.21	-0.15		0.57
17 POLITICALLY NAT. MODERN	0.71	0.21	0.24	0.07	0.57
18 LARGE DISTANCE FROM US	0.49	0.36	0.03	0.25	0.43
19 LARGE DISTANCE FROM USSR	0.24	-0.51	0.55	-0.40	0.78
20 LARGE DISTANCE FROM CHINA	-0.31	-0.48	0.53	-0.44	0.81
21 MANY MEN UNDER ARMS	-0.26	-0.08	-0.66	0.10	0.83
COLUMN SQ SUM	5.13	4.95	3.86	1.96	

Rotated factor loadings matrix/3.0

	F1	F2	ROW SQ SUM
01 MANY INHABITANTS PER PHYSICIAN	0.83	0.08	0.70
02 BIG GROSS NATIONAL PRODUCT	-0.93	-0.09	0.87
03 LARGE PERCENTAGE INCREASE IN POPULATION	0.55	-0.26	0.38
04 LARGE PERCENTAGE CHRISTIANS OF TOTAL POPULATION	-0.42	-0.57	0.50
05 BIG POPULATION PER SQUARE KILOMETER	-0.11	0.44	0.21
06 BIG AREA	-0.32	-0.24	0.16
07 LARGE PERCENTAGE EXPORT OF TOTAL TO US	-0.02	-0.88	0.77
08 LARGE PERCENTAGE INPORT OF TOTAL TO US	0.15	-0.85	0.74
09 LARGE NUMBER OF PERSONS IN UN MISSIONS	-0.62	-0.14	0.40
10 RECENTLY ENTERED UN	0.56	0.51	0.57
11 ALLIED WITH USSR	0.17	0.75	0.59
12 LARGE DOLLAR INPORT FROM US	-0.46	-0.67	0.66
13 LARGE DOLLAR EXPORT TO US	-0.49	-0.72	0.76
14 SMALL POPULATION	0.58	-0.05	0.34
15 SMALL PER CAPITA GROSS NATIONAL PRODUCT	0.79	0.10	0.63
16 NON-WESTERN	0.85	0.15	0.74
17 POLITICALLY NAT. MODERN	0.70	0.02	0.50
18 LARGE DISTANCE FROM US	0.43	0.43	0.37
19 LARGE DISTANCE FROM USSR	0.49	-0.70	0.73
20 LARGE DISTANCE FROM CHINA	0.55	-0.72	0.52
21 MANY MEN UNDER ARMS	-0.74	0.12	0.56
COLUMN SQ SUM	6.50	5.21	

Rotated factor loadings matrix/0.0

	F1	F2	F3	F4	F5	F6	F7	F8	F9	F10	F11	F12	F13	F14	F15	F16	F17	F18	F19	F20	F21	ROW SQ SUM
01 MANY INHABITANTS PER PHYSICIAN	0.88	0.03	0.20	0.01	0.10	0.09	-0.03	-0.17	-0.15	0.02	0.02	0.31	0.10	0.02	-0.06	-0.12	-0.01	-0.02	-0.00	-0.01	0.03	1.00
02 BIG GROSS NATIONAL PRODUCT	-0.56	-0.27	-0.67	-0.06	-0.12	-0.17	0.01	0.03	0.17	-0.03	0.07	0.03	0.03	0.07	-0.01	0.14	-0.03	-0.02	-0.04	-0.00	-0.01	1.00
03 LARGE PERCENTAGE INCREASE IN POPULATION	0.42	-0.03	0.06	-0.06	0.02	0.03	0.01	0.07	-0.00	-0.89	0.14	0.01	0.01	-0.01	-0.07	-0.01	0.00	0.00	0.00	-0.00	0.00	1.00
04 LARGE PERCENTAGE CHRISTIANS OF TOTAL POPULATION	-0.54	-0.31	0.15	0.05	-0.42	0.07	-0.09	-0.02	0.07	0.09	0.11	-0.00	0.42	-0.07	-0.02	-0.07	0.00	0.00	-0.00	0.00	0.00	1.00
05 BIG POPULATION PER SQUARE KILOMETER	0.01	0.07	-0.08	0.96	-0.04	-0.03	-0.05	0.04	0.04	-0.22	0.04	-0.00	0.01	0.02	0.01	0.00	0.22	-0.00	-0.01	-0.01	0.00	1.00
06 BIG AREA	0.01	-0.17	-0.59	-0.74	-0.05	-0.04	-0.03	0.10	0.08	-0.04	0.00	0.04	0.02	0.01	0.01	0.06	-0.00	-0.00	-0.02	0.00	0.00	1.00
07 LARGE PERCENTAGE EXPORT OF TOTAL TO US	-0.03	-0.72	0.09	-0.10	-0.16	0.29	0.18	0.05	0.05	0.25	-0.44	0.06	-0.05	-0.03	0.06	-0.03	0.01	-0.01	0.00	0.00	0.00	1.00
08 LARGE PERCENTAGE INPORT OF TOTAL TO US	0.25	-0.69	0.01	-0.21	-0.05	0.36	-0.05	0.08	0.05	-0.20	-0.09	-0.03	0.02	0.01	-0.36	-0.01	0.01	-0.01	0.00	0.00	0.00	1.00
09 LARGE NUMBER OF PERSONS IN UN MISSIONS	-0.26	-0.14	-0.32	-0.10	-0.17	0.10	0.07	0.04	0.88	-0.06	0.00	-0.01	-0.00	-0.00	0.00	-0.01	-0.00	0.00	-0.00	0.00	0.00	1.00
10 RECENTLY ENTERED UN	0.35	0.39	0.17	0.02	0.13	-0.79	-0.07	-0.15	-0.13	0.00	0.02	0.03	0.01	-0.01	-0.01	0.01	-0.00	-0.00	0.00	0.00	0.00	1.00
11 ALLIED WITH USSR	0.12	0.52	0.03	0.10	0.12	-0.16	-0.78	-0.05	-0.05	0.02	0.11	0.02	0.00	0.01	0.01	0.00	0.00	-0.00	0.01	-0.00	0.00	1.00
12 LARGE DOLLAR INPORT FROM US	-0.17	-0.90	-0.22	-0.09	-0.07	0.04	-0.17	0.16	0.10	0.03	0.05	0.14	-0.01	-0.01	0.01	-0.01	-0.09	0.01	-0.00	-0.00	0.00	1.00
13 LARGE DOLLAR EXPORT TO US	-0.28	-0.89	-0.22	-0.09	-0.06	-0.02	0.05	-0.05	0.13	-0.08	0.08	-0.01	-0.03	-0.00	-0.01	0.09	-0.02	0.00	0.00	0.00	0.00	1.00
14 SMALL POPULATION	0.04	0.12	0.96	0.03	0.09	-0.02	0.05	0.21	0.04	-0.20	-0.08	0.01	0.13	0.00	-0.01	-0.13	0.04	0.00	-0.01	0.00	0.00	1.00
15 SMALL PER CAPITA GROSS NATIONAL PRODUCT	0.91	0.05	0.03	0.09	-0.10	0.08	-0.04	-0.06	-0.03	-0.18	0.14	-0.11	-0.01	-0.01	-0.03	-0.01	0.04	-0.15	-0.00	0.01	0.00	1.00
16 NON-WESTERN	0.69	0.22	0.14	-0.10	0.12	0.12	-0.06	-0.08	-0.03	-0.18	0.14	0.02	0.01	-0.01	-0.01	-0.01	0.00	0.04	0.00	-0.00	0.00	1.00
17 POLITICALLY NAT. MODERN	0.54	0.07	0.19	-0.03	0.12	0.10	0.10	0.80	-0.09	-0.01	0.02	0.00	0.00	-0.01	0.01	0.01	0.00	0.00	0.00	-0.00	0.00	1.00
18 LARGE DISTANCE FROM US	0.15	-0.21	-0.16	-0.06	0.23	-0.16	0.08	0.08	0.08	-0.18	0.02	0.06	-0.03	-0.03	0.00	0.02	0.03	0.00	0.19	-0.02	0.00	1.00
19 LARGE DISTANCE FROM USSR	0.15	-0.23	-0.16	0.01	0.84	0.22	-0.09	0.15	-0.06	0.04	0.06	-0.36	-0.03	-0.06	-0.32	-0.08	-0.06	0.05	0.02	0.00	0.01	1.00
20 LARGE DISTANCE FROM CHINA	-0.10	-0.24	0.22	-0.17	-0.28	0.04	-0.04	-0.15	-0.11	0.08	0.06	-0.08	0.05	-0.03	-0.02	0.05	0.19	-0.01	-0.14	0.00	0.01	1.00
21 MANY MEN UNDER ARMS	-0.22	-0.24	-0.80	0.08	0.19	-0.08	-0.21	0.06	0.06	-0.21	0.08	0.04	-0.05	-0.08	0.06	-0.08	-0.01	-0.37	-0.01	-0.00	0.00	1.00
COLUMN SQ SUM	3.96	3.45	2.84	1.64	1.26	0.98	0.90	0.74	0.97	1.92	1.14	0.27	0.18	0.20	0.18	0.14	0.06	0.06	0.05	0.02	0.03	

TABLE 4
Factor Scores Computed by Formula F = ZA

Unrotated

01	ALBANIA	11.34	2.99
02	ARGENTINA	-7.22	-2.21
03	AUSTRALIA	-6.50	-1.83
04	BELGIUM	-8.56	-1.66
05	BRAZIL	-7.86	-4.51
06	BULGARIA	3.30	-7.54
07	CAMBODIA	14.25	-0.22
08	CANADA	-12.15	-3.08
09	CEYLON	10.62	-0.67
10	CHILE	-0.65	-5.23
11	COLOMBIA	-5.09	-5.81
12	COSTA RICA	0.41	-7.55
13	CZECHOSLOVAKIA	-2.30	7.83
14	DENMARK	-5.52	-1.76
15	ECUADOR	1.21	-7.81
16	FRANCE	-11.26	-2.27
17	INDIA	-3.19	2.60
18	IRAN	2.85	0.67
19	IRAQ	6.35	1.64
20	LEBANON	4.69	1.68
21	LIBERIA	7.85	-5.99
22	LIBYA	9.63	-2.74
23	MEXICO	-7.13	-6.64
24	MOROCCO	8.07	-0.58
25	NEPAL	14.00	-2.60
26	NETHERLANDS	-5.96	2.06
27	NEW ZEALAND	-3.40	-2.24
28	NORWAY	-4.31	1.46
29	PANAMA	-0.18	-8.06
30	POLAND	-3.80	5.71
31	SWEDEN	-5.30	2.95
32	THAILAND	5.90	-1.20
33	TUNESIA	9.08	1.68
34	TURKEY	-3.00	1.40
35	USSR	-5.21	8.42
36	UNITED STATES	-5.97	8.48
37	YUGOSLAVIA	-2.01	3.63
38	KENYA	9.59	-1.30
39	TAIWAN	1.61	-1.60

Rotated

01	ALBANIA	8.16	8.42
02	ARGENTINA	-5.04	-5.62
03	AUSTRALIA	-4.62	-4.92
04	BELGIUM	-8.19	-3.00
05	BRAZIL	-4.40	-7.92
06	BULGARIA	-1.07	8.16
07	CAMBODIA	12.31	7.17
08	CANADA	-8.85	-8.92
09	CEYLON	8.85	5.89
10	CHILE	-0.93	-7.85
11	COLOMBIA	-4.25	-6.60
12	COSTA RICA	-1.35	-7.60
13	CZECHOSLOVAKIA	4.25	5.31
14	DENMARK	-5.66	-1.34
15	ECUADOR	5.08	-6.06
16	FRANCE	-10.81	-3.87
17	INDIA	-4.08	0.58
18	IRAN	2.09	2.05
19	IRAQ	4.59	4.68
20	LEBANON	3.15	3.87
21	LIBERIA	9.82	-1.06
22	LIBYA	9.66	2.63
23	MEXICO	-2.67	-9.37
24	MOROCCO	7.21	3.67
25	NEPAL	10.60	9.53
26	NETHERLANDS	-6.17	-1.32
27	NEW ZEALAND	-1.75	-3.67
28	NORWAY	-4.44	-0.97
29	PANAMA	4.01	-6.99
30	POLAND	-6.20	2.92
31	SWEDEN	-6.06	-0.21
32	THAILAND	5.67	2.02
33	TUNESIA	6.91	6.13
34	TURKEY	-3.30	-0.35
35	USSR	-8.81	4.52
36	UNITED STATES	-9.49	4.17
37	YUGOSLAVIA	-3.60	2.07
38	KENYA	8.88	3.84
39	TAIWAN	2.20	-0.53

TABLE 5
Factor Scores Computed by Formula F = ZA, Standardized

Unrotated

01	ALBANIA	1.57	0.67
02	ARGENTINA	-1.00	-0.49
03	AUSTRALIA	-0.90	-0.41
04	BELGIUM	-1.18	0.37
05	BRAZIL	-1.09	-1.01
06	BULGARIA	0.46	1.69
07	CAMBODIA	1.97	-0.05
08	CANADA	-1.68	-0.69
09	CEYLON	1.47	0.10
10	CHILE	-0.67	-1.39
11	COLOMBIA	-0.70	-1.30
12	COSTA RICA	0.06	-1.69
13	CZECHOSLOVAKIA	-0.32	1.75
14	DENMARK	-0.76	0.39
15	ECUADOR	0.17	-1.75
16	FRANCE	-1.56	0.51
17	INDIA	-0.44	0.58
18	IRAN	0.39	0.15
19	IRAQ	0.88	0.37
20	LEBANON	0.65	0.38
21	LIBERIA	1.09	-1.34
22	LIBYA	1.33	-0.61
23	MEXICO	-0.98	-1.48
24	MOROCCO	1.11	-0.13
25	NEPAL	1.93	-0.60
26	NETHERLANDS	-0.82	0.46
27	NEW ZEALAND	-0.47	-0.50
28	NORWAY	-0.59	0.33
29	PANAMA	-0.02	-1.80
30	POLAND	-0.52	1.28
31	SWEDEN	-0.73	0.66
32	THAILAND	0.81	-0.27
33	TUNESIA	1.25	0.38
34	TURKEY	-0.41	0.31
35	USSR	-0.72	1.88
36	UNITED STATES	-0.82	1.90
37	YUGOSLAVIA	-0.28	0.81
38	KENYA	1.32	-0.29
39	TAIWAN	0.22	-0.36

Rotated

01	ALBANIA	1.23	1.57
02	ARGENTINA	-0.76	-1.05
03	AUSTRALIA	-0.70	-0.92
04	BELGIUM	-1.24	-0.56
05	BRAZIL	-0.66	-1.48
06	BULGARIA	-0.16	1.53
07	CAMBODIA	1.86	1.34
08	CANADA	-1.33	-1.67
09	CEYLON	1.34	1.10
10	CHILE	-0.14	-1.47
11	COLOMBIA	-0.20	-1.42
12	COSTA RICA	0.64	-1.17
13	CZECHOSLOVAKIA	-0.91	1.03
14	DENMARK	-0.85	-0.25
15	ECUADOR	0.77	-1.13
16	FRANCE	-1.63	-0.72
17	INDIA	-0.62	0.11
18	IRAN	0.32	0.38
19	IRAQ	0.69	0.68
20	LEBANON	0.48	0.72
21	LIBERIA	1.48	-0.20
22	LIBYA	1.46	0.49
23	MEXICO	-0.40	-1.75
24	MOROCCO	1.09	0.69
25	NEPAL	1.60	1.76
26	NETHERLANDS	-0.93	-0.18
27	NEW ZEALAND	-0.26	-0.61
28	NORWAY	-0.67	-0.18
29	PANAMA	0.61	-1.31
30	POLAND	-0.94	0.55
31	SWEDEN	-0.92	-0.04
32	THAILAND	0.86	0.38
33	TUNESIA	1.04	1.15
34	TURKEY	-0.50	-0.07
35	USSR	-1.33	0.84
36	UNITED STATES	-1.44	0.78
37	YUGOSLAVIA	-0.54	0.39
38	KENYA	1.34	-0.72
39	TAIWAN	0.33	-0.10

TABLE 6
Correlations of Factor Scores Computed by Formula F = ZA with Original Scores

Unrotated

	1	2	3	4	5	6	7	8	9	10	11	12	13	14	15	16	17	18	19	20	21
VAR 1	0.76	-0.84	0.34	-0.66	0.14	-0.39	-0.47	-0.31	-0.60	0.74	0.54	-0.74	-0.79	0.47	0.72	0.80	0.61	0.59	0.05	-0.33	-0.57
VAR 2	-0.36	0.40	-0.51	0.27	0.44	-0.04	-0.74	-0.80	0.20	0.15	0.55	-0.33	-0.36	-0.35	-0.32	-0.31	-0.35	0.15	-0.86	-0.64	0.49

Rotated

	1	2	3	4	5	6	7	8	9	10	11	12	13	14	15	16	17	18	19	20	21
VAR 1	0.83	-0.93	0.49	-0.52	-0.03	-0.36	-0.18	-0.01	-0.64	0.64	0.31	-0.58	-0.61	0.56	0.79	0.86	0.70	0.50	0.35	-0.08	-0.70
VAR 2	0.27	-0.30	-0.13	-0.65	0.41	-0.30	-0.86	-0.79	-0.28	0.62	0.77	-0.76	-0.81	0.08	0.28	0.34	0.18	0.52	-0.57	-0.69	-0.05

Original matrix A

```
 0.76  -0.36
-0.84   0.40
 0.34  -0.51
-0.66  -0.27
 0.14   0.44
-0.39  -0.04
-0.47  -0.74
-0.31  -0.80
-0.60   0.20
 0.74   0.15
 0.54   0.55
-0.74  -0.33
-0.79  -0.36
 0.47  -0.35
 0.72  -0.32
 0.80  -0.31
 0.61  -0.35
 0.59   0.15
 0.05  -0.86
-0.33  -0.64
-0.57   0.49
```

TABLE 7
Application of Formula $F = ZA\,(A'\,A)^{-1}$, Unrotated Case

A transposed

```
0.76  -0.84   0.34  -0.66   0.14  -0.39  -0.47  -0.31  -0.60   0.74   0.54  -0.74  -0.79   0.47
0.72   0.80   0.61   0.55   0.05  -0.33  -0.57
-0.36   0.40  -0.51  -0.27   0.44  -0.04  -0.74  -0.80   0.20   0.15   0.55  -0.33  -0.36  -0.35
-0.32  -0.31  -0.35   0.15  -0.86  -0.64   0.49
```

A transposed ° A

```
 7.23  -0.00
-0.00   4.47
```

Inverted matrix

```
0.14   0.0
0.0    0.22
```

Original matrix A

```
 0.83   0.08
-0.93  -0.09
 0.55  -0.26
-0.42  -0.57
-0.11   0.44
-0.32  -0.24
-0.02  -0.88
 0.15  -0.85
-0.62  -0.14
 0.56   0.51
 0.17   0.75
-0.46  -0.67
-0.49  -0.72
 0.58  -0.05
 0.79   0.10
 0.65   0.15
 0.70   0.02
 0.43   0.43
 0.49  -0.70
 0.05  -0.72
-0.74   0.12
```

TABLE 8
Application of Formula $F = ZA\,(A'\,A)^{-1}$, Rotated Case

A transposed

```
0.83  -0.93   0.55  -0.42  -0.11  -0.32  -0.02   0.15  -0.62   0.56   0.17  -0.46  -0.49   0.58
0.79   0.85   0.70   0.43   0.49   0.05  -0.74
0.08  -0.09  -0.26  -0.57   0.44  -0.24  -0.88  -0.85  -0.14   0.51   0.75  -0.67  -0.72  -0.05
0.10   0.15   0.02   0.43  -0.70  -0.72   0.12
```

A transposed ° A

```
6.50   1.22
1.22   5.21
```

Inverted matrix

```
 0.16  -0.04
-0.04   0.20
```

TABLE 9
Factor Scores Computed by Formula $F = ZA(A'A)^{-1}$

	Unrotated				Rotated	
01	ALBANIA	1.57	0.67	01	ALBANIA	1.00	1.38
02	ARGENTINA	-1.00	-0.49	02	ARGENTINA	-0.60	-0.94
03	AUSTRALIA	-0.90	-0.41	03	AUSTRALIA	-0.56	-0.81
04	BELGIUM	-1.18	0.37	04	BELGIUM	-1.21	-0.29
05	BRAZIL	-1.09	-1.01	05	BRAZIL	-0.41	-1.43
06	BULGARIA	0.46	1.69	06	BULGARIA	-0.48	1.68
07	CAMBODIA	1.97	-0.05	07	CAMBODIA	1.71	0.97
08	CANADA	-1.68	-0.69	08	CANADA	-1.08	-1.46
09	CEYLON	1.47	0.10	09	CEYLON	1.20	0.85
10	CHILE	-0.67	-1.39	10	CHILE	0.15	-1.54
11	COLOMBIA	-0.70	-1.30	11	COLOMBIA	0.07	-1.48
12	COSTA RICA	0.06	-1.69	12	COSTA RICA	0.92	-1.42
13	CZECHOSLOVAKIA	-0.32	1.75	13	CZECHOSLOVAKIA	-1.18	1.33
14	DENMARK	-0.76	0.39	14	DENMARK	-0.86	-0.06
15	ECUADOR	0.17	-1.75	15	ECUADOR	1.05	-1.41
16	FRANCE	-1.56	0.51	16	FRANCE	-1.59	-0.37
17	INDIA	-0.44	0.58	17	INDIA	-0.68	0.27
18	IRAN	0.39	0.15	18	IRAN	0.26	0.33
19	IRAQ	0.88	0.37	19	IRAQ	0.56	0.77
20	LEBANON	0.65	0.38	20	LEBANON	0.36	0.66
21	LIBERIA	1.09	-1.34	21	LIBERIA	1.62	-0.58
22	LIBYA	1.33	-0.61	22	LIBYA	1.46	0.16
23	MEXICO	-0.98	-1.48	23	MEXICO	-0.07	-1.78
24	MOROCCO	1.11	-0.13	24	MOROCCO	1.02	0.46
25	NEPAL	1.93	0.60	25	NEPAL	1.35	1.51
26	NETHERLANDS	-0.82	0.46	26	NETHERLANDS	-0.94	-0.03
27	NEW ZEALAND	-0.47	-0.50	27	NEW ZEALAND	-0.14	-0.67
28	NORWAY	-0.59	0.33	28	NORWAY	-0.68	-0.03
29	PANAMA	-0.02	-1.80	29	PANAMA	0.91	-1.56
30	POLAND	-0.52	1.28	30	POLAND	-1.11	0.82
31	SWEDEN	-0.73	0.66	31	SWEDEN	-0.97	0.19
32	THAILAND	0.81	-0.27	32	THAILAND	0.84	0.19
33	TUNESIA	1.25	0.38	33	TUNESIA	0.88	0.97
34	TURKEY	-0.41	0.31	34	TURKEY	-0.52	0.05
35	USSR	-0.72	1.88	35	USSR	-1.59	1.24
36	UNITED STATES	-0.82	1.90	36	UNITED STATES	-1.69	1.20
37	YUGOSLAVIA	-0.28	0.81	37	YUGOSLAVIA	-0.66	0.55
38	KENYA	1.32	-0.29	38	KENYA	1.28	0.44
39	TAIWAN	0.22	-0.36	39	TAIWAN	0.38	-0.19

TABLE 10
Correlations of Factor Scores Computed by Formula $F = ZA(A'A)^{-1}$ with Original Scores

Unrotated

		1	2	3	4	5	6	7	8	9
VAR	1	0.76	-0.84	0.34	-0.66	0.14	-0.39	-0.47	-0.31	-0.60
VAR	2	-0.36	0.40	-0.51	-0.27	0.44	-0.04	-0.74	-0.80	0.20

10	11	12	13	14	15	16	17	18	19	20	21
0.74	0.54	-0.74	-0.79	0.47	0.72	0.80	0.61	0.59	0.05	-0.33	-0.57
0.15	0.55	-0.33	-0.36	-0.35	-0.32	-0.31	-0.35	0.15	-0.86	-0.64	0.49

Rotated

		1	2	3	4	5	6	7	8	9
VAR	1	0.83	-0.93	0.55	-0.42	-0.11	-0.32	-0.02	0.15	-0.62
VAR	2	0.08	-0.09	-0.26	-0.57	0.44	-0.24	-0.88	-0.85	-0.14

10	11	12	13	14	15	16	17	18	19	20	21
0.56	0.18	-0.46	-0.49	0.58	0.79	0.85	0.70	0.43	0.49	0.05	-0.74
0.51	0.75	-0.67	-0.72	-0.05	0.10	0.15	0.02	0.43	-0.70	-0.72	0.12

TABLE 11
Simple Correlation of Rotated Factor Scores (1.0 Eigenvalue) with Radios per 1,000 Population

		R	R SQUARED
1 .	2	-0.88210	0.77810
1 .	3	-0.17706	0.03135
1 .	4	-0.09188	0.00844
1 .	5	-0.04679	0.00219

Total 0.82008

Where 1 equals radios per 1,000 population and 2-5 equals factor scores

TABLE 12
Multiple Regression of Rotated Factor Scores (1.0 Eigenvalue) with Radios per 1,000 Population

Beta Coefficients

F1	F2	F3	F4
-0.8821	-0.1771	-0.0919	-0.0468

Multiple correlation

0.9056

Variance explained

.82

Where F1 to F4 are sets of factor scores calculated from the original 21 variables

TABLE 13
Simple Correlation and Multiple Regression of Ten Correlated Predictors with Radios per 1,000 Population

Simple Correlation

R

		R	Code
1.	2	-0.82974	01. MANY RADIOS PER 1000 POPULATION.
1.	3	0.61529	02. MANY INHABITANTS PER PHYSICIAN
1.	4	-0.65214	03. BIG GROSS NATIONAL PRODUCT
1.	5	0.08526	04. LARGE PERCENTAGE INCREASE IN POPULATION
1.	6	0.14225	05. BIG AREA
1.	7	-0.07788	06. LARGE PERCENTAGE EXPORT OF TOTAL TO US
1.	8	-0.29979	07. LARGE PERCENTAGE INPORT OF TOTAL TO US
1.	9	-0.11377	08. ALLIED WITH USSR
1.	10	-0.85256	09. SMALL POPULATION
1.	11	0.31450	10. SMALL PER CAPITA GROSS NATIONAL PRODUCT
			11. MANY MEN UNDER ARMS

Multiple regression

Beta coefficients

2.	3.	4.	5.	6.	7.	8.	9.	10.	11.
-0.3813	0.1211	-0.3530	0.0257	0.0990	0.0078	-0.1436	0.1143	-0.2537	-0.0494

Multiple correlation

0.9314

TABLE 14
Simple and Canonical Correlation of Ten Correlated Predictors with Ten Correlated Dependent Variables

Simple correlation

pair	value		pair	value		pair	value
1. 2	-0.68735		5. 6	0.82134		10. 11	0.31450
1. 3	0.48135		5. 7	-0.60464		10. 12	-0.14712
1. 4	-0.10872		5. 8	0.00781		10. 13	0.48103
1. 5	-0.04344		5. 9	-0.00603		10. 14	0.58285
1. 6	0.13991		5. 10	-0.07457		10. 15	-0.77020
1. 7	0.18896		5. 11	0.14225		10. 16	0.12876
1. 8	0.24032		5. 12	0.02820		10. 17	-0.25954
1. 9	0.84367		5. 13	-0.48047		10. 18	0.20251
1. 10	-0.49461		5. 14	-0.32433		10. 19	0.00056
1. 11	-0.82974		5. 15	-0.19686		10. 20	-0.20115
1. 12	0.79672		5. 16	-0.57372		11. 12	-0.84585
1. 13	-0.28694		5. 17	-0.13443		11. 13	0.25844
1. 14	-0.37577		5. 18	-0.37464		11. 14	0.32225
1. 15	0.70075		5. 19	-0.53646		11. 15	-0.61178
1. 16	0.24444		5. 20	-0.22832		11. 16	-0.42275
1. 17	0.82767		6. 7	-0.54980		11. 17	-0.83085
1. 18	0.08788		6. 8	-0.06656		11. 18	-0.17986
1. 19	0.20995		6. 9	0.20824		11. 19	-0.25905
1. 20	0.67755		6. 10	-0.06322		11. 20	-0.81052
2. 3	-0.38999		6. 11	-0.07788		12. 13	-0.23310
2. 4	0.52693		6. 12	0.25030		12. 14	-0.19255
2. 5	0.10398		6. 13	-0.38382		12. 15	0.45684
2. 6	-0.04554		6. 14	-0.36072		12. 16	0.36307
2. 7	-0.22400		6. 15	-0.06399		12. 17	0.84496
2. 8	-0.74039		6. 16	-0.43379		12. 18	0.19198
2. 9	-0.63651		6. 17	0.13605		12. 19	0.18797
2. 10	0.76989		6. 18	-0.36455		12. 20	0.75114
2. 11	0.61529		6. 19	-0.41621		13. 14	0.69975
2. 12	-0.51624		6. 20	0.01915		13. 15	-0.35659
2. 13	0.31651		7. 8	0.07268		13. 16	-0.32523
2. 14	0.35321		7. 9	0.20597		13. 17	-0.17028
2. 15	-0.90713		7. 10	-0.12456		13. 18	-0.39244
2. 16	-0.27821		7. 11	-0.29979		13. 19	0.22467
2. 17	-0.57481		7. 12	0.10701		13. 20	-0.17112
2. 18	-0.14728		7. 13	0.17892		14. 15	-0.37624
2. 19	-0.21794		7. 14	0.11646		14. 16	0.40086
2. 20	-0.53322		7. 15	0.31306		14. 17	-0.20533
3. 4	0.05427		7. 16	0.62285		14. 18	0.55406
3. 5	0.25169		7. 17	0.30242		14. 19	0.31851
3. 6	0.38654		7. 18	0.59785		14. 20	-0.17898
3. 7	-0.14152		7. 19	0.70414		15. 16	0.31085
3. 8	0.09128		7. 20	0.37925		15. 17	0.60669
3. 9	0.55104		8. 9	0.10403		15. 18	0.16283
3. 10	-0.23033		8. 10	-0.78665		15. 19	0.36435
3. 11	-0.65214		8. 11	-0.11377		15. 20	0.63835
3. 12	0.54323		8. 12	-0.02441		16. 17	0.47084
3. 13	-0.16354		8. 13	-0.29291		16. 18	0.72269
3. 14	-0.27076		8. 14	-0.34550		16. 19	0.63178
3. 15	0.28423		8. 15	0.62022		16. 20	0.57945
3. 16	0.05817		8. 16	-0.09192		17. 18	0.24214
3. 17	0.40889		8. 17	-0.05298		17. 19	0.35762
3. 18	-0.04700		8. 18	-0.08896		17. 20	0.78404
3. 19	-0.10652		8. 19	-0.03867		18. 19	0.77801
3. 20	0.43287		8. 20	-0.03657		18. 20	0.40583
4. 5	0.14225		9. 10	-0.27553		19. 20	0.53972
4. 6	0.26259		9. 11	-0.85256			
4. 7	-0.19738		9. 12	0.90654			
4. 8	-0.65710		9. 13	-0.22623			
4. 9	-0.09802		9. 14	-0.31776			
4. 10	0.41932		9. 15	0.59841			
4. 11	0.08526		9. 16	0.37957			
4. 12	-0.02779		9. 17	0.83487			
4. 13	0.15748		9. 18	0.22020			
4. 14	-0.00058		9. 19	0.25043			
4. 15	-0.46311		9. 20	0.74944			
4. 16	-0.16310						
4. 17	-0.01548						
4. 18	-0.22654						
4. 19	-0.05799						
4. 20	0.04056						

```
CANONICAL CORRELATION      0.98351

COEFFICIENTS FOR LEFT HAND VARIABLES (1st set)
-0.20493    0.28599    -0.03668    0.13511    0.06812    0.00018    -0.01843    0.03887
-0.37960    0.14811

COEFFICIENTS FOR RIGHT HAND VARIABLES (2nd set)
 0.22028   -0.49927    -0.11880    0.20726   -0.67779   -0.04484     0.05415   -0.17044
 0.03090    0.38831
```

First set

01. MANY INHABITANTS PER PHYSICIAN
02. BIG GROSS NATIONAL PRODUCT
03. LARGE PERCENTAGE INCREASE IN POPULATION
04. BIG AREA
05. LARGE PERCENTAGE EXPORT OF TOTAL TO US
06. LARGE PERCENTAGE INPORT OF TOTAL TO US
07. ALLIED WITH USSR
08. SMALL POPULATION
09. SMALL PER CAPITA GROSS NATIONAL PRODUCT
10. MANY MEN UNDER ARMS

Second set

01. MANY RADIOS PER 1000 POPULATION.
02. LARGE NUMBER OF INHABITANTS PER HOSPITAL BED.
03. LARGE DEFENSE EXPENDITURES AS A % OF GNP.
04. LARGE % MILITARY PERSONNEL AS A % OF POP AGED 15-64.
05. VERY LOW INTERNATIONAL FINANCIAL STATUS.
06. CENSORSHIP.
07. VERY LOW NEWSPAPER CIRCULATION PER 1000 POPULATION.
08. NO EFFECTIVE CONSTITUTIONAL LIMITATIONS.
09. OPPOSITION GROUPS NOT TOLERATED.
10. TRADITIONAL BUREAUCRACY.

TABLE 15
Simple Correlation of Canonical Variate Scores with Original Variables

Correlation of first set with their canonical variate scores

1	2	3	4	5	6	7	8	9	10
-0.29	0.57	0.18	0.62	0.74	0.73	-0.72	-0.54	-0.20	0.46

Correlation of second set with their canonical variate scores

1	2	3	4	5	6	7	8	9	10
0.18	-0.32	-0.74	-0.74	0.28	-0.61	-0.27	-0.55	-0.28	-0.21

TABLE 16
Canonical Correlation of Factor Scores

Canonical correlation   0.93283

Coefficients for left-hand variables (first set)

0.99268	0.11449	-0.01179	0.03643

Coefficients for right-hand variables (second set)

-0.98752	-0.09801	0.12330

First set = factor scores from 21 variable analyses

01	ALBANIA	0.13	1.66	1.20	0.33
02	ARGENTINA	-0.61	-0.41	-0.33	-1.72
03	AUSTRALIA	-0.47	-0.54	-0.34	-1.04
04	BELGIUM	-1.55	-0.73	0.36	1.53
05	BRAZIL	0.16	-1.21	-0.83	-1.45
06	BULGARIA	-0.98	1.80	0.17	0.37
07	CAMBODIA	1.58	1.06	0.58	0.11
08	CANADA	-0.63	-1.26	-0.88	-1.31
09	CEYLON	1.04	0.60	0.65	1.08
10	CHILE	-0.28	-1.06	0.67	-1.19
11	COLOMBIA	0.24	-1.35	-0.03	-0.77
12	COSTA RICA	0.00	-1.16	1.76	0.01
13	CZECHOSLOVAKIA	-1.39	1.13	-0.24	0.95
14	DENMARK	-1.41	-0.36	0.67	1.40
15	ECUADOR	0.34	-1.05	1.44	-0.50
16	FRANCE	-1.19	-0.69	-0.91	0.41
17	INDIA	1.13	-0.71	-2.35	1.06
18	IRAN	1.36	-0.02	-1.34	-0.01
19	IRAQ	0.94	0.72	-0.45	-0.09
20	LEBANON	-0.04	0.27	0.80	1.67
21	LIBERIA	1.03	-0.35	1.41	-0.09
22	LIBYA	1.08	0.81	0.67	-1.44
23	MEXICO	0.07	-1.65	0.01	-0.79
24	MOROCCO	0.75	0.88	0.44	-0.82
25	NEPAL	1.45	1.43	0.12	0.43
26	NETHERLANDS	-0.94	-0.64	0.00	1.75
27	NEW ZEALAND	-0.90	-0.27	0.93	-0.52
28	NORWAY	-1.32	0.20	0.55	-0.05
29	PANAMA	0.13	-1.32	1.60	-0.07
30	POLAND	-0.89	0.51	-0.68	0.76
31	SWEDEN	-1.44	0.43	0.16	-0.26
32	THAILAND	1.72	-0.19	-0.79	0.30
33	TUNESIA	0.11	1.38	1.02	-0.23
34	TURKEY	0.75	-0.62	-1.64	0.67
35	USSR	-0.84	1.45	-1.99	-1.30
36	UNITED STATES	-1.26	1.40	-1.55	-0.95
37	YUGOSLAVIA	-0.37	0.22	-0.56	0.73
38	KENYA	0.98	0.96	0.52	-1.12
39	TAIWAN	1.50	-1.30	-0.81	2.15

Second set = factor scores from 10 variable analyses

01	ALBANIA	-0.13	0.36	2.02
02	ARGENTINA	0.21	-0.24	-0.87
03	AUSTRALIA	1.37	-0.26	-0.72
04	BELGIUM	1.00	0.03	-0.64
05	BRAZIL	-0.60	-0.07	-0.79
06	BULGARIA	0.85	0.20	2.16
07	CAMBODIA	-1.48	0.43	1.03
08	CANADA	1.22	0.48	-0.77
09	CEYLON	-0.64	-1.08	-0.63
10	CHILE	0.07	-0.01	-0.80
11	COLOMBIA	-0.28	-0.73	-0.94
12	COSTA RICA	0.51	-2.14	-0.10
13	CZECHOSLOVAKIA	1.25	0.26	2.10
14	DENMARK	1.39	-0.21	-0.49
15	ECUADOR	-0.69	-0.79	-0.83
16	FRANCE	0.65	1.61	-0.72
17	INDIA	-1.42	0.07	-1.73
18	IRAN	-1.27	0.84	0.68
19	IRAQ	-1.08	0.88	0.61
20	LEBANON	-0.48	-0.84	-0.46
21	LIBERIA	-0.33	-1.23	1.13
22	LIBYA	-0.68	-0.42	0.17
23	MEXICO	-0.57	-0.11	-0.52
24	MOROCCO	-0.73	-0.27	-0.12
25	NEPAL	-1.36	-1.07	1.11
26	NETHERLANDS	0.91	0.46	-0.65
27	NEW ZEALAND	1.51	-0.68	-0.39
28	NORWAY	1.36	0.10	-0.48
29	PANAMA	0.60	-2.16	-0.09
30	POLAND	0.80	0.17	1.52
31	SWEDEN	1.57	0.23	-0.38
32	THAILAND	-1.11	-0.07	0.98
33	TUNESIA	-0.40	-0.97	0.27
34	TURKEY	-0.99	0.80	-0.90
35	USSR	0.69	1.46	1.33
36	UNITED STATES	1.21	1.38	-0.89
37	YUGOSLAVIA	-0.06	1.26	1.18
38	KENYA	-1.52	-0.78	-0.15
39	TAIWAN	-1.33	3.06	-1.23

TABLE 17
Simple Correlation of Canonical Variate Scores with Original Scores

Correlation of first set with their canonical variate scores	Correlation of second set with their canonical variate scores

Canonical variate scores (first set)

1	2	3	4
0.99	0.11	−0.01	0.04

Canonical variate scores (second set)

1	2	3
−0.99	−0.10	0.12

#	Country (first set)	Value	#	Country (second set)	Value
01	ALBANIA	0.32	01	ALBANIA	0.34
02	ARGENTINA	−0.71	02	ARGENTINA	−0.29
03	AUSTRALIA	−0.56	03	AUSTRALIA	−1.41
04	BELGIUM	−1.57	04	BELGIUM	−1.07
05	BRAZIL	−0.02	05	BRAZIL	0.51
06	BULGARIA	−0.75	06	BULGARIA	−0.59
07	CAMBODIA	1.69	07	CAMBODIA	1.54
08	CANADA	−0.81	08	CANADA	−1.35
09	CEYLON	1.14	09	CEYLON	0.66
10	CHILE	−0.45	10	CHILE	−0.17
11	COLOMBIA	0.06	11	COLOMBIA	0.23
12	COSTA RICA	−0.15	12	COSTA RICA	−0.31
13	CZECHOSLOVAKIA	−1.21	13	CZECHOSLOVAKIA	−1.00
14	DENMARK	−1.40	14	DENMARK	−1.41
15	ECUADOR	0.18	15	ECUADOR	0.66
16	FRANCE	−1.24	16	FRANCE	−0.89
17	INDIA	1.11	17	INDIA	1.18
18	IRAN	1.36	18	IRAN	1.26
19	IRAQ	1.02	19	IRAQ	1.06
20	LEBANON	0.05	20	LEBANON	0.50
21	LIBERIA	0.96	21	LIBERIA	0.59
22	LIBYA	1.11	22	LIBYA	0.74
23	MEXICO	−0.15	23	MEXICO	0.51
24	MOROCCO	0.82	24	MOROCCO	0.74
25	NEPAL	1.62	25	NEPAL	1.58
26	NETHERLANDS	−0.94	26	NETHERLANDS	−1.02
27	NEW ZEALAND	−0.96	27	NEW ZEALAND	−1.47
28	NORWAY	−1.30	28	NORWAY	−1.41
29	PANAMA	−0.04	29	PANAMA	−0.40
30	POLAND	−0.79	30	POLAND	−0.62
31	SWEDEN	−1.39	31	SWEDEN	−1.62
32	THAILAND	1.71	32	THAILAND	1.22
33	TUNESIA	0.25	33	TUNESIA	0.53
34	TURKEY	0.72	34	TURKEY	0.79
35	USSR	−0.69	35	USSR	−0.66
36	UNITED STATES	−1.10	36	UNITED STATES	−1.44
37	YUGOSLAVIA	−0.31	37	YUGOSLAVIA	0.08
38	KENYA	1.04	38	KENYA	1.56
39	TAIWAN	1.42	39	TAIWAN	0.87

TABLE 18
Unrotated Factor Loadings (Six Principal Components)

VARIABLE	1	2	3	4	5	6
1	−0.34	−0.69	0.14	−0.02	−0.39	−0.19
2	−0.23	−0.19	−0.13	−0.05	−0.53	0.59
3	0.86	0.32	−0.11	−0.14	−0.04	−0.01
4	−0.58	−0.46	0.32	−0.36	0.10	0.00
5	0.80	0.16	0.21	−0.24	−0.17	0.04
6	0.80	−0.01	0.22	−0.02	−0.09	−0.06
7	0.41	0.30	0.21	−0.03	−0.12	0.07
8	0.58	−0.06	−0.17	−0.46	−0.01	−0.08
9	−0.37	0.18	0.40	0.40	0.28	0.35
10	−0.64	0.14	0.29	0.44	0.18	0.26
11	0.57	0.18	0.03	−0.47	−0.55	0.01
12	0.16	0.48	0.60	0.31	0.19	0.15
13	−0.54	0.04	−0.28	0.04	−0.41	0.37
14	0.11	−0.42	−0.21	0.38	0.19	0.07
15	−0.33	0.17	−0.34	−0.17	0.27	−0.57
16	−0.33	−0.28	−0.01	0.01	−0.55	0.54
17	0.15	−0.11	0.32	−0.46	0.52	−0.40
18	−0.88	−0.20	−0.06	−0.23	0.08	0.01
19	−0.01	−0.51	0.63	−0.34	−0.22	−0.07
20	−0.38	−0.43	−0.29	0.25	0.07	−0.02
21	−0.15	−0.57	−0.21	0.01	−0.32	−0.18
22	0.37	−0.55	−0.35	−0.17	0.08	0.14
23	0.83	−0.06	−0.18	−0.03	0.27	0.09
24	−0.06	0.49	−0.07	−0.19	−0.20	−0.17
25	−0.74	−0.26	0.01	0.0	0.05	−0.02
26	−0.86	−0.24	−0.14	0.17	−0.11	0.03
27	−0.84	−0.21	−0.19	0.26	0.04	0.02
28	0.22	−0.05	−0.10	0.32	−0.37	−0.45
29	0.09	−0.05	0.40	−0.28	0.19	0.43
30	0.34	0.11	0.41	−0.21	0.20	0.28
31	0.03	0.21	0.16	−0.17	0.24	0.35
32	0.32	−0.24	−0.05	−0.32	−0.13	0.36
33	−0.04	−0.26	0.08	−0.13	0.28	0.16
34	−0.26	−0.03	0.19	−0.54	−0.14	0.24
35	−0.23	0.40	0.52	0.14	−0.14	−0.07
36	−0.19	0.38	0.69	−0.08	−0.19	0.02
37	−0.34	−0.13	0.38	−0.31	0.19	0.13
38	0.38	0.05	−0.67	0.05	−0.07	0.11
39	0.44	−0.44	−0.51	0.03	0.18	0.08
40	−0.63	0.05	0.51	−0.30	−0.17	−0.01
41	−0.68	0.07	0.43	−0.16	−0.04	−0.03
42	−0.46	−0.51	0.02	−0.34	0.21	0.09
43	−0.38	−0.50	−0.04	−0.44	0.14	0.09
44	−0.32	−0.47	−0.05	−0.33	0.19	0.03
45	−0.24	−0.45	−0.19	0.31	0.13	0.09
46	0.22	0.50	−0.40	0.45	−0.07	0.04
47	0.76	0.24	−0.26	0.03	−0.04	0.04
48	−0.86	−0.00	0.00	0.03	0.03	0.06
49	0.89	0.18	−0.00	−0.09	−0.08	0.01
50	0.64	0.41	−0.32	0.41	−0.04	0.01
51	0.85	0.23	0.09	0.09	−0.13	0.03
52	0.48	−0.68	0.21	0.09	−0.04	0.01
53	0.99	0.33	−0.11	−0.10	0.02	−0.08
54	0.16	0.16	−0.32	0.09	0.20	−0.06
55	0.06	0.38	0.37	0.24	0.19	0.07
56	0.44	0.06	−0.08	−0.32	0.06	−0.05
57	0.56	0.25	−0.44	−0.22	0.09	0.38
58	0.85	0.27	−0.12	−0.20	−0.01	0.01
59	0.64	0.42	−0.31	0.54	−0.06	0.02
60	−0.23	0.56	−0.23	−0.10	−0.19	−0.28
61	0.46	−0.80	0.23	0.24	−0.03	−0.04
62	0.51	0.22	0.50	−0.08	−0.02	0.13
63	0.70	−0.51	0.28	0.14	0.01	−0.06
64	0.67	−0.58	−0.06	0.31	0.05	0.05
65	0.64	−0.63	−0.05	0.28	0.13	−0.05
66	0.52	0.00	0.44	0.02	−0.04	−0.04
67	−0.20	0.07	−0.16	0.52	0.20	0.24
68	0.84	−0.14	−0.13	0.17	−0.01	−0.03
69	−0.59	0.51	−0.23	−0.06	0.02	−0.07
70	−0.75	0.01	−0.10	0.23	0.13	0.12
71	−0.69	0.16	−0.42	0.08	−0.04	0.08
72	0.40	−0.54	−0.02	0.19	0.16	0.04
73	−0.14	−0.13	0.45	0.41	−0.41	−0.38
74	0.80	−0.53	0.13	0.24	0.08	−0.09
75	0.40	0.52	0.33	−0.01	−0.09	−0.00
76	−0.35	−0.29	−0.44	−0.00	0.10	−0.18
77	−0.32	0.50	−0.39	0.32	0.16	0.13
78	−0.55	0.17	0.17	0.40	−0.18	−0.34
79	0.17	−0.12	−0.08	0.32	−0.35	0.17
80	0.71	−0.51	0.20	0.17	0.05	−0.01
81	0.65	−0.53	0.24	0.31	−0.05	−0.14
82	−0.45	0.15	0.20	−0.26	0.00	−0.31
83	−0.64	0.53	−0.16	−0.09	−0.05	−0.15
84	0.84	0.13	−0.02	−0.05	0.05	−0.14
85	−0.35	0.42	−0.65	−0.10	0.11	0.03
86	−0.61	0.40	−0.50	−0.18	−0.11	0.07
87	−0.04	0.75	0.10	−0.34	−0.18	−0.04
88	0.14	0.48	0.24	0.28	0.24	0.12
89	−0.06	0.53	0.24	0.40	0.42	0.17
91	−0.35	−0.66	0.45	−0.34	−0.11	−0.05

TABLE 19

Factor I
(28.1% of variance accounted for in 91 original variables)

Low Newspaper Circulation (.99) Low Per Capita Gross National Product (.89) Few Radios Per 1,000 Population (.88) Large Number of Inhabitants Per Physician (.86) Large Agricultural Population (.86) Small Percentage of Primary and Secondary School Pupils as a Percentage of Population Age 5–19 (.86) Non-Western (.85) Underdeveloped Economic Development Status (.85) Small Percentage Literate of Population Age 15 and over (.84) Negligible Interest Articulation by Associational Groups (.84) Traditional Bu·reaucracy (.84) Votes for Self-Determination (.83) Large Number of Inhabitants Per Hospital Bed (.80) Negligible Interest Aggregation by Legislature (.80) Large Percentage of Labor Force Employed in Agriculture (.80) Non-Urban (.76) Considerable Interest Articulation by Non-Associational Groups (.75) Few Students Enrolled in Higher Education per 100,000 Population (.74) Negligible Horizontal Power Distribution (.71) Current Regime Unrepresentative (.70) Frequent Interest Articulation by Anomic Groups (.69) Few Exports to United States (.68) Non-Competitive Electoral System (.67) Ineffective Legislature (.65) Censorship of the Press (.64) Not Politically Modern (.64) Group Opposition Not Tolerated (.64) Strong Executive (.64) Small Percentage of Christians as a Percentage of Total Population (.64) Few Imports from the United States (.63) Police Politically Significant (.61) Considerable Interest Articulation by Institutional Groups (.59) Large Percentage of Moslems as a Percentage of Population (.58) Small Gross National Product (.58) Asian Region (.57) Late Date of Independence (.56) Considerable Leadership Charisma (.55) Not Western European Region (.54) Low Political Inculturation (.52) Unstable Government (.51)

Factor II
(13.9% of variance accounted for in 91 original variables)

Effective Constitutional Limitations (.80) Not Communist Bloc (.75) Small Percentage Military Personnel as a Percentage of Total Population (.69) Long Distance from USSR (.68) Free Press (.68) Few Men Under Arms (.66) Opposition Groups Tolerated (.63) Competitive Electoral System (.58) Small Percentage Military Personnel as a Per-

Table 19 continued

centage of Population Age 15–64 (.57) Non-Mobilizational System Style (.56) Votes with West (.55) Considerable Interest Articulation by Political Parties (.54) Effective Legislature (.53) Long Distance from China (.53) Weak Executive (.53) Unstable Party System (.52) Representative Character Regime (.51) Considerable Horizontal Power Distribution (.51) Few Exports to the USSR (.51) Few Expenditures on Defense as a Percentage of Gross National Product (.51) Limited Interest Articulation by Institutional Groups (.51) Non-Elitist Political Leadership (.50) Small Population (.50) Few Imports from the USSR (.50)

Factor III
(8.9% of variance accounted for in 91 original variables)

Large Percentage of Total Imports from U.S. (.69) Entered the United Nations Early (.67) Considerable Political Participation by Military (.65) Latin American Region (.60) Large Percentage of Total Exports to U.S. (.52) Large Imports from U.S. (.51) Allied with U.S. (.51) Police Politically Significant (.50) Government Unstable (.50)

Factor IV
(6.6% of variance accounted for in 91 original variables)

Not Belligerent (.54) Negligible Sectionalism (.52)

Factor V
(4.0% of variance accounted for in 91 original variables)

Not Asian Region (.55) Small Population Per Square Kilometer (.55) Small Population Per 1,000 Hectares of Agricultural Land (.53) Big Area (.52)

Factor VI
(3.7% of variance accounted for in 91 original variables)

Large Population Per 1,000 Hectares of Agricultural Land (.59) Not Anglo-Saxon Region (.57) Big Population Per Square Kilometer (.54)

TABLE 20

Unrotated Factor Scores (Six Principal Components)

#	Country	F1	F2	F3	F4	F5	F6
1	AFGHANISTAN	1.03	-0.50	0.46	-0.76	-1.10	-1.23
2	ALBANIA	0.70	-1.61	-0.87	1.49	-0.25	-0.38
3	ALGERIA	0.62	-0.43	0.05	-0.50	0.76	-0.19
4	ARGENTINA	-0.89	0.01	1.90	1.92	-0.04	-0.37
5	AUSTRALIA	-1.66	0.15	-0.23	-0.71	1.44	-1.92
6	AUSTRIA	-1.54	-0.12	-0.56	0.04	0.34	0.66
7	BELGIUM	-1.69	-0.06	0.12	-0.31	-0.63	0.91
8	BOLIVIA	0.11	1.36	0.87	0.63	1.07	-0.29
9	BRAZIL	-0.50	0.52	2.29	1.10	1.10	-0.45
10	BULGARIA	-0.07	-2.17	-0.40	1.15	0.22	-0.09
11	BURMA	0.76	-0.51	0.05	-1.52	0.03	-0.12
12	BURUNDI	0.71	-0.73	-0.73	0.82	-0.45	0.71
13	CAMBODIA	1.17	-0.50	-0.43	-0.42	-0.57	0.31
14	CAMEROUN	-1.73	0.00	-0.72	-0.39	0.76	-0.04
15	CANADA	-1.73	0.31	-0.41	-1.07	1.40	-1.42
16	CENTRAL AFRICAN REP	1.18	0.43	0.31	0.43	0.44	0.06
17	CEYLON	0.13	0.55	-0.48	-0.78	-1.10	0.25
18	CHAD	1.16	0.55	-0.75	0.01	-0.04	-0.06
19	CHILE	-0.87	0.78	1.45	0.12	0.04	0.32
20	CHINA, PR	0.76	-1.87	-0.12	-0.20	0.18	-0.08
21	COLOMBIA	-0.52	0.91	1.87	0.40	1.12	0.31
22	CONGO (BRA)	0.99	-0.30	-0.91	-0.28	0.90	0.21
23	CONGO (LEO)	0.87	0.63	0.36	-0.50	0.40	-0.60
24	COSTA RICA	-0.60	-1.38	0.59	1.15	0.61	1.05
25	CUBA	-0.32	0.51	0.78	2.05	1.77	
26	CYPRUS	-0.18	0.35	-0.97	0.12	-0.84	-0.06
27	CZECHOSLOVAKIA	-0.48	-2.27	-0.16	0.86	0.95	0.58
28	DAHOMEY	1.23	0.41	-0.84	0.25	0.31	0.30
29	DENMARK	-1.75	0.01	-0.48	-0.11	-0.76	-0.01
30	DOMINICAN REPUBLIC	-0.54	0.95	0.99	1.05	-0.70	0.64
31	ECUADOR	-0.13	1.09	1.69	0.63	0.12	-0.10
32	EL SALVADOR	0.19	-0.70	1.57	2.03	-0.91	0.77
33	ETHIOPIA	0.97	-0.21	0.68	-0.27	0.08	-1.14
34	FINLAND	-1.41	-0.20	-1.05	-0.36	-0.29	-0.51
35	FRANCE	-1.40	-0.82	0.56	-0.99	0.18	0.81
36	GABON	0.85	0.38	-0.49	0.52	0.60	0.18
37	EAST GERMANY	-0.41	-2.07	-0.28	1.41	0.27	0.71
38	WEST GERMANY	-0.36	0.05	-0.45	-0.77	0.02	0.58
39	GHANA	-0.93	0.0	0.05	-0.30	0.79	0.97
40	GREECE	-0.94	0.93	1.92	-1.09	-1.09	-0.28
41	GUATEMALA	0.18	0.93	1.04	-0.13	-0.23	0.96
42	GUINEA	1.19	-0.48	0.03	-0.34	0.34	0.35
43	HAITI	0.75	0.42	1.41	1.58	-0.07	0.42
44	HONDURAS	0.14	1.05	1.21	1.31	-0.02	0.79
45	HUNGARY	-0.33	-2.20	-0.08	0.94	1.31	1.67
46	ICELAND	-1.25	0.74	-0.86	0.59	0.42	-0.87
47	INDIA	-0.03	0.14	0.57	-3.22	0.09	1.34
48	INDONESIA	0.80	-0.51	0.71	-1.68	0.28	1.06
49	IRAN	1.21	-0.34	1.18	-1.41	-0.58	-2.22
50	IRAQ	0.56	0.25	0.40	-1.67	-0.31	-0.58
51	IRELAND	-1.39	0.45	-0.71	-0.80	-0.17	-0.01
52	ISRAEL	-0.96	-0.59	-0.71	-0.61	-1.58	0.43
53	ITALY	-1.62	-0.15	-0.85	-0.55	-0.11	0.78
54	IVORY COAST	0.98	-0.25	0.37	0.20	0.58	0.44
55	JAMAICA	-0.50	1.23	-0.63	0.56	-0.39	1.19
56	JAPAN	-1.42	-0.16	0.07	-1.21	-0.96	-0.15
57	JORDAN	0.85	-0.78	0.46	-0.75	-0.89	-0.04
58	NORTH KOREA	0.59	-1.79	-0.70	0.46	-0.42	0.10
59	SOUTH KOREA	0.17	-0.34	0.97	-0.14	-2.63	0.35
60	LAOS	1.21	0.25	0.71	1.18	-0.13	-1.17
61	LEBANON	-0.13	0.35	0.40	0.09	-1.03	0.89
62	LIBERIA	0.66	0.47	0.15	1.06	-1.03	-1.14
63	LIBYA	-0.33	0.34	-0.48	0.00	0.65	-1.03
64	LUXEMBOURG	-1.59	0.40	-0.48	-0.69	-0.82	0.59
65	MALAGASY REP	0.49	1.14	-0.87	-0.20	0.70	-0.78
66	MALI	1.15	0.36	-0.82	-0.22	0.94	0.26
67	MAURITANIA	1.08	0.69	-0.66	-0.05	0.48	0.08
68	MEXICO	-0.59	0.51	-0.58	0.26	0.58	-0.42
69	MONGOLIA	0.71	-1.28	-0.22	1.28	0.14	-1.15
70	MOROCCO	1.21	-0.17	-0.44	0.49	0.49	0.08
71	NEPAL	1.75	-0.30	-0.09	0.46	-1.16	-0.98
72	NETHERLANDS	-1.52	0.41	-0.40	-0.37	-0.92	-0.29
73	NEW ZEALAND	-1.21	0.53	-0.55	1.30	-0.43	-1.09
74	NICARAGUA	0.15	0.53	-1.55	0.38	0.27	-0.05
75	NIGER	1.23	0.52	-0.81	-0.24	-0.27	-0.08
76	NIGERIA	0.60	0.69	-0.08	-1.05	0.39	-0.65
77	NORWAY	-1.68	-0.04	-0.72	-0.32	-0.77	0.77
78	PAKISTAN	0.72	-0.32	1.28	-2.14	-0.26	0.58
79	PANAMA	-0.34	1.36	1.48	1.45	0.06	-0.53
80	PARAGUAY	0.33	0.23	1.25	1.73	-0.28	-0.30
81	PERU	-0.18	0.71	1.92	0.06	0.75	0.49
82	PHILIPPINES	-0.57	0.81	-0.62	-0.81	-0.80	1.04
83	POLAND	-0.42	-2.22	0.36	0.89	0.53	-0.26
84	PORTUGAL	-0.32	-0.72	0.77	1.56	-1.13	0.02
85	ROMANIA	-0.10	-2.24	-0.07	0.71	0.39	0.59
86	RWANDA	0.68	0.47	-1.06	1.02	-0.33	-1.47
87	SAUDI ARABIA	0.88	-0.20	-0.39	-0.14	-0.54	0.29
88	SENEGAL	1.03	0.45	-0.72	-0.14	0.48	-0.19
89	SIERRA LEONE	0.82	0.24	-0.79	-0.39	-0.34	-1.27
90	SOMALIA	0.96	1.16	-0.79	0.02	0.03	0.09
91	SOUTH AFRICA	-0.03	0.43	-0.72	0.17	-0.14	-0.16
92	SPAIN	-0.46	-0.05	-0.46	0.03	0.03	-0.66
93	SUDAN	-0.38	-1.09	-0.72	0.71	-0.50	-0.61
94	SWEDEN	-1.06	0.17	1.17	0.26	1.03	-0.13
95	SWITZERLAND	-0.17	-0.33	-0.66	-0.78	-0.12	-0.05
96	SYRIA	1.51	0.24	-0.17	-0.48	-0.13	-1.32
97	THAILAND	-0.53	-0.37	-0.73	-0.69	0.48	-0.18
98	TOGO	0.62	-0.23	-1.28	-1.39	-0.48	0.78
99	TRINIDAD	1.00	0.46	-0.72	-0.32	-0.23	0.55
100	TUNISIA	-0.56	1.22	-0.62	0.56	-0.81	-1.31
101	TURKEY	0.35	-0.24	-0.71	-0.32	-0.31	-0.17
102	UGANDA	-0.48	-0.14	0.80	-1.27	-1.18	1.90
103	U.S.S.R.	0.52	0.81	-0.62	-0.66	-0.66	-0.03
104	U.A.R.	-0.33	-2.45	0.17	-0.06	1.64	-0.65
105	UNITED KINGDOM	0.27	-1.21	0.80	-1.64	-0.12	0.14
106	UNITED STATES	-1.88	-0.56	-0.02	-1.02	-0.31	-0.77
107	UPPER VOLTA	-1.56	-0.46	-0.15	-0.59	1.44	-0.59
108	URUGUAY	1.24	0.50	-1.06	0.17	0.24	0.63
109	VENEZUELA	-1.05	1.07	0.14	0.66	0.67	0.17
110	NORTH VIETNAM	-0.73	0.77	1.33	0.42	1.33	-0.69
111	SOUTH VIETNAM	0.69	-1.60	-0.88	0.34	-0.38	-0.56
112	YEMEN	0.77	-0.46	0.46	0.46	-1.65	-0.37
113	YUGOSLAVIA	0.86	0.12	-0.54	-0.09	-0.51	-0.00
114	BOTSWANA	-0.01	-0.55	0.26	-0.05	0.41	0.60
115	GAMBIA	0.45	0.22	-1.42	0.17	0.96	0.29
116	KUWAIT	0.51	0.74	0.50	0.23	-0.46	0.60
117	LESOTHO	-0.11	-0.04	-1.48	-0.15	-0.17	0.85
118	MALAWI	0.33	0.80	-0.87	0.23	0.24	1.98
119	MALTA	-0.76	0.55	-1.28	0.35	0.32	-0.60
120	TANZANIA	-0.53	0.81	-0.03	0.90	-0.84	0.50
121	ZAMBIA	0.23	0.71	-0.07	0.17	1.09	0.60
122	KENYA	0.39	0.52	-0.93	-0.20	0.78	-0.53
123	W. SAMOA	0.45	0.48	-0.80	-0.80	0.50	1.36
124	SINGAPORE	0.76	-0.98	0.94	-0.38	-0.54	-0.36
125	GUYANA	-0.56	0.38	0.30	0.94	-0.54	-0.63
126	MALAYSIA	-0.32	0.35	0.47	0.09	-1.12	-0.55
127	MALDIVE ISLANDS	-0.13	0.70	-0.40	0.45	-0.39	1.26
128	BARBADOS	-0.54	0.44	0.40	-0.60	-0.55	1.26
129	TAIWAN	0.56	-1.41	-1.02	0.25	-2.48	1.36

TABLE 21
Factor Loadings for 91 Variables and 19 Factors

	1	2	3	4	5	6	7	8	9	10	11	12	13	14	15	16	17	18	19
VAR 1	-.45	-.28	0.04	-.24	-.11	-.12	0.06	-.17	0.05	0.71	0.02	0.20	-.01	-.19	0.02	0.06	0.01	0.03	-.08
VAR 2	-.16	0.05	0.02	-.05	-.90	0.09	-.01	0.14	-.05	0.10	0.06	-.04	-.04	-.08	0.11	0.02	0.05	0.06	0.05
VAR 3	0.91	-.10	-.07	0.16	0.04	0.11	-.02	0.10	-.01	-.11	-.03	-.04	0.04	-.06	-.01	-.05	0.06	0.05	0.08
VAR 4	-.52	-.02	0.16	-.74	-.08	0.06	0.09	0.03	0.01	0.06	-.01	-.02	0.03	-.05	0.14	0.01	-.04	-.01	0.04
VAR 5	0.83	-.23	0.11	-.06	-.03	0.04	0.0	-.01	-.12	0.01	0.05	-.06	0.16	-.10	-.13	-.16	0.11	-.01	0.05
VAR 6	0.71	-.43	0.06	0.01	0.03	0.10	0.03	-.12	0.14	0.01	-.05	-.12	0.04	-.10	-.19	-.09	0.06	0.10	-.17
VAR 7	0.40	-.02	0.31	0.29	0.03	0.13	0.10	-.01	0.07	0.13	0.24	-.22	0.28	-.10	0.04	-.27	0.15	-.29	-.04
VAR 8	0.58	-.12	-.05	-.03	0.18	0.19	0.08	-.05	-.24	0.02	-.13	0.26	0.01	-.39	0.0	0.15	-.14	-.13	-.27
VAR 9	-.36	0.04	0.22	0.03	-.12	-.04	0.17	-.02	0.04	-.16	-.05	-.02	-.04	0.75	-.09	-.06	0.07	-.13	-.03
VAR 10	-.61	0.14	0.21	0.09	-.14	-.09	0.03	-.10	0.06	-.23	-.10	0.02	-.10	0.52	-.03	-.13	0.06	0.04	-.10
VAR 11	0.72	0.05	0.20	-.02	-.27	-.01	0.13	-.08	-.00	0.02	-.05	-.13	0.15	-.56	-.04	0.03	-.02	0.14	-.10
VAR 12	0.18	-.02	0.23	0.15	0.18	0.02	0.11	-.35	0.06	-.17	-.03	0.07	0.35	0.56	-.26	-.34	0.31	0.05	-.15
VAR 13	-.39	0.35	-.17	0.06	-.58	0.05	-.08	-.07	-.10	0.11	-.22	0.47	0.17	0.13	0.04	0.09	0.02	0.05	0.12
VAR 14	-.09	-.31	-.38	0.08	-.02	-.05	0.27	-.02	0.04	0.08	0.01	0.15	-.70	0.04	-.10	-.17	-.10	-.03	-.13
VAR 15	-.30	0.42	-.18	0.15	0.68	0.01	0.06	0.09	0.07	0.08	-.06	0.03	-.01	-.39	0.14	0.01	-.01	0.18	-.06
VAR 16	-.26	0.01	0.02	-.13	-.90	-.12	0.09	-.06	-.02	0.01	-.00	0.02	-.06	-.09	0.05	0.00	-.10	-.04	0.00
VAR 17	0.18	-.12	0.08	-.62	0.63	0.14	0.01	0.08	-.07	-.02	-.05	-.07	0.08	0.02	0.06	-.07	0.01	0.01	0.09
VAR 18	-.91	0.20	-.00	0.00	0.01	-.06	-.05	-.08	-.04	0.09	0.07	0.11	-.09	0.10	0.01	-.00	-.10	0.07	-.04
VAR 19	-.09	-.16	-.05	-.20	-.02	0.05	-.03	-.00	-.28	0.76	-.03	-.23	-.10	-.17	0.03	0.17	0.12	-.06	-.17
VAR 20	-.43	-.08	-.06	-.04	-.12	0.13	-.14	0.05	0.17	0.18	0.12	-.02	-.79	-.02	0.03	0.15	-.07	0.04	-.06
VAR 21	-.24	-.18	-.21	-.06	-.07	0.05	0.06	-.06	0.09	0.87	0.06	0.18	-.08	-.10	0.01	0.08	-.11	0.02	0.02
VAR 22	0.15	-.43	-.36	-.01	-.09	0.10	-.04	0.08	-.24	-.09	0.09	0.17	-.06	-.34	0.26	0.09	-.23	-.01	-.30
VAR 23	0.66	-.38	-.22	0.18	0.11	0.21	0.12	0.22	-.05	-.02	-.05	-.05	-.13	0.09	0.03	0.11	-.07	-.21	-.18
VAR 24	0.21	0.45	0.21	0.06	0.03	-.01	-.01	0.07	0.12	-.01	0.05	-.10	-.02	-.07	-.07	0.14	0.05	0.60	0.02
VAR 25	-.80	0.14	0.02	-.18	-.01	0.21	-.13	0.02	0.14	0.21	0.00	-.11	0.12	-.04	-.02	0.09	0.11	-.16	-.05
VAR 26	-.85	0.25	-.01	-.02	-.18	-.08	0.00	-.02	0.11	0.18	0.13	-.01	-.14	-.05	0.07	0.05	-.07	-.10	0.06
VAR 27	-.87	0.19	-.05	0.00	-.12	0.12	-.07	0.01	0.21	0.08	-.05	-.11	-.17	0.03	0.10	0.05	-.06	0.00	0.13
VAR 28	0.20	-.15	0.18	0.28	-.01	-.14	-.02	-.08	0.72	0.18	-.02	0.08	-.22	-.33	0.02	0.07	-.13	-.18	-.05
VAR 29	0.09	-.03	0.09	-.26	-.09	0.12	0.79	0.02	-.18	0.07	0.05	-.11	-.06	0.19	0.05	-.11	0.10	0.00	-.18
VAR 30	0.27	-.12	0.02	-.14	0.03	-.03	0.28	-.10	-.27	-.21	-.03	-.69	0.13	0.12	-.14	-.03	0.11	0.05	-.10
VAR 31	0.10	0.15	0.04	-.21	-.11	0.25	0.07	-.15	0.14	-.48	0.16	-.23	-.00	0.03	-.17	0.14	0.22	-.46	-.20
VAR 32	0.21	-.19	-.10	0.01	-.10	0.05	0.07	-.23	-.78	0.14	0.05	-.08	-.08	-.22	0.03	0.03	0.03	-.18	0.09
VAR 33	-.13	-.09	-.10	-.01	0.07	0.05	0.17	0.05	0.00	0.10	0.05	-.07	-.07	0.02	0.19	0.01	0.07	-.04	-.80
VAR 34	0.30	-.02	0.17	-.07	-.07	0.04	0.13	-.02	-.57	0.22	-.06	-.07	0.34	-.08	0.15	0.04	-.01	-.01	-.26
VAR 35	-.07	0.14	0.77	0.01	-.03	-.07	-.02	0.00	0.17	-.16	-.07	0.0	0.03	0.14	-.12	-.17	0.02	0.04	0.11
VAR 36	-.02	0.16	0.82	-.08	-.02	-.14	0.13	-.06	-.05	-.04	0.00	-.09	0.11	0.09	-.17	-.12	0.13	0.00	-.05
VAR 37	-.21	0.09	0.35	-.48	0.04	0.11	0.14	-.14	-.11	-.08	0.18	0.28	-.13	0.02	-.00	-.06	0.05	-.24	-.28
VAR 38	0.35	0.03	-.49	0.31	-.17	0.14	0.15	0.19	0.09	0.05	0.07	-.08	0.00	-.05	0.14	0.13	-.34	0.06	0.34
VAR 39	0.20	-.39	-.55	0.12	-.03	0.06	0.07	0.18	-.05	-.10	0.07	0.11	-.14	-.21	0.13	0.12	-.33	-.17	-.15
VAR 40	-.41	0.26	0.65	-.41	-.10	-.06	0.08	-.05	-.02	0.04	-.06	0.09	0.15	0.03	0.04	0.09	0.03	0.05	-.03
VAR 41	-.45	0.26	0.61	-.40	-.09	-.06	0.04	0.00	0.10	-.11	-.05	0.15	-.01	0.12	0.12	0.03	-.03	0.02	0.06
VAR 42	-.45	-.03	-.15	-.55	-.06	0.08	-.06	-.06	-.17	-.09	-.02	-.10	-.15	-.12	0.35	0.06	0.01	0.18	-.10
VAR 43	-.32	0.02	-.26	-.57	-.10	0.04	0.00	-.11	-.19	-.03	-.06	-.04	-.13	-.17	0.31	0.04	-.02	0.22	-.21
VAR 44	-.29	-.05	-.10	-.34	0.02	-.05	-.01	0.01	-.04	0.05	0.03	-.00	-.02	-.05	0.80	0.12	0.00	0.00	-.11
VAR 45	-.21	-.05	-.17	-.23	-.12	0.05	0.03	0.07	-.03	0.05	0.03	0.10	0.07	-.08	0.81	0.03	-.11	-.04	-.11
VAR 46	0.14	0.17	-.11	0.86	0.07	0.05	-.11	0.02	-.03	-.08	-.08	0.05	-.07	0.02	-.13	-.02	0.08	0.07	-.08
VAR 47	0.71	-.16	-.13	0.25	-.03	0.04	-.11	0.23	-.06	-.19	0.15	0.00	-.01	-.02	-.14	0.15	-.01	0.03	0.18
VAR 48	-.79	0.36	0.08	-.08	-.05	-.11	0.03	0.02	-.10	-.01	-.02	0.06	0.07	0.07	0.10	0.03	-.04	-.10	0.13
VAR 49	0.88	-.25	-.08	0.10	-.00	0.04	-.03	-.01	0.00	-.03	0.01	-.22	0.07	0.02	-.09	0.08	0.04	0.04	0.10
VAR 50	0.55	-.04	-.28	0.68	0.07	-.07	-.05	-.02	0.06	-.11	0.03	-.04	-.06	0.10	-.14	-.01	0.09	0.03	0.04
VAR 51	0.78	-.29	-.00	0.30	0.01	0.0	0.06	-.10	0.05	-.06	-.08	-.11	0.08	0.02	-.06	-.01	0.15	0.06	-.01
VAR 52	0.16	-.79	-.10	-.13	0.00	-.21	0.12	-.15	-.12	0.13	0.28	0.03	-.02	-.06	0.08	0.12	-.05	0.07	-.04
VAR 53	0.97	-.19	-.08	0.24	0.20	0.01	0.01	0.19	-.19	-.03	-.02	0.14	0.11	0.08	-.15	0.09	0.00	-.09	0.16
VAR 54	0.14	0.06	-.16	0.17	-.03	0.00	0.03	0.79	0.15	-.10	-.05	-.01	-.11	0.08	0.18	-.10	-.04	0.05	0.08
VAR 55	0.09	0.07	0.20	0.12	0.10	-.12	0.13	0.12	-.03	-.21	0.04	-.00	0.04	0.14	-.10	-.80	0.09	-.13	0.05
VAR 56	0.42	-.08	0.02	-.14	-.01	0.16	-.01	0.67	0.00	-.07	0.09	0.08	0.12	-.20	-.16	0.06	0.02	0.06	-.18
VAR 57	0.54	0.09	-.20	0.35	-.08	0.33	0.25	0.15	-.24	-.07	0.13	-.16	0.03	-.04	0.02	0.27	-.18	-.20	-.01
VAR 58	0.87	-.09	-.08	0.21	0.11	0.10	0.12	0.09	-.04	0.03	0.11	-.08	0.10	-.05	0.03	0.09	0.05	-.07	0.01
VAR 59	0.69	0.06	-.07	0.34	0.02	0.03	-.06	0.25	-.07	-.10	0.14	0.08	-.11	0.01	-.08	0.07	0.09	-.10	0.33
VAR 60	0.03	0.46	0.26	0.00	0.10	-.17	-.08	-.00	0.08	-.02	-.65	0.05	0.19	0.10	-.10	0.01	0.21	0.04	0.09
VAR 61	0.04	-.96	-.10	-.14	-.07	-.13	-.09	-.00	-.01	0.09	0.01	-.04	-.04	-.08	0.05	0.03	-.01	-.04	-.03
VAR 62	0.50	-.11	0.12	0.00	0.01	-.19	0.47	-.04	0.10	-.04	0.07	-.23	0.03	0.10	0.04	-.05	0.52	-.08	-.14
VAR 63	0.32	-.86	-.04	-.08	0.05	-.05	-.06	0.08	-.09	0.03	-.08	-.04	-.02	-.07	-.03	0.01	0.20	-.04	0.06
VAR 64	0.24	-.89	-.18	0.11	-.02	0.19	-.07	0.11	-.01	0.11	0.07	0.12	-.04	0.02	-.07	0.08	-.10	0.03	0.06
VAR 65	0.25	-.86	-.31	-.00	0.03	0.05	-.12	0.06	0.06	0.02	0.08	0.08	-.02	-.02	0.06	-.04	-.10	-.01	0.03
VAR 66	0.32	-.39	0.11	0.02	0.06	-.19	0.29	0.34	-.07	0.12	-.25	0.01	0.27	0.13	-.22	0.00	0.24	-.11	-.07
VAR 67	-.33	-.03	-.03	0.50	-.19	0.17	0.07	-.38	0.16	-.09	0.18	0.04	0.03	-.05	-.01	-.16	0.03	0.29	0.19
VAR 68	0.65	-.51	-.21	0.23	0.00	-.07	-.01	0.11	0.02	0.02	0.01	-.03	-.06	0.03	0.01	0.02	-.13	-.05	0.03
VAR 69	-.27	0.71	0.14	0.09	0.06	0.15	-.22	-.01	-.04	-.06	-.19	0.07	-.17	0.04	0.07	-.13	0.03	0.07	0.20
VAR 70	-.67	0.32	-.02	0.04	-.00	-.04	-.14	-.14	-.03	0.04	0.42	0.15	-.05	0.27	0.02	-.04	-.01	0.12	-.02
VAR 71	-.48	0.52	-.04	0.13	-.13	0.09	-.45	-.06	-.06	0.07	0.25	0.16	-.05	0.15	0.03	0.08	-.16	0.08	-.01
VAR 72	0.07	-.72	-.12	-.06	0.03	0.48	-.23	-.03	0.08	0.13	-.12	-.04	0.01	0.01	0.02	-.05	-.03	-.20	0.13
VAR 73	-.20	-.23	0.20	0.03	-.10	-.97	-.04	-.03	0.19	-.01	-.13	-.00	0.02	0.02	0.03	-.02	0.06	0.05	0.10
VAR 74	0.49	-.77	-.02	0.11	0.11	0.08	0.06	0.04	0.11	0.03	-.15	-.07	-.13	0.02	0.08	-.01	0.02	-.04	0.12
VAR 75	0.50	0.18	0.07	0.17	0.12	-.32	0.42	-.05	0.07	0.02	0.03	0.00	0.10	0.18	-.09	-.07	0.57	-.00	-.13
VAR 76	-.29	0.11	-.24	-.13	0.02	-.04	-.06	-.03	0.14	0.05	0.12	0.03	-.08	-.08	0.07	0.04	-.82	-.00	0.02
VAR 77	-.01	0.70	-.01	0.06	0.05	0.54	0.08	-.00	0.06	-.08	-.19	0.01	-.13	0.01	0.11	0.17	0.03	0.12	-.02
VAR 78	-.48	0.27	0.12	0.12	0.02	-.50	-.00	-.08	0.47	0.02	-.09	-.14	0.00	0.06	-.10	-.06	0.01	-.03	0.04
VAR 79	-.05	-.35	0.10	0.31	-.22	0.16	0.12	-.22	0.01	-.10	-.21	0.27	-.08	-.34	-.15	-.19	0.25	0.20	0.07
VAR 80	0.36	-.85	-.02	-.01	0.05	0.05	0.05	0.04	-.03	0.06	-.09	0.01	-.09	-.00	0.04	0.04	0.05	0.04	-.02
VAR 81	0.30	-.86	0.0	0.07	-.04	-.07	0.03	-.05	0.02	0.14	-.07	-.04	-.15	0.03	0.05	-.08	0.02	-.02	-.03
VAR 82	-.12	0.43	0.26	-.33	0.14	-.31	-.19	-.00	-.08	0.20	-.18	-.05	-.08	0.15	0.10	-.15	-.16	-.04	-.08
VAR 83	-.27	0.83	0.04	0.02	-.00	-.15	-.04	0.02	0.25	-.02	-.05	-.05	-.05	0.04	-.00	-.01	0.00	-.09	-.03
VAR 84	0.83	-.23	-.19	0.05	0.17	-.06	0.05	0.03	0.10	-.02	0.02	0.06	-.08	0.04	-.02	0.05	0.17	-.06	0.03
VAR 85	-.06	0.67	-.21	0.13	-.03	0.41	-.14	0.15	0.10	0.01	-.11	0.15	-.16	0.09	0.01	0.04	-.24	0.04	0.17
VAR 86	-.26	0.80	-.09	0.07	-.14	0.11	-.16	0.08	-.05	0.09	0.15	0.08	-.00	-.02	0.08	-.02	-.15	-.06	0.24
VAR 87	0.30	0.61	0.32	0.07	0.09	0.04	-.02	-.02	-.18	-.06	-.16	0.14	0.26	0.04	-.14	0.10	0.22	0.15	0.10
VAR 88	0.44	0.06	-.22	0.11	0.10	-.04	0.02	0.18	-.12	0.03	0.07	0.08	0.07	-.28	0.10	0.54	0.03	-.16	0.09
VAR 89	0.22	0.16	0.36	0.39	0.20	0.01	0.13	0.15	0.02	-.27	0.19	-.19	0.08	0.44	-.06	-.10	0.17	-.03	0.20
VAR 90	-.06	0.12	0.29	0.36	0.21	0.09	0.15	0.09	-.01	-.31	-.06	-.04	0.02	0.60	-.05	-.17	0.01	0.14	-.01
VAR 91	-.37	-.26	0.15	-.73	-.11	-.07	0.09	-.12	-.06	0.38	0.03	-.11	0.00	-.10	0.05	0.01	0.04	-.03	-.01

TABLE 22

Factor I
Underdeveloped
21.1% of variance accounted for in 91 original variables

Very Low Newspaper Circulation per 1,000 Population (.96), Few Radios per 1,000 Population (.91), Large Number of Inhabitants per Physician (.90), Large Percentage Illiterate in Population—age 15 and over (.87), Very Low Per Capita Gross National Product (.87), Non-Western (.86), Small Percentage of Primary and Secondary School Pupils as a Percentage of Population—age 5–19 (.85), Large Number of Inhabitants per Hospital Bed (.83), Traditional Bureaucracy (.83), Few Students Enrolled in Higher Education per 100,000 Population (.80), Large Agricultural Population (.78), Very Low Economic Development Status (.77), Asian Region (.72), Low Urbanization (.71), Large Percentage of Labor Force Employed in Agriculture (.70), Not Politically Modern (.68), Considerable Interest Articulation by Non-Associational Groups (.67), Votes for Self-Determination (.65), Negligible Interest Articulation by Associational Groups (.65), Few Christians as a Per Cent of Total Population (.61), Large Percentage of Moslems as a Per Cent of Population (.57), Very Low International Financial Status (.55), Late Date of Independence (.53), Small Gross National Product (.52), Unstable Government (.50), Large Percentage Annual Rate of Increase in Population (.40).

Factor II
Democracy
14.9% of variance accounted for in 91 original variables

Effective Constitutional Limitations (.96), Current Electoral System is Competitive (.89), Current Regime as Representative (.86), Freedom of Group Opposition (.86), Considerable Horizontal Power Distribution (.85), Effective Current Legislature (.85), Weak Executive (.83), Police Not Politically Significant (.80), Free Press (.78), Considerable Interest Aggregation by Legislature (.77), Limited Interest Articulation by Institutional Groups (.71), Non-Elitist Political Leadership (.70), Military Neutral in Political Affairs (.67), Non-Communist (.61), Infrequent Interest Articulation by Anomic Groups (.52), Considerable Interest Articulation by Associational Groups (.50), Bicameral Legislature (.43), Votes with West in U.N. (.42), Low Political Inculturation (.39), Power Distributed Vertically (.35).

Table 22 continued

Factor III
U.S. Relations
5.5% of variance accounted for in 91 original variables

Large Percentage Imports of Total Imports from U.S. (.82), Large Percentage Exports of Total Exports to U.S. (.77), Large Imports from the United States (.65), Large Exports to the United States (.61), Allied with the United States (.55), Long Time in U.S. (.48).

Factor IV
Smallness
6.9% of variance accounted for in 91 original variables

Small Population (.86), Small Gross National Product (.74), Small Number of Men Under Arms (.72), Very Low International Financial Status (.68), Small Area in Square Kilometers (.62), Small Imports from the U.S.S.R. (.57), Small Exports to the U.S.S.R. (.55), Negligible Sectionalism (.49), Few Permanent Missions at U.N. (.47).

Factor V
Diffusion
4.0% of variance accounted for in 91 original variables

Small Population per 1,000 Hectares of Agricultural Land (.90), Small Population per Square Kilometer (.89), Anglo-Saxon Region (.67), Big Area in Square Kilometers (.62), Not Western European Region (.57).

Factor VI
Executive Leadership
3.4% of variance accounted for in 91 original variables

Considerable Interest Aggregation by Executive (.96), Non-Elitist Political Leadership (.54), Pronounced Leadership Charisma (.49).

Factor VII
Turmoil
2.4% of variance accounted for in 91 original variables

Much Turmoil (.78).

Factor VIII
Religious & Linguistic Heterogeneity
2.5% of variance accounted for in 91 original variables

Religious Heterogeneity (.79), Linguistic Heterogeneity (.66).

Table 22 continued

Factor IX
Peaceful
3.0% of variance accounted for in 91 original variables

Little War-Making (.78), Large Percentage Annual Increase of Per Cent Literate of Population 15 and over (.72), Not Belligerent (.56).

Factor X
Militarism
3.5% of variance accounted for in 91 original variables

Large Percentage of Military Personnel as a Percentage of Population Age 15–64 (.86), Large Defense Expenditures Percentage of Gross National Product (.76), Large Percentage of Military Personnel as a Percentage of Total Population (.71), Little Subversion (.47).

Factor XI
Mobilization
1.7% of variance accounted for in 91 original variables

Mobilizational Systems Style (.65).

Factor XII
Internal Peace
2.0% of variance accounted for in 91 original variables

Little Revolutionary Activity (.68).

Factor XIII
Voting Intensity
2.5% of variance accounted for in 91 original variables

Few Votes in National Elections as a Percentage of Voting Age Population (.78), Not Eastern European Region (.70).

Factor XIV
Catholicism
3.8% of variance accounted for in 91 original variables

Large Percentage of Roman Catholics as a Percentage of Total Population (.74), Long Distance from China (.60), Not Asian Region (.56), Latin American Region (.56), Large Percentage of Christians as a Percentage of Total Population (.51), Long Distance from the U.S.S.R. (.43).

Table 22 continued

Factor XV
Communist China Economic Relations
2.6% of variance accounted for in 91 original variables

Large Imports from Communist China (.81), Large Exports to Communist China (.80).

Factor XVI
Racial Homogeneity
2.0% of variance accounted for in 91 original variables

Racial Homogeneity (.80), Long Distance from U.S. (.53).

Factor XVII
Personalissimo
2.7% of variance accounted for in 91 original variables

Personalissimo (.81), Party System Unstable (.56), Government Unstable (.51).

Factor XVIII
U.N. Supranationalism
1.7% of variance accounted for in 91 original variables

Votes for U.N. Supranationalism (.60).

Factor XIX
Friendly Diplomatic Activity
2.3% of variance accounted for in 91 original variables

Friendly Diplomatic Activity (.80).

TABLE 23
Factor Scores Using Formula $F = ZA(A'A)^{-1}$

Factor scores

AFGHANISTAN	1.51	-0.50	0.20	-1.08	0.04	-1.48	-1.90	-0.62	-1.14	-0.16
	-0.34	1.08	-0.29	-1.43	-0.34	-1.75	-0.49	0.79	-0.26	
ALBANIA	-0.09	-1.50	-0.55	1.33	-0.36	-0.34	-0.59	-0.10	1.22	0.88
	0.58	-0.20	-0.66	-0.45	0.94	-0.48	-0.90	-1.18	-1.15	
ALGERIA	0.22	-0.65	-0.50	-0.72	1.11	0.51	0.94	-0.53	-0.24	-0.32
	1.13	1.15	0.36	-0.07	-1.78	0.63	-0.76	1.16	-0.38	
ARGENTINA	-1.20	-0.04	0.41	-0.94	1.87	-0.30	1.45	-0.58	0.57	-0.50
	0.47	-1.50	0.61	1.31	0.80	2.04	1.68	-0.20	-0.48	
AUSTRALIA	-1.50	0.77	1.35	-0.21	2.41	0.11	-1.21	0.57	0.01	-0.04
	0.63	-0.73	-0.10	-0.31	1.90	1.17	-0.09	-1.25	1.26	
AUSTRIA	-0.93	1.02	-1.10	-0.60	-0.73	-0.24	-0.47	-1.18	-0.03	-0.80
	-0.64	-0.02	-0.64	0.73	0.51	0.24	-0.36	0.03	-0.17	
BELGIUM	-1.33	0.86	-0.05	-1.17	-1.51	-0.12	-0.18	1.02	-0.87	0.20
	-1.18	0.92	0.20	0.74	-1.99	0.60	-0.12	-0.34	-0.08	
BOLIVIA	0.24	0.45	0.60	0.23	1.04	0.16	0.67	0.71	0.19	-0.39
	0.44	-0.63	-0.74	0.71	-1.02	-1.44	0.42	1.92	-0.31	
BRAZIL	0.01	0.41	0.69	-1.67	1.09	-0.92	0.95	-0.34	0.08	0.09
	0.23	-0.89	0.77	1.38	0.18	-0.85	0.40	-0.45	0.30	
BULGARIA	-1.17	-1.65	-1.27	-0.48	-0.15	-0.22	0.11	-0.18	0.41	-0.01
	0.57	0.45	-1.02	-0.57	-1.48	0.19	-0.53	-0.48	-0.29	
BURMA	1.04	0.06	-0.54	-1.56	-0.13	-0.62	-1.01	-0.32	-0.10	-0.78
	1.39	-1.72	-0.65	-1.16	0.14	0.99	1.01	-1.03	0.04	
BURUNDI	0.70	-0.05	0.33	1.47	-0.60	1.11	-0.16	-0.61	0.45	0.13
	-0.48	-0.62	-0.26	0.59	0.13	0.78	-0.27	-0.28	0.66	
CAMBODIA	1.41	-0.43	-0.28	0.15	-0.28	0.62	-1.26	-0.91	-0.63	1.14
	0.50	0.58	-0.83	-0.10	1.82	-0.44	1.77	-0.62	0.75	
CAMEROUN	1.13	0.52	-0.57	-0.08	0.20	0.78	-0.52	0.77	0.24	0.01
	-0.76	-0.33	-0.57	0.98	-0.45	1.01	0.19	-0.13	-0.43	
CANADA	-1.31	0.70	1.25	-1.17	1.53	0.74	-1.52	1.53	0.03	0.68
	-0.12	-0.26	0.63	0.58	1.02	0.01	-0.14	-0.21	-0.08	
CENTRAL AFRICAN RE	0.54	-0.75	-0.34	0.85	0.47	1.61	-0.07	0.45	0.20	-0.02
	-1.29	0.26	0.44	0.38	-0.56	0.88	0.47	0.00	0.41	
CEYLON	0.62	0.71	-0.14	-0.03	-1.41	-1.34	0.54	0.96	0.38	-1.37
	-0.60	-0.23	-0.28	-1.21	2.20	0.55	0.94	0.14	0.86	
CHAD	0.88	-0.05	-1.15	0.23	0.39	1.42	-0.43	1.21	-0.04	-0.04
	-1.73	-0.04	-0.25	0.11	-0.77	-0.79	0.74	-0.16	0.05	
CHILE	-0.64	0.42	1.38	0.37	0.42	-0.29	1.19	-0.46	-0.62	0.75
	-0.16	0.25	0.78	1.35	0.65	-0.07	0.47	0.30	0.52	
CHINA, PR	0.03	-1.35	-2.26	-1.96	-0.30	-0.14	0.10	-0.41	-0.29	-0.45
	-0.22	-2.24	-0.41	-1.52	-1.36	-1.82	-0.72	0.77	1.60	
COLOMBIA	-0.14	0.23	1.21	-1.11	0.49	0.24	0.14	-0.63	0.17	-0.89
	-0.06	-0.38	-0.39	1.52	-1.17	-0.81	0.21	-1.09	0.71	
CONGO (BRA)	0.39	-0.40	-1.09	0.67	0.42	1.24	-0.04	1.08	-0.15	-0.22
	-1.56	-0.08	-0.29	0.74	-0.44	1.35	0.99	0.33	0.15	
CONGO (LEO)	1.02	0.11	0.45	-0.42	0.77	-0.13	1.48	0.99	0.40	-0.04
	-0.57	0.79	-0.67	0.44	-0.50	1.46	0.43	-0.39	0.46	
COSTA RICA	-0.43	0.52	0.91	1.12	-0.21	0.92	-1.12	-1.64	0.73	-1.70
	0.29	-0.99	-0.28	0.81	-0.05	0.07	1.10	-0.29	-1.11	
CUBA	-0.79	-1.05	-1.29	-0.68	-0.19	0.56	0.63	-1.15	0.14	-1.42
	1.44	0.95	-0.00	1.28	1.59	-1.61	0.55	-0.53	-0.93	
CYPRUS	-0.55	0.55	-0.86	0.95	-0.45	-0.19	1.05	0.74	0.26	-0.21
	-1.11	0.34	0.33	-1.26	-1.52	0.74	-0.30	0.83	-1.12	
CZECHOSLOVAKIA	-1.53	-1.60	-0.82	-0.55	-0.60	-0.25	0.24	1.66	0.24	-0.01
	0.94	1.06	-0.83	0.42	-0.24	0.25	-0.14	-0.68	-1.55	
DAHOMEY	0.80	-0.40	-0.87	0.74	-0.12	1.52	-0.31	1.03	0.31	0.14
	-1.24	-0.25	0.29	0.50	-0.51	0.58	0.69	0.84	-0.17	
DENMARK	-1.39	0.87	0.06	0.16	-0.48	0.37	-1.15	-0.98	0.35	-0.26
	0.15	0.70	0.53	-0.76	0.32	0.29	0.47	1.67	-1.16	
DOMINICAN REPUBLIC	0.21	0.62	1.77	0.86	-0.78	0.44	-0.10	-1.46	0.49	0.96
	-0.49	0.06	-0.86	1.05	-0.07	-1.82	-0.48	-1.38	-0.24	
ECUADOR	-0.19	0.22	1.07	0.42	0.01	-0.87	0.21	1.67	0.05	0.17
	-0.37	-0.50	0.37	0.62	-0.63	-1.82	1.11	-0.09	-0.96	
EL SALVADOR	-0.35	-0.96	1.04	1.34	-0.88	-1.01	-0.61	-0.39	-0.05	-0.90
	0.00	0.18	0.75	0.79	-0.00	-0.91	1.67	0.03	0.72	
ETHIOPIA	1.30	-0.73	0.51	-0.97	0.19	-1.35	-1.53	1.06	-0.17	-1.03
	-0.51	1.74	-0.57	-0.24	0.89	-1.07	0.51	0.49	-0.44	
FINLAND	-1.32	0.89	-0.69	-0.13	0.20	0.36	0.26	-0.41	0.35	-0.51
	-0.29	0.65	-0.16	-1.73	0.08	-0.10	-0.05	1.34	-0.05	
FRANCE	-1.01	0.41	0.51	-1.17	0.12	1.01	1.31	-1.11	-1.39	1.04
	-0.73	1.76	-1.53	0.31	1.30	-0.14	1.34	-0.54	1.22	
GABON	0.22	-0.64	0.54	1.14	0.47	1.48	0.01	1.29	-0.14	0.03
	-0.89	-0.07	-0.78	0.56	-0.30	1.48	0.75	0.10	0.34	
EAST GERMANY	-1.69	-1.89	-0.78	0.24	-0.87	-0.69	1.40	0.50	0.34	-0.85
	-0.22	-0.77	-0.68	0.05	1.32	0.77	-0.63	0.49	1.20	
WEST GERMANY	-0.94	1.03	-0.21	-0.90	1.02	0.02	0.19	0.51	0.03	0.62
	-1.02	-1.02	-0.30	0.73	1.68	0.58	-0.62	-0.09	0.65	
GHANA	0.68	-1.10	0.73	-0.18	0.26	1.10	-0.03	0.56	-0.17	-1.63
	0.98	0.90	-0.03	-0.05	1.40	0.28	0.22	0.80	-0.08	
GREECE	-0.62	0.81	-0.71	-0.39	-0.18	0.13	-0.09	-1.41	0.49	0.86
	-0.39	0.96	-0.34	-0.59	-0.83	0.26	2.08	0.78	-0.37	
GUATEMALA	-0.17	-0.42	1.11	0.62	-0.80	-0.69	0.74	1.05	0.13	-0.76
	0.55	-0.40	0.88	0.59	-0.87	-1.53	1.05	1.22	-1.12	
GUINEA	0.58	-1.15	0.97	1.00	0.11	1.47	-0.53	0.90	-0.23	-0.07
	0.71	-0.16	-0.14	-0.10	-0.48	0.29	-0.99	0.60	-0.40	
HAITI	0.62	-0.98	1.60	0.88	-1.00	-0.54	0.31	-0.17	-0.21	-0.48
	-0.85	-0.83	-1.31	0.88	-0.44	-1.45	-0.52	0.50	0.09	
HONDURAS	-0.26	-0.26	1.19	1.50	0.14	-0.92	0.23	-0.76	-1.19	-1.13
	0.15	-0.76	-0.29	0.22	-0.13	-1.07	1.04	0.46	-0.19	
HUNGARY	-1.65	-1.68	-1.56	-0.09	-0.55	-0.22	1.07	-0.14	-1.16	-0.81
	0.12	-0.84	-1.09	0.48	0.43	-0.07	-0.17	-0.02	-0.44	
ICELAND	-1.44	0.98	0.45	1.54	1.42	0.02	-0.93	-0.93	-0.70	-0.43
	-0.19	-1.85	-0.90	-1.07	-0.58	-0.00	-0.61	1.27	-0.45	
INDIA	1.13	1.09	0.78	-2.83	-1.25	1.03	-0.24	-0.22	-1.27	-1.22
	-0.28	-1.40	0.12	-0.88	-1.38	0.27	-1.69	-0.44	-1.32	
INDONESIA	0.86	-0.44	1.03	-1.87	-0.63	1.14	0.74	0.45	-0.34	-1.36
	0.88	-0.48	-1.77	-1.21	-1.13	1.36	0.80	-0.23	-0.44	
IRAN	0.98	-0.23	0.26	-2.02	1.18	-0.88	-0.08	-0.35	0.59	0.88
	-0.76	0.15	-0.50	-0.92	-0.72	0.80	1.02	1.49	-0.28	
IRAQ	0.57	-0.03	-0.18	-0.41	0.44	-0.29	1.19	-0.04	0.17	1.16
	-0.90	0.79	1.61	-1.02	0.85	-0.25	0.25	-0.70	-2.56	

Table 23 continued

Factor scores

IRELAND	-1.13	0.83	-0.47	0.54	-0.14	0.12	-0.43	-1.34	0.28	-0.38
	-1.11	0.33	0.34	0.48	-0.53	0.26	-1.10	0.32	0.15	
ISRAEL	-1.29	0.79	-0.91	0.98	-0.32	0.29	1.80	0.18	-1.11	1.03
	1.15	0.32	-0.70	-2.45	-1.47	1.01	-0.51	-0.45	0.29	
ITALY	-0.55	0.98	-0.16	-1.77	-1.00	0.09	1.09	-0.91	0.24	0.43
	-0.68	1.19	-0.99	1.51	0.63	0.42	0.16	0.51	0.79	
IVORY COAST	0.83	-0.55	0.59	0.49	0.21	1.86	0.07	0.68	-0.04	-0.17
	0.12	0.43	-1.45	0.46	0.11	0.87	0.47	1.57	0.51	
JAMAICA	-0.03	1.05	0.89	1.24	-1.13	0.13	0.87	0.40	-0.64	-0.76
	-0.53	0.04	-0.95	-0.10	0.09	-1.53	-1.55	0.02	0.47	
JAPAN	-0.53	0.96	0.60	-1.57	-1.00	0.43	0.35	-0.14	1.08	-0.60
	-0.50	0.05	-0.57	-1.53	1.85	-0.40	0.89	0.99	1.59	
JORDAN	0.66	-0.31	-0.21	0.75	0.22	-1.34	1.53	-1.24	-1.58	1.98
	-1.71	-0.19	-0.46	-0.13	0.63	0.82	-0.03	-1.18	-1.26	
NORTH KOREA	-0.00	-1.34	-1.92	-0.10	-0.62	0.26	0.05	-1.03	1.05	0.98
	0.51	-0.95	1.27	-0.31	1.37	-0.88	-0.52	0.74	0.61	
SOUTH KOREA	0.27	-0.31	0.45	-0.29	-1.88	-0.01	0.94	-0.62	1.35	2.05
	-0.16	-1.21	1.83	-0.35	-0.68	0.09	0.12	-0.39	0.67	
LAOS	1.35	-0.08	-0.02	-0.02	0.09	-0.87	-0.42	1.00	0.23	1.95
	-0.39	-0.39	-0.60	-0.02	-0.82	0.12	1.60	0.33	0.16	
LEBANON	-0.59	0.29	-0.48	0.54	-1.16	-0.61	1.16	0.15	0.56	-0.87
	-0.78	0.99	0.74	-0.99	-1.34	1.10	1.94	-1.69	-1.78	
LIBERIA	0.39	-0.92	1.49	0.89	0.21	-0.95	-1.92	1.51	-0.26	-0.28
	-0.03	1.01	-0.34	0.29	-1.06	0.87	-0.38	1.31	-0.47	
LIBYA	1.10	0.70	-0.01	-0.10	1.31	-0.80	-1.09	-1.11	-0.93	0.52
	-0.49	0.41	-0.94	0.93	0.42	0.97	-0.91	-0.92	-0.70	
LUXEMBOURG	-1.02	1.05	0.08	0.64	-0.95	-0.48	-0.41	-1.15	-0.64	0.04
	-0.94	0.45	-0.93	0.67	-0.33	0.32	-0.59	1.10	-0.36	
MALAGASY REP	1.36	1.19	-0.26	0.43	0.78	-0.47	-0.47	-0.19	-0.10	0.38
	-1.31	-0.69	-1.17	1.22	0.59	0.43	-1.16	-1.39	0.35	
MALI	0.89	-0.11	-1.10	0.38	0.37	1.78	-0.48	0.91	-0.02	-0.07
	-0.74	0.47	0.64	-0.34	0.61	-2.01	0.72	-0.65	0.04	
MAURITANIA	0.92	0.10	-0.15	0.71	0.67	1.50	-0.42	0.44	-0.85	0.31
	0.35	0.23	0.03	-0.65	-0.42	-1.74	0.20	-1.01	0.54	
MEXICO	0.06	0.05	1.64	-0.97	0.68	1.25	-0.00	-0.82	0.86	0.93
	-0.03	0.34	1.14	1.63	-0.27	-1.84	-1.77	-0.37	1.24	
MONGOLIA	-0.33	-1.65	-0.96	0.80	0.94	0.09	-0.88	-1.32	1.09	0.00
	0.33	-0.49	0.96	-0.95	0.16	-0.53	-1.04	-0.20	0.17	
MOROCCO	1.04	0.39	0.24	-0.63	0.24	1.02	-0.06	-0.83	-0.03	0.00
	-0.03	-0.10	0.33	-0.16	0.99	-0.29	-1.31	0.97	-1.44	
NEPAL	1.18	-0.49	-1.51	-0.73	-0.51	-1.47	-0.40	0.70	1.06	-0.07
	0.41	1.22	0.18	-0.71	-0.80	-0.72	1.52	0.90	0.93	
NETHERLANDS	-0.93	1.02	0.23	-0.65	-1.48	-0.12	-1.10	0.47	0.04	0.73
	-0.62	0.61	-0.76	0.32	0.71	0.12	0.33	0.40	-0.39	
NEW ZEALAND	-1.98	0.46	1.19	1.07	1.62	0.62	-0.27	0.37	-0.01	-0.29
	0.41	-0.24	-0.19	-1.00	0.34	1.21	0.26	0.04	1.26	
NICARAGUA	0.01	-0.55	1.76	1.35	0.26	-1.03	-1.44	-0.57	-1.13	0.34
	0.03	-0.34	-1.44	0.75	-0.24	-1.46	1.12	0.13	-0.87	
NIGER	1.01	-0.07	-0.89	0.59	0.48	1.57	-0.48	1.15	0.29	0.13
	-1.39	0.14	0.79	-0.34	1.19	-1.68	0.83	-0.79	0.47	
NIGERIA	1.61	0.71	0.23	-1.19	0.12	-1.12	0.60	0.96	0.21	-0.69
	-0.31	1.16	-0.10	0.82	0.05	0.71	-1.78	0.02	-0.12	
NORWAY	-1.15	1.08	0.05	-0.14	0.17	-0.05	-1.23	-0.78	-0.11	0.23
	-0.22	0.30	-0.55	-1.21	0.36	-0.21	0.19	1.71	0.06	
PAKISTAN	0.85	-0.67	1.82	-1.05	-0.41	0.01	0.84	0.36	-1.38	-0.39
	0.14	-1.33	1.58	-0.10	1.41	1.41	-1.04	0.75	0.34	
PANAMA	-0.63	0.12	1.17	1.04	-0.03	-0.23	0.17	-0.73	0.99	-1.77
	0.45	0.37	0.28	-0.03	-0.52	0.61	2.00	-1.96	-0.17	
PARAGUAY	-0.43	-1.23	1.00	1.31	0.83	-0.84	-0.13	-1.09	0.04	0.29
	0.41	-1.59	0.83	1.06	-0.60	0.70	-0.13	1.45	0.77	
PERU	-0.30	-0.13	1.44	-0.37	0.89	-0.58	0.59	1.25	-0.70	-0.25
	0.03	-0.33	0.03	0.56	-0.40	-1.26	0.80	0.88	-0.00	
PHILIPPINES	0.24	0.88	1.50	-1.18	-1.28	1.41	-1.17	-0.64	1.45	-0.68
	-0.27	-0.48	0.92	-0.50	-1.54	0.33	-0.59	-2.96	-0.03	
POLAND	-1.13	-1.70	-0.38	-0.71	-0.69	-0.05	0.55	-1.28	-0.17	-0.61
	0.13	0.40	-1.35	0.35	0.53	0.08	-0.52	-0.79	0.14	
PORTUGAL	-0.73	-1.36	0.20	0.07	-0.71	-0.88	-1.32	-1.15	0.27	0.27
	-0.83	0.82	1.74	1.32	-1.01	0.57	-1.02	0.31	0.65	
ROMANIA	-0.88	-1.73	-1.05	-0.73	-0.39	0.06	-0.21	-0.81	1.23	-0.44
	0.43	-0.07	-0.01	-0.31	0.58	0.09	-0.47	0.16	-0.80	
RWANDA	0.32	0.01	-1.04	1.52	-0.45	0.08	0.47	-0.42	0.14	0.00
	-0.75	-0.84	-0.47	0.38	-0.01	0.91	0.31	-0.60	1.06	
SAUDI ARABIA	0.79	-0.70	0.70	-0.51	1.51	-1.32	-0.32	-1.18	-1.43	-0.63
	-0.12	2.88	-0.04	-1.18	-0.64	0.49	-0.25	-1.28	1.66	
SENEGAL	0.86	-0.35	-0.09	0.47	0.32	1.20	-0.00	-0.07	-0.18	-0.59
	-0.04	0.11	-0.22	-0.12	0.01	0.60	-0.27	0.77	0.21	
SIERRA LEONE	1.46	1.00	-0.45	0.48	-0.15	-0.27	0.50	0.29	0.46	-0.00
	-0.03	0.50	0.33	-0.10	-0.21	-0.24	-0.82	0.23	-0.58	
SOMALIA	0.94	0.18	-1.03	0.42	1.11	-0.84	0.14	-0.97	0.10	-0.17
	-1.22	0.26	-0.22	-0.34	-0.80	1.42	-0.32	-0.33	0.32	
SOUTH AFRICA	-1.24	-0.46	1.10	0.06	1.67	-0.63	1.07	1.79	-0.65	-1.76
	-0.78	0.04	1.55	-1.72	-0.99	-0.82	-1.27	-1.08	0.85	
SPAIN	-1.00	-1.57	0.43	-0.88	-0.32	-0.08	0.93	-0.26	0.71	-0.03
	-0.94	1.67	2.08	0.77	-0.84	0.61	-1.19	0.38	0.94	
SUDAN	0.96	-0.41	-0.89	-0.90	0.91	-0.71	0.27	0.30	0.19	-1.46
	0.04	0.63	0.90	-0.77	1.30	-1.08	0.25	-0.60	-0.00	
SWEDEN	-1.29	1.02	0.01	-0.33	0.27	0.07	-0.51	-0.48	-0.02	0.07
	-0.14	0.50	-0.45	-0.88	0.99	0.20	-0.07	1.45	-0.78	
SWITZERLAND	-1.18	1.02	-0.29	0.00	-1.02	-0.88	-1.16	1.89	0.19	-0.62
	-1.45	-0.68	1.44	0.19	1.11	0.51	-0.95	-0.07	-0.95	
SYRIA	0.10	-0.04	-0.45	0.60	0.45	-0.59	1.02	-0.91	-1.11	0.69
	-1.19	-0.76	1.39	-1.45	1.20	-0.14	1.21	-0.35	-2.10	
THAILAND	0.94	-0.51	0.73	-1.16	-0.35	-1.08	-0.60	-0.79	1.54	0.60
	0.73	-0.95	-0.24	-0.97	-0.33	0.52	1.76	0.85	1.45	
TOGO	0.77	-0.13	-0.64	1.04	-0.04	0.12	-0.15	0.40	0.03	-0.38
	-0.86	-0.13	-0.01	0.31	0.29	0.58	0.21	-0.18	-0.45	
TRINIDAD	-0.17	1.15	1.18	1.52	-1.00	-0.02	1.08	0.72	-0.24	-0.30
	-0.18	0.46	-0.88	-0.75	0.06	-1.49	-1.28	-0.81	0.43	
TUNISIA	0.45	-0.24	-0.13	0.30	0.22	1.58	0.52	-2.01	0.10	-0.60
	-0.08	0.53	0.15	-0.26	0.63	0.43	-1.29	1.75	-0.73	

Table 23 continued

Factor scores

TURKEY	0.51	0.76	0.81	-1.60	0.31	0.19	0.49	-1.76	1.40	0.89
	0.22	0.92	-0.15	-1.55	-0.91	-0.50	-0.29	1.05	-1.03	
UGANDA	1.31	0.86	-0.19	0.01	-0.15	-1.06	0.27	1.02	0.16	-0.20
	-0.19	0.51	0.28	1.06	1.70	0.77	-0.73	-1.07	0.31	
U.S.S.R.	-1.08	-1.45	0.05	-1.15	1.65	0.04	-0.16	1.35	0.94	1.30
	0.67	0.50	-0.87	-0.42	0.49	-0.54	-1.00	-1.87	-1.40	
U.A.R.	-0.12	-1.02	0.56	-0.24	-0.56	1.16	-0.56	-1.41	-2.26	-0.11
	0.44	0.07	1.71	-0.33	1.99	0.36	0.85	-0.22	-1.30	
UNITED KINGDOM	-1.23	0.85	-0.64	-1.05	-0.63	-0.05	-0.81	0.39	-0.47	0.36
	-0.63	0.20	-0.44	-0.29	1.51	-0.18	0.10	-0.21	0.19	
UNITED STATES	-1.45	1.16	-2.48	-2.04	1.09	0.48	-0.58	0.44	-0.60	1.45
	-0.44	-0.71	0.21	0.24	-1.86	-2.80	0.43	-0.80	0.37	
UPPER VOLTA	0.80	-0.26	-1.06	0.79	0.12	1.57	-0.26	0.93	0.24	0.09
	-0.90	-0.23	0.40	-0.05	-0.40	0.12	0.83	0.91	0.36	
URUGUAY	-1.03	0.84	0.14	0.52	1.36	0.11	-0.63	-0.50	0.09	-0.33
	-0.31	-0.02	-0.25	0.50	-0.62	0.51	1.19	0.24	0.38	
VENEZUELA	-0.24	0.57	1.88	0.02	1.74	0.25	1.51	-0.59	0.84	0.70
	1.96	0.14	-1.06	1.15	-0.11	-0.73	-0.55	0.40	-0.79	
NORTH VIETNAM	-0.01	-1.34	-1.56	0.08	-0.89	0.33	0.32	-0.16	0.60	0.64
	1.10	-0.79	1.07	-0.19	1.50	-0.04	-0.17	0.43	1.16	
SOUTH VIETNAM	0.17	-0.97	0.41	0.53	-0.81	0.02	1.02	0.29	0.64	1.63
	0.63	-1.12	0.68	-0.38	-0.34	0.64	0.79	0.15	1.57	
YEMEN	0.44	0.01	-1.45	0.53	0.74	-1.29	-0.71	-1.04	-1.42	-0.21
	-0.99	0.09	0.58	-1.09	-1.32	0.19	0.07	-1.56	0.35	
YUGOSLAVIA	-0.39	-1.41	0.38	-1.14	-0.36	-0.43	-1.09	1.86	0.21	0.84
	0.29	0.0	-1.15	0.35	-0.32	0.50	-0.77	0.07	-1.90	
BOTSWANA	0.84	0.74	-1.66	0.28	0.73	-1.00	-0.84	0.41	-0.40	0.24
	1.61	-0.46	-0.29	1.54	0.32	0.88	-0.34	0.67	0.38	
GAMBIA	0.92	1.07	-1.84	0.61	-0.30	-0.98	-0.94	-0.38	0.07	0.42
	1.69	-0.68	0.15	0.25	-0.83	0.38	-0.01	-0.23	-0.02	
KUWAIT	-0.73	-0.06	0.14	1.21	0.51	0.73	1.14	-0.43	-0.80	-0.09
	1.73	1.37	1.59	-1.07	-0.31	0.80	-0.99	-0.47	0.11	
LESOTHO	0.81	1.05	-1.51	1.04	-0.18	-1.05	-0.60	0.09	-0.04	0.49
	1.65	-0.81	0.08	1.36	0.80	0.54	-0.22	-0.10	-0.10	
MALAWI	0.47	0.59	-1.16	0.77	-0.20	0.57	-0.01	0.93	-0.18	-0.26
	2.37	0.67	0.38	-0.11	-1.08	-0.83	0.20	0.32	-0.19	
MALTA	-0.91	0.72	-0.68	1.38	-1.67	-0.38	0.09	-0.25	-0.88	-0.56
	1.23	0.31	0.36	0.65	-1.51	1.49	-0.24	0.06	-0.62	
TANZANIA	1.05	0.70	-0.93	-0.41	0.70	0.80	0.49	-0.24	-0.09	-0.97
	1.35	1.67	0.06	0.45	1.28	0.40	0.73	0.35	1.09	
ZAMBIA	0.48	0.78	-1.15	-0.06	0.53	-0.59	-0.64	1.10	-0.36	0.48
	2.65	-0.03	0.39	1.10	-0.43	0.83	0.01	-0.33	0.27	
KENYA	1.27	0.86	-0.51	-0.48	-0.07	-0.13	0.67	-0.01	0.15	-0.19
	2.34	1.41	0.13	1.02	1.11	0.93	-0.59	0.10	0.29	
W. SAMOA	0.13	0.19	-0.48	1.65	-0.54	0.63	0.23	0.49	0.39	0.45
	0.23	-0.88	0.44	0.28	-0.43	0.15	-0.07	0.24	0.48	
SINGAPORE	-0.65	0.92	-0.28	1.37	-1.15	0.41	0.04	1.21	-0.51	-0.07
	2.04	-0.32	0.11	-1.43	1.46	-1.44	0.89	-0.55	0.39	
GUYANA	0.34	0.70	0.48	1.32	0.86	-0.09	1.58	0.39	0.10	0.06
	-0.05	-0.51	-1.01	-0.72	0.06	-1.38	-0.64	-0.35	0.48	
MALAYSIA	0.42	1.07	0.46	0.10	-0.65	-0.52	0.17	1.93	0.78	1.37
	0.41	-0.75	0.74	-0.57	0.19	-0.62	-0.54	-0.72	0.67	
MALDIVE ISLANDS	0.47	0.48	-1.25	1.15	0.34	0.23	-0.09	-1.06	0.11	0.74
	-0.36	-0.35	0.50	-0.31	-0.38	0.72	-0.39	-0.75	0.85	
BARBADOS	-0.28	0.90	0.22	1.83	-1.27	-0.69	-0.01	0.17	-0.95	-0.38
	0.39	-0.69	-1.09	0.36	-0.17	-1.44	-2.00	-0.02	-0.19	
TAIWAN	0.06	-0.54	2.30	-0.09	-1.76	0.97	-1.38	0.67	-1.83	2.27
	2.73	1.10	0.55	-0.73	-1.23	0.43	1.11	-0.51	0.46	

TABLE 24
Intercorrelations of Factor Scores

X	Y	R		X	Y	R		X	Y	R		X	Y	R		X	Y	R
1.	2	0.00590		3.	4	0.00951		5.	6	0.01378		8.	9	0.00865		12.	13	-0.02146
1.	3	-0.07413		3.	5	0.03549		5.	7	-0.01380		8.	10	-0.00996		12.	14	-0.04064
1.	4	-0.01388		3.	6	0.02611		5.	8	-0.02060		8.	11	0.01028		12.	15	-0.01847
1.	5	0.04396		3.	7	0.03158		5.	9	-0.02627		8.	12	-0.01733		12.	16	0.03654
1.	6	0.00129		3.	8	0.03090		5.	10	0.00031		8.	13	-0.00683		12.	17	0.01555
1.	7	-0.08404		3.	9	-0.11193		5.	11	-0.00465		8.	14	0.01025		12.	18	-0.02003
1.	8	0.08024		3.	10	-0.02060		5.	12	0.03407		8.	15	-0.03778		12.	19	-0.06378
1.	9	0.01298		3.	11	0.02962		5.	13	-0.00758		8.	16	-0.03820		13.	14	-0.07504
1.	10	0.04954		3.	12	0.03417		5.	14	0.03180		8.	17	0.02549		13.	15	0.01105
1.	11	-0.00217		3.	13	-0.05631		5.	15	-0.00403		8.	18	-0.04624		13.	16	0.01325
1.	12	0.04896		3.	14	0.06316		5.	16	0.01595		8.	19	0.00225		13.	17	-0.02522
1.	13	-0.00006		3.	15	-0.03501		5.	17	0.01305		9.	10	0.04400		13.	18	-0.04888
1.	14	0.03528		3.	16	-0.00840		5.	18	-0.03303		9.	11	-0.00830		13.	19	0.02195
1.	15	-0.00156		3.	17	-0.00840		5.	19	0.02789		9.	12	-0.07936		14.	15	0.06581
1.	16	0.00541		3.	18	-0.00567		6.	7	0.01904		9.	13	0.10823		14.	16	0.06113
1.	17	0.03985		3.	19	-0.00629		6.	8	0.01477		9.	14	0.04745		14.	17	-0.05869
1.	18	-0.01693		4.	5	-0.03651		6.	9	0.06061		9.	15	0.00750		14.	18	-0.00634
1.	19	0.04670		4.	6	0.04234		6.	10	0.00198		9.	16	-0.00903		14.	19	0.03153
2.	3	0.05944		4.	7	0.01102		6.	11	0.00059		9.	17	0.03509		15.	16	-0.00481
2.	4	0.02051		4.	8	0.02844		6.	12	0.01667		9.	18	0.06005		15.	17	0.03407
2.	5	-0.00124		4.	9	-0.10237		6.	13	0.01900		9.	19	0.07478		15.	18	-0.01881
2.	6	-0.00978		4.	10	-0.01843		6.	14	-0.01931		10.	11	0.01551		15.	19	0.05197
2.	7	-0.03301		4.	11	0.00998		6.	15	0.02303		10.	12	-0.04543		16.	17	0.00874
2.	8	0.00998		4.	12	-0.12355		6.	16	-0.00071		10.	13	0.04681		16.	18	0.04996
2.	9	-0.07003		4.	13	0.01057		6.	17	0.01668		10.	14	0.02819		16.	19	0.00954
2.	10	0.04183		4.	14	0.06221		6.	18	0.03345		10.	15	-0.03835		17.	18	0.03178
2.	11	-0.06756		4.	15	-0.04035		6.	19	0.04648		10.	16	-0.00595		17.	19	0.03261
2.	12	-0.01305		4.	16	-0.00786		7.	8	-0.01763		10.	17	0.02692		18.	19	-0.00305
2.	13	-0.10260		4.	17	0.01039		7.	9	0.03868		10.	18	-0.03658				
2.	14	0.02757		4.	18	-0.00161		7.	10	-0.04187		10.	19	0.04116				
2.	15	0.01134		4.	19	0.00081		7.	11	-0.00723		11.	12	0.02017				
2.	16	0.00325						7.	12	0.01818		11.	13	0.02099				
2.	17	-0.02829						7.	13	0.04733		11.	14	-0.00661				
2.	18	-0.05116						7.	14	-0.03097		11.	15	0.03139				
2.	19	0.00426						7.	15	0.01105		11.	16	-0.00258				
								7.	16	0.04857		11.	17	0.01429				
								7.	17	-0.05239		11.	18	0.02564				
								7.	18	-0.02053		11.	19	0.03210				
								7.	19	0.04658								

TABLE 25
Correlation of Raw Factor Scores with Original Variables

Factor scores

	1	2	3	4	5	6	7	8	9	10	11	12	13	14	15	16	17	18	19
VAR 1	−0.45	−0.28	0.00	−0.30	−0.11	−0.12	0.09	−0.24	0.10	0.56	0.04	0.17	0.07	−0.23	−0.00	0.07	0.01	0.03	−0.09
VAR 2	−0.14	0.05	0.01	−0.04	−0.86	0.07	−0.01	0.14	−0.04	0.08	0.06	−0.06	−0.04	−0.07	0.11	0.01	0.06	0.07	0.04
VAR 3	0.89	−0.10	−0.14	0.18	0.07	0.12	−0.10	0.19	0.01	−0.05	−0.04	0.00	0.02	−0.01	−0.02	−0.04	0.11	0.03	0.14
VAR 4	−0.50	−0.02	0.19	−0.74	−0.07	0.04	0.14	−0.04	0.09	0.02	−0.00	0.06	0.04	−0.09	0.15	−0.00	−0.07	−0.01	0.04
VAR 5	0.83	−0.21	0.06	−0.06	−0.01	0.03	−0.06	0.05	−0.08	0.04	0.05	−0.00	0.15	−0.11	−0.13	−0.17	0.13	−0.02	0.08
VAR 6	0.64	−0.39	−0.01	−0.00	0.04	0.09	−0.03	−0.06	0.19	0.03	−0.03	−0.08	0.08	−0.09	−0.22	−0.09	0.09	0.13	−0.15
VAR 7	0.39	−0.04	0.29	0.30	0.04	0.14	0.06	0.04	0.07	0.18	0.25	−0.22	0.27	−0.11	0.04	−0.29	0.16	−0.30	0.00
VAR 8	0.57	−0.09	−0.13	−0.08	0.19	0.17	0.05	−0.02	−0.22	−0.01	−0.10	0.32	0.06	−0.38	−0.04	0.15	−0.09	−0.14	−0.25
VAR 9	−0.36	0.03	0.31	0.08	−0.10	−0.03	0.18	−0.01	0.01	−0.14	−0.06	−0.06	−0.09	0.71	−0.05	−0.03	0.01	−0.12	−0.05
VAR 10	−0.62	0.13	0.31	0.12	−0.13	−0.08	0.07	−0.11	−0.01	−0.24	−0.11	−0.02	−0.15	0.49	−0.01	−0.11	−0.00	0.06	−0.12
VAR 11	0.53	0.01	0.15	−0.01	−0.22	−0.02	0.09	−0.04	−0.01	0.02	−0.03	−0.11	0.17	−0.49	−0.06	0.02	−0.01	0.12	−0.10
VAR 12	0.13	−0.04	0.22	0.14	0.17	0.02	0.09	−0.27	0.07	−0.12	−0.00	0.04	0.31	0.48	−0.21	−0.26	0.24	0.02	−0.10
VAR 13	−0.27	0.30	−0.13	0.05	−0.52	0.06	−0.08	−0.08	−0.08	0.10	−0.22	0.42	0.12	0.13	0.06	0.09	0.00	0.03	0.09
VAR 14	−0.08	−0.24	−0.32	0.04	−0.03	−0.06	0.25	−0.02	0.02	0.04	0.00	0.13	−0.61	0.02	−0.10	−0.15	−0.08	−0.01	−0.13
VAR 15	−0.23	0.33	−0.13	0.11	0.59	0.03	0.05	0.07	0.03	0.06	−0.07	0.02	−0.03	−0.32	0.13	0.01	0.00	0.15	−0.02
VAR 16	−0.28	0.00	0.02	−0.12	−0.86	−0.12	0.13	−0.07	0.01	−0.02	0.01	−0.00	−0.02	−0.12	0.05	−0.01	−0.10	−0.00	−0.03
VAR 17	0.21	−0.10	0.08	−0.62	0.64	0.12	−0.02	0.07	0.01	−0.02	−0.03	0.03	0.05	0.02	0.07	−0.06	0.01	−0.02	0.11
VAR 18	−0.89	0.18	0.07	0.01	−0.03	−0.07	0.03	−0.17	−0.07	0.04	0.07	0.02	−0.10	0.06	0.03	0.02	−0.13	0.06	−0.08
VAR 19	−0.05	−0.12	−0.06	−0.18	−0.02	0.03	−0.05	−0.01	−0.21	0.68	−0.03	−0.18	−0.05	−0.14	0.02	0.16	0.13	−0.05	−0.15
VAR 20	−0.38	−0.05	−0.07	−0.05	−0.13	0.10	−0.11	0.02	0.12	0.13	0.09	−0.06	−0.71	0.00	0.02	0.15	−0.04	0.08	−0.07
VAR 21	−0.19	−0.14	−0.22	−0.07	−0.07	0.05	0.06	−0.06	0.09	0.77	0.05	0.17	−0.04	−0.08	−0.01	0.06	−0.09	0.03	0.01
VAR 22	0.15	−0.36	−0.37	−0.04	−0.10	0.05	−0.05	0.06	−0.21	−0.14	0.08	0.19	−0.01	−0.34	0.21	0.06	−0.19	−0.03	−0.34
VAR 23	0.60	−0.32	−0.23	0.15	0.08	0.18	0.04	0.25	−0.03	−0.02	−0.07	0.02	−0.07	0.07	0.03	0.08	−0.04	−0.22	−0.16
VAR 24	0.19	0.39	0.20	0.05	0.02	0.01	−0.03	0.07	0.11	0.04	0.04	−0.12	−0.05	−0.04	−0.07	0.14	0.05	0.59	0.04
VAR 25	−0.72	0.12	0.02	−0.14	−0.02	0.20	−0.10	−0.02	0.12	0.18	−0.02	−0.16	0.11	−0.05	−0.00	0.08	0.10	−0.14	−0.05
VAR 26	−0.82	0.23	0.01	−0.03	−0.20	−0.08	0.06	−0.10	0.05	0.12	0.12	−0.10	−0.12	−0.05	0.07	0.04	−0.08	−0.07	0.02
VAR 27	−0.80	0.18	−0.04	0.01	−0.14	0.16	−0.03	−0.03	0.18	0.07	−0.05	−0.18	−0.15	0.04	0.10	0.03	−0.05	0.03	0.11
VAR 28	0.09	−0.09	0.11	0.18	−0.02	−0.07	−0.00	−0.04	0.46	0.13	0.01	0.05	−0.12	−0.21	−0.00	0.03	−0.09	−0.14	−0.05
VAR 29	0.08	−0.03	0.09	−0.16	−0.07	0.07	0.67	0.01	−0.14	0.07	0.03	−0.09	−0.05	0.17	0.06	−0.06	0.05	−0.02	−0.14
VAR 30	0.20	−0.09	0.04	−0.09	0.03	−0.04	0.25	−0.08	−0.23	−0.17	−0.01	−0.58	0.10	0.09	−0.09	−0.01	0.07	0.02	−0.08
VAR 31	0.08	0.11	0.04	−0.14	−0.09	0.19	0.06	−0.11	0.15	−0.41	0.12	−0.19	−0.00	0.04	−0.13	0.12	0.17	−0.38	−0.14
VAR 32	0.16	−0.13	−0.07	0.02	−0.08	−0.01	0.06	−0.19	−0.70	0.10	0.03	−0.06	−0.08	−0.25	0.04	0.01	0.01	−0.18	0.04
VAR 33	−0.09	−0.07	−0.08	−0.01	0.05	0.03	0.15	0.04	0.01	0.07	0.04	−0.04	−0.06	0.02	0.16	0.01	0.04	−0.04	−0.69
VAR 34	0.22	−0.01	0.14	−0.04	−0.06	0.01	0.12	−0.03	−0.50	0.20	−0.05	−0.06	0.28	−0.11	0.13	0.03	−0.03	−0.03	−0.22
VAR 35	−0.09	0.13	0.80	0.04	−0.01	−0.04	−0.01	0.05	0.07	−0.12	−0.07	−0.01	−0.07	0.13	−0.13	−0.18	−0.03	0.02	0.09
VAR 36	−0.04	0.15	0.85	−0.05	0.01	−0.10	0.12	−0.01	−0.10	−0.01	0.01	−0.03	0.01	0.07	−0.16	−0.13	0.09	−0.02	−0.04
VAR 37	−0.20	0.09	0.38	−0.47	0.07	0.08	0.13	−0.13	−0.12	−0.10	0.19	0.37	−0.17	−0.01	0.01	−0.07	0.02	−0.23	−0.27
VAR 38	0.34	0.03	−0.52	0.30	−0.19	0.14	0.15	0.18	0.15	0.06	0.08	−0.17	0.09	−0.01	0.13	0.13	−0.29	0.06	0.33
VAR 39	0.21	−0.41	−0.61	0.09	−0.04	0.04	0.07	0.15	0.01	−0.13	0.08	0.09	−0.04	−0.21	0.12	0.12	−0.30	−0.14	−0.13
VAR 40	−0.41	0.25	0.69	−0.41	−0.06	−0.06	0.13	−0.07	−0.05	0.03	−0.05	0.15	0.04	0.01	0.03	0.07	−0.02	0.04	−0.04
VAR 41	−0.44	0.26	0.66	−0.38	−0.06	−0.06	0.09	−0.02	0.02	−0.11	−0.06	0.16	−0.09	0.10	0.12	0.00	−0.09	0.02	0.04
VAR 42	−0.42	−0.03	−0.13	−0.56	−0.07	0.06	−0.02	−0.12	−0.09	−0.14	−0.02	−0.05	−0.15	−0.13	0.38	0.07	0.02	0.19	−0.11
VAR 43	−0.30	0.02	−0.25	−0.58	−0.10	0.02	0.01	−0.16	−0.12	−0.08	−0.06	0.01	−0.11	−0.18	0.33	0.05	0.00	0.22	−0.21
VAR 44	−0.27	−0.04	−0.11	−0.37	0.01	−0.05	0.01	−0.05	0.02	0.01	0.04	0.03	0.00	−0.04	0.80	0.12	0.03	0.00	−0.10
VAR 45	−0.20	−0.05	−0.19	−0.26	−0.13	0.03	0.05	0.02	0.03	0.01	0.04	0.11	0.10	−0.07	0.82	0.03	−0.07	−0.04	−0.09
VAR 46	0.13	0.16	−0.11	0.87	0.05	0.06	−0.13	0.04	−0.13	−0.06	−0.09	−0.04	−0.07	0.04	−0.14	−0.01	0.08	0.07	−0.11
VAR 47	0.69	−0.14	−0.18	0.24	−0.02	0.05	−0.17	0.28	−0.04	−0.13	0.14	0.01	0.01	0.01	−0.15	0.13	0.03	0.03	0.17
VAR 48	−0.78	0.35	0.17	−0.07	−0.06	−0.10	0.10	−0.04	−0.13	−0.03	−0.02	0.03	0.02	0.05	0.12	0.03	−0.09	−0.12	0.08
VAR 49	0.86	−0.24	−0.16	0.11	0.04	0.04	−0.09	0.07	0.04	0.02	0.02	−0.16	0.07	0.05	−0.10	0.07	0.10	0.05	0.14
VAR 50	0.53	−0.05	−0.28	0.68	0.06	−0.05	−0.10	0.05	0.00	−0.06	0.01	−0.10	−0.06	0.12	−0.14	−0.00	0.11	0.04	0.05
VAR 51	0.76	−0.28	−0.06	0.31	0.03	0.02	−0.01	−0.03	0.08	−0.02	−0.08	−0.10	0.09	0.03	−0.08	−0.01	0.17	0.05	0.01
VAR 52	0.16	−0.77	−0.15	−0.16	0.01	−0.18	0.10	−0.16	−0.03	0.08	0.31	0.07	0.05	−0.07	0.07	0.12	−0.01	0.11	−0.02
VAR 53	0.88	−0.13	−0.12	0.16	0.14	0.01	−0.08	0.19	−0.07	0.03	−0.02	0.11	0.08	0.14	−0.12	0.09	0.02	−0.07	0.11
VAR 54	0.14	0.06	−0.15	0.16	−0.05	0.00	0.02	0.77	0.13	−0.07	−0.05	−0.06	−0.10	0.10	0.17	−0.11	−0.02	0.02	0.10
VAR 55	0.07	0.07	0.25	0.15	0.11	−0.08	0.07	0.18	−0.09	−0.17	0.04	−0.02	−0.01	0.11	−0.09	−0.80	0.06	−0.15	0.05
VAR 56	0.42	−0.08	−0.02	−0.14	−0.01	0.15	−0.05	0.69	0.02	−0.07	0.10	0.12	0.13	−0.20	−0.18	0.05	0.06	0.05	−0.18
VAR 57	0.53	0.08	−0.21	0.36	−0.06	0.31	0.20	0.21	−0.17	−0.03	0.14	−0.11	0.05	−0.01	0.04	0.28	−0.14	−0.20	0.04
VAR 58	0.85	−0.10	−0.14	0.22	0.14	0.10	0.04	0.17	−0.01	0.07	0.10	−0.01	0.11	−0.03	0.03	0.10	0.11	−0.09	0.09
VAR 59	0.66	0.06	−0.10	0.37	0.04	0.06	−0.10	0.33	−0.10	−0.02	0.13	0.07	−0.10	0.06	−0.08	0.08	0.12	−0.11	0.33
VAR 60	0.03	0.47	0.30	0.02	0.11	−0.14	−0.11	0.00	0.02	0.02	−0.65	0.03	0.07	0.10	−0.11	0.00	0.19	−0.01	0.07
VAR 61	0.04	−0.91	−0.16	−0.16	−0.06	−0.11	−0.07	−0.02	0.07	0.04	0.07	0.01	0.04	−0.09	0.02	0.02	0.02	0.03	−0.02
VAR 62	0.44	−0.11	0.10	0.01	0.03	−0.17	0.43	−0.02	0.14	−0.01	0.08	−0.19	0.04	0.08	0.05	−0.06	0.49	−0.08	−0.09
VAR 63	0.35	−0.78	−0.16	−0.07	0.07	−0.09	−0.07	0.12	−0.02	0.04	0.08	0.00	0.08	−0.08	−0.02	0.01	0.21	−0.02	0.09
VAR 64	0.21	−0.81	−0.22	0.09	−0.01	0.20	−0.06	0.08	0.07	0.07	0.10	0.08	0.08	0.02	−0.07	0.08	−0.05	0.07	0.06
VAR 65	0.24	−0.84	−0.35	−0.03	0.04	0.05	−0.11	0.05	0.12	−0.00	0.10	0.06	0.08	−0.03	0.06	−0.04	−0.05	0.05	0.04
VAR 66	0.30	−0.36	0.09	0.01	0.07	−0.16	0.25	0.37	−0.03	0.12	−0.24	0.03	0.25	0.09	−0.22	−0.00	0.23	−0.12	−0.06
VAR 67	−0.33	−0.05	−0.02	0.48	−0.20	0.19	0.10	−0.37	0.11	−0.10	0.17	−0.03	0.04	−0.05	−0.01	−0.14	0.03	0.31	0.19
VAR 68	0.64	−0.50	−0.29	0.21	0.01	−0.05	−0.06	0.15	0.07	0.04	0.01	−0.01	0.02	0.05	−0.00	0.03	−0.06	−0.03	0.06
VAR 69	−0.26	0.70	0.18	0.10	0.05	0.14	−0.22	−0.03	−0.10	−0.03	−0.20	0.03	−0.23	0.06	0.07	−0.13	0.00	0.03	0.17
VAR 70	−0.66	0.30	0.07	0.06	−0.02	−0.05	−0.10	−0.19	−0.11	0.02	0.40	0.09	−0.10	0.24	0.05	−0.02	−0.04	0.12	−0.05
VAR 71	−0.45	0.47	0.03	0.15	−0.14	0.07	−0.39	−0.09	−0.12	0.06	0.24	0.10	−0.09	0.15	0.05	0.09	−0.19	0.06	−0.06
VAR 72	0.07	−0.65	−0.15	−0.06	0.03	0.46	−0.16	−0.04	0.16	0.11	−0.10	−0.05	0.09	−0.01	0.03	−0.06	−0.03	−0.15	0.15
VAR 73	−0.13	−0.16	0.14	0.01	−0.08	−0.83	−0.02	−0.05	0.08	−0.04	−0.11	−0.06	0.01	0.03	0.02	−0.01	0.01	0.04	0.07
VAR 74	0.46	−0.73	−0.10	0.08	0.11	0.11	0.03	0.04	0.15	0.02	−0.15	−0.06	−0.00	0.03	0.06	−0.01	0.06	−0.01	0.14
VAR 75	0.42	0.14	0.08	0.17	0.12	−0.26	0.34	−0.02	0.07	0.05	0.04	0.01	0.06	0.15	−0.06	−0.05	0.55	−0.01	−0.07
VAR 76	−0.25	0.10	−0.23	−0.13	−0.01	−0.07	0.01	−0.06	0.11	0.04	0.09	−0.00	−0.03	−0.05	0.07	0.04	−0.79	−0.01	0.01
VAR 77	−0.01	0.66	0.03	0.08	0.04	0.49	0.10	−0.02	0.00	−0.07	−0.19	−0.02	−0.17	0.03	0.12	0.18	−0.00	0.10	−0.01
VAR 78	−0.45	0.26	0.13	0.11	−0.02	−0.48	0.02	−0.08	0.33	0.03	−0.13	−0.21	−0.01	0.07	−0.12	−0.08	−0.03	−0.04	0.00
VAR 79	−0.08	−0.34	0.07	0.30	−0.21	0.18	0.14	−0.20	−0.02	−0.13	−0.20	0.24	−0.04	−0.37	−0.17	−0.18	0.25	0.22	0.03
VAR 80	0.39	−0.80	−0.16	0.01	0.05	0.03	0.01	0.09	0.04	0.07	0.07	0.03	0.02	−0.00	0.04	0.06	0.08	0.06	0.03
VAR 81	0.32	−0.80	−0.14	0.08	−0.04	−0.07	−0.01	−0.01	0.08	0.15	0.12	−0.04	−0.02	0.04	0.04	−0.07	0.04	0.02	0.02
VAR 82	−0.11	0.42	0.26	−0.33	0.13	−0.32	−0.21	−0.02	−0.11	0.20	−0.18	−0.03	−0.14	0.17	0.10	−0.16	−0.17	−0.07	−0.08
VAR 83	−0.25	0.79	0.09	0.04	−0.02	−0.15	−0.03	0.02	0.15	0.01	−0.09	−0.12	−0.10	0.06	−0.00	−0.03	−0.04	−0.11	−0.06
VAR 84	0.81	−0.22	−0.24	0.05	0.18	−0.05	−0.02	0.10	0.10	0.01	0.00	0.09	−0.06	0.08	−0.03	0.06	0.21	−0.05	0.10
VAR 85	−0.06	0.63	−0.16	0.15	−0.04	0.37	−0.10	0.13	0.05	0.02	−0.12	0.10	−0.20	0.12	0.03	0.06	−0.23	0.02	0.16
VAR 86	−0.25	0.76	−0.04	0.08	−0.14	0.09	−0.13	0.06	−0.11	0.10	0.13	0.03	−0.05	0.01	0.09	−0.01	−0.16	−0.09	0.20
VAR 87	0.30	0.63	0.36	0.14	0.10	0.04	−0.05	0.03	−0.22	−0.01	−0.18	0.11	0.14	0.05	−0.14	0.10	0.22	0.09	0.11
VAR 88	0.44	0.05	−0.29	0.08	0.10	−0.04	0.01	0.19	−0.08	0.02	0.07	0.10	0.11	−0.23	0.07	0.54	0.09	−0.16	0.13
VAR 89	0.22	0.15	0.42	0.44	0.22	0.03	0.09	0.19	−0.05	−0.22	0.19	−0.22	0.02	0.46	−0.03	−0.10	0.14	−0.04	0.22
VAR 90	−0.07	0.11	0.36	0.40	0.22	0.11	0.13	0.12	−0.08	−0.30	−0.06	−0.08	−0.07	0.60	−0.03	−0.16	−0.03	0.13	−0.02
VAR 91	−0.35	−0.25	0.14	−0.76	−0.09	−0.09	0.10	−0.17	0.07	0.26	0.04	0.00	0.04	−0.14	0.07	0.00	0.03	−0.00	−0.01

TABLE 26
Communalities of Variables in Rank Order

VARIABLE 44	0.89		VARIABLE 1	1.00
VARIABLE 5	0.88		VARIABLE 11	1.00
VARIABLE 29	0.88		VARIABLE 12	1.00
VARIABLE 46	0.88		VARIABLE 13	1.00
VARIABLE 55	0.88		VARIABLE 15	1.00
VARIABLE 60	0.88		VARIABLE 53	1.00
VARIABLE 70	0.88		VARIABLE 61	1.00
VARIABLE 84	0.88		VARIABLE 64	1.00
VARIABLE 45	0.87		VARIABLE 73	1.00
VARIABLE 83	0.87		VARIABLE 75	1.00
VARIABLE 89	0.87		VARIABLE 28	0.99
VARIABLE 8	0.86		VARIABLE 20	0.98
VARIABLE 9	0.86		VARIABLE 21	0.98
VARIABLE 10	0.86		VARIABLE 91	0.97
VARIABLE 30	0.86		VARIABLE 27	0.96
VARIABLE 31	0.86		VARIABLE 63	0.95
VARIABLE 38	0.86		VARIABLE 65	0.95
VARIABLE 51	0.86		VARIABLE 74	0.95
VARIABLE 59	0.86		VARIABLE 16	0.94
VARIABLE 6	0.85		VARIABLE 18	0.94
VARIABLE 36	0.85		VARIABLE 3	0.93
VARIABLE 48	0.85		VARIABLE 14	0.93
VARIABLE 78	0.85		VARIABLE 26	0.93
VARIABLE 39	0.84		VARIABLE 49	0.93
VARIABLE 87	0.84		VARIABLE 50	0.93
VARIABLE 90	0.84		VARIABLE 62	0.93
VARIABLE 68	0.83		VARIABLE 72	0.93
VARIABLE 69	0.83		VARIABLE 2	0.92
VARIABLE 47	0.82		VARIABLE 57	0.92
VARIABLE 54	0.82		VARIABLE 77	0.92
VARIABLE 56	0.82		VARIABLE 86	0.92
VARIABLE 66	0.81		VARIABLE 4	0.91
VARIABLE 22	0.80		VARIABLE 19	0.91
VARIABLE 42	0.80		VARIABLE 23	0.91
VARIABLE 43	0.80		VARIABLE 40	0.91
VARIABLE 67	0.79		VARIABLE 58	0.91
VARIABLE 79	0.79		VARIABLE 71	0.91
VARIABLE 33	0.78		VARIABLE 81	0.91
VARIABLE 7	0.77		VARIABLE 17	0.90
VARIABLE 35	0.77		VARIABLE 32	0.90
VARIABLE 37	0.75		VARIABLE 41	0.90
VARIABLE 88	0.75		VARIABLE 52	0.90
VARIABLE 34	0.74		VARIABLE 76	0.90
VARIABLE 24	0.73		VARIABLE 80	0.90
VARIABLE 82	0.70		VARIABLE 85	0.90
			VARIABLE 25	0.89

TABLE 27
Correlation of Ranked Factor Scores with Original Variables

Factor scores ranked

	1	2	3	4	5	6	7	8	9	10	11	12	13	14	15	16	17	18	19
VAR 1	-0.44	-0.23	0.01	-0.32	-0.13	-0.07	0.09	-0.26	0.16	0.51	0.07	0.18	0.03	-0.23	0.01	0.02	-0.03	0.01	-0.05
VAR 2	-0.15	0.11	-0.01	-0.03	-0.85	0.06	0.00	0.15	-0.02	0.03	0.02	-0.02	-0.06	-0.07	0.10	0.00	0.06	0.10	0.06
VAR 3	0.88	-0.18	-0.14	0.18	0.14	0.06	-0.09	0.22	-0.04	-0.03	-0.06	-0.02	0.07	-0.04	-0.04	0.01	0.13	0.03	0.15
VAR 4	-0.48	0.03	0.22	-0.75	-0.13	0.09	0.15	-0.05	0.11	0.00	0.03	0.07	0.01	-0.09	0.15	-0.06	-0.07	-0.00	0.05
VAR 5	0.82	-0.30	0.05	-0.04	0.06	-0.02	-0.04	0.07	-0.07	0.01	0.04	-0.01	0.18	-0.13	-0.16	-0.13	0.15	0.00	0.09
VAR 6	0.62	-0.44	-0.02	0.02	0.08	0.06	-0.02	-0.04	0.20	0.02	0.03	-0.07	0.09	-0.10	-0.24	-0.11	0.08	0.15	-0.13
VAR 7	0.36	-0.10	0.28	0.32	0.06	0.12	0.05	0.05	0.06	0.21	0.25	-0.24	0.30	-0.11	0.04	-0.26	0.16	-0.31	0.06
VAR 8	0.57	-0.15	-0.15	-0.06	0.25	0.13	0.04	0.00	-0.20	-0.01	-0.12	0.32	0.06	-0.39	-0.06	0.18	-0.09	-0.15	-0.23
VAR 9	-0.37	0.04	0.31	0.07	-0.13	-0.04	0.18	-0.02	0.00	-0.11	-0.05	-0.08	-0.11	0.73	-0.03	-0.02	0.01	-0.09	-0.04
VAR 10	-0.63	0.16	0.30	0.10	-0.17	-0.07	0.08	-0.12	-0.01	-0.21	-0.10	-0.02	-0.17	0.50	0.02	-0.13	0.00	0.07	-0.15
VAR 11	0.52	-0.03	0.14	0.03	-0.17	-0.04	0.10	-0.03	-0.01	-0.02	-0.04	-0.10	0.17	-0.49	-0.08	0.04	-0.00	0.13	-0.05
VAR 12	0.10	-0.10	0.22	0.13	0.20	0.02	0.11	-0.28	0.11	-0.09	0.00	0.01	0.35	0.48	-0.22	-0.24	0.25	0.04	-0.09
VAR 13	-0.26	0.34	-0.14	0.02	-0.49	0.07	-0.08	-0.10	-0.03	0.10	-0.30	0.41	0.11	0.14	0.07	0.08	-0.00	0.04	0.09
VAR 14	-0.08	-0.19	-0.32	0.05	-0.03	-0.05	0.25	-0.00	-0.01	0.05	0.00	0.15	-0.61	0.04	-0.07	-0.12	-0.11	-0.03	-0.15
VAR 15	-0.21	0.32	-0.10	0.09	0.55	0.07	0.02	0.06	0.00	0.09	-0.06	0.04	-0.02	-0.30	0.15	0.00	-0.00	0.10	-0.02
VAR 16	-0.29	0.08	0.02	-0.11	-0.90	-0.09	0.15	-0.07	0.05	-0.08	-0.01	0.02	-0.06	-0.11	0.05	-0.04	-0.11	0.03	-0.01
VAR 17	0.23	-0.14	0.12	-0.62	0.64	0.11	-0.02	0.07	-0.02	0.01	0.01	0.03	0.06	0.00	0.06	-0.05	0.04	-0.03	0.10
VAR 18	-0.88	0.26	0.08	-0.01	-0.08	-0.01	0.01	-0.19	-0.06	0.05	0.07	0.04	-0.14	0.07	0.06	-0.03	-0.14	0.03	-0.09
VAR 19	-0.04	-0.12	-0.07	-0.17	-0.02	0.04	-0.10	-0.02	-0.16	0.64	-0.04	-0.17	-0.09	-0.15	0.03	0.16	0.11	-0.08	-0.11
VAR 20	-0.36	0.01	-0.07	-0.05	-0.18	0.14	-0.11	0.03	0.06	0.13	0.12	-0.03	-0.70	0.01	0.06	0.12	-0.05	0.08	-0.13
VAR 21	-0.18	-0.09	-0.23	-0.10	-0.05	0.08	0.03	-0.06	0.14	0.74	0.05	0.18	-0.09	-0.09	0.02	0.04	-0.10	-0.01	0.05
VAR 22	0.16	-0.33	-0.39	-0.06	-0.08	0.03	-0.04	0.06	-0.19	-0.14	0.10	0.21	-0.02	-0.37	0.22	0.03	-0.17	-0.09	-0.30
VAR 23	0.60	-0.35	-0.27	0.15	0.14	0.11	0.04	0.27	-0.02	-0.03	-0.06	-0.00	-0.04	0.04	0.03	0.13	-0.02	-0.23	-0.14
VAR 24	0.18	0.36	0.21	0.07	0.05	0.01	-0.01	0.08	0.09	0.03	0.02	-0.08	-0.02	-0.02	-0.07	0.15	0.07	0.59	-0.00
VAR 25	-0.72	0.18	0.03	-0.14	-0.08	0.24	-0.14	-0.04	0.16	0.18	-0.01	-0.14	0.10	-0.06	0.00	0.04	0.09	-0.12	-0.04
VAR 26	-0.82	0.31	0.03	-0.05	-0.27	-0.01	0.04	-0.12	0.06	0.11	0.12	-0.07	-0.15	-0.04	0.09	-0.00	-0.09	-0.08	0.03
VAR 27	-0.79	0.27	-0.03	-0.01	-0.19	0.19	-0.07	-0.04	0.16	0.09	-0.04	-0.14	-0.17	0.05	0.12	-0.01	-0.05	0.03	0.10
VAR 28	0.09	-0.08	0.08	0.15	-0.04	-0.08	-0.02	-0.05	0.45	0.14	0.04	0.07	-0.14	-0.22	-0.01	-0.00	-0.15	-0.06	-0.05
VAR 29	0.07	-0.05	0.09	-0.13	-0.06	0.06	0.67	0.03	-0.11	0.06	0.03	-0.09	-0.06	0.18	0.06	-0.07	0.07	-0.06	-0.10
VAR 30	0.19	-0.11	0.08	-0.03	0.04	-0.08	0.24	-0.07	-0.23	-0.20	0.03	-0.57	0.12	0.09	-0.09	0.01	0.09	0.02	-0.10
VAR 31	0.08	0.10	0.05	-0.10	-0.10	0.17	0.06	-0.11	0.13	-0.40	0.13	-0.18	0.04	0.03	-0.13	0.12	0.15	-0.33	-0.16
VAR 32	0.16	-0.15	-0.07	0.04	-0.05	-0.02	0.05	-0.20	-0.61	0.04	-0.00	-0.07	-0.10	-0.22	0.04	0.02	0.02	-0.21	0.04
VAR 33	-0.09	-0.05	-0.09	-0.00	0.03	0.02	0.14	0.03	0.06	0.08	0.06	-0.03	-0.07	0.00	0.18	-0.03	0.05	-0.06	-0.65
VAR 34	0.21	-0.04	0.11	-0.02	-0.02	-0.00	0.13	-0.03	-0.40	0.14	-0.08	-0.05	0.25	-0.12	0.13	0.03	-0.00	-0.05	-0.15
VAR 35	-0.11	0.11	0.81	0.05	-0.03	-0.03	0.03	0.04	0.06	-0.12	-0.04	-0.04	-0.08	0.16	-0.14	-0.19	-0.02	0.05	0.07
VAR 36	-0.06	0.09	0.86	-0.04	-0.00	-0.10	0.15	-0.03	-0.08	-0.04	0.05	-0.06	-0.02	0.09	-0.18	-0.13	0.10	0.01	-0.04
VAR 37	-0.20	0.08	0.39	-0.47	0.04	0.10	0.13	-0.13	-0.08	-0.12	0.20	0.38	-0.15	0.01	0.04	-0.08	0.04	-0.22	-0.29
VAR 38	0.34	0.05	-0.53	0.29	-0.15	0.13	0.15	0.20	0.11	0.05	0.03	-0.17	0.09	-0.03	0.13	0.18	-0.29	0.05	0.33
VAR 39	0.22	-0.36	-0.63	0.08	-0.03	0.01	0.06	0.15	0.02	-0.13	0.08	0.09	-0.05	-0.22	0.13	0.11	-0.31	-0.16	-0.11
VAR 40	-0.41	0.24	0.69	-0.42	-0.11	-0.03	0.13	-0.10	-0.03	0.01	-0.04	0.14	-0.01	0.03	0.03	0.02	-0.02	0.06	-0.05
VAR 41	-0.43	0.28	0.66	-0.39	-0.12	-0.04	0.11	-0.05	0.03	-0.10	-0.05	0.15	-0.13	0.12	0.12	-0.04	-0.09	0.03	0.02
VAR 42	-0.40	0.04	-0.10	-0.55	-0.12	0.09	-0.00	-0.12	-0.08	-0.11	0.01	-0.02	-0.17	-0.12	0.38	0.02	0.02	0.15	-0.12
VAR 43	-0.28	0.07	-0.23	-0.58	-0.12	0.05	0.02	-0.16	-0.11	-0.06	-0.06	0.05	-0.15	-0.17	0.34	-0.00	-0.00	0.17	-0.23
VAR 44	-0.24	0.00	-0.12	-0.41	-0.05	-0.03	0.03	-0.05	0.05	0.03	0.07	0.04	-0.02	-0.06	0.80	0.04	0.03	-0.02	-0.07
VAR 45	-0.18	0.01	-0.20	-0.31	-0.16	0.05	0.07	0.02	0.07	0.01	0.05	0.12	0.07	-0.10	0.81	-0.04	-0.08	-0.06	-0.05
VAR 46	0.12	0.13	-0.15	0.87	0.10	0.04	-0.15	0.05	-0.15	-0.04	-0.12	-0.05	-0.04	0.06	-0.13	0.03	0.08	0.03	-0.11
VAR 47	0.69	-0.19	-0.20	0.26	0.03	-0.00	-0.16	0.30	-0.07	-0.13	0.11	-0.02	0.06	-0.01	-0.17	0.19	0.04	0.03	0.15
VAR 48	-0.77	0.41	0.18	-0.08	-0.13	-0.05	0.07	-0.06	-0.11	-0.02	-0.04	0.03	-0.01	0.08	0.14	0.01	-0.10	-0.13	0.07
VAR 49	0.85	-0.31	-0.17	0.14	0.12	-0.01	-0.08	0.10	0.03	0.02	0.01	-0.17	0.11	0.03	-0.13	0.12	0.12	0.05	0.14
VAR 50	0.52	-0.10	-0.31	0.69	0.12	-0.09	-0.11	0.07	-0.04	-0.03	-0.01	-0.10	-0.02	0.12	-0.15	0.06	0.09	0.03	0.03
VAR 51	0.73	-0.35	-0.06	0.33	0.09	-0.03	0.00	-0.01	0.08	0.00	-0.04	-0.12	0.12	0.03	-0.11	0.03	0.18	0.06	0.01
VAR 52	0.15	-0.76	-0.15	-0.18	0.00	-0.19	0.11	-0.15	0.01	0.04	0.34	0.04	0.02	-0.07	0.05	0.11	-0.04	0.12	-0.02
VAR 53	0.87	-0.21	-0.14	0.16	0.22	-0.05	-0.07	0.20	-0.08	0.05	-0.07	0.08	0.10	0.13	-0.14	0.16	0.03	-0.07	0.12
VAR 54	0.15	0.07	-0.13	0.15	-0.07	-0.03	0.04	0.78	0.08	-0.07	-0.08	-0.09	-0.09	0.09	0.15	-0.05	0.00	0.03	0.10
VAR 55	0.05	0.03	0.27	0.16	0.12	-0.08	0.10	0.17	-0.13	-0.15	0.05	-0.04	0.01	0.11	-0.08	-0.73	0.10	-0.18	0.04
VAR 56	0.43	-0.13	-0.04	-0.14	0.03	0.10	-0.02	0.70	-0.01	-0.07	0.07	0.15	0.16	-0.18	-0.20	0.08	0.08	0.07	-0.17
VAR 57	0.53	0.04	-0.25	0.36	-0.01	0.28	0.19	0.22	-0.15	-0.05	0.10	-0.12	0.09	-0.02	0.04	0.33	-0.13	-0.21	0.06
VAR 58	0.84	-0.18	-0.17	0.23	0.19	0.05	0.04	0.19	-0.03	0.07	0.06	-0.06	0.15	-0.04	0.02	0.15	0.13	-0.11	0.12
VAR 59	0.65	-0.00	-0.12	0.37	0.09	0.02	-0.09	0.35	-0.14	-0.01	0.06	0.01	-0.05	0.05	-0.10	0.15	0.14	-0.13	0.34
VAR 60	0.03	0.42	0.30	0.04	0.11	-0.15	-0.10	-0.01	0.05	0.04	-0.68	0.01	0.07	0.11	-0.12	0.01	0.20	0.01	0.09
VAR 61	0.04	-0.87	-0.17	-0.18	-0.08	-0.13	-0.06	-0.02	0.10	-0.00	0.15	0.00	-0.00	-0.11	0.01	0.00	-0.01	0.04	-0.01
VAR 62	0.43	-0.18	0.09	0.02	0.07	-0.19	0.42	0.00	0.16	-0.03	0.10	-0.20	0.06	0.07	0.03	-0.06	0.47	-0.07	-0.03
VAR 63	0.35	-0.80	-0.18	-0.07	0.08	-0.14	-0.07	0.12	-0.01	0.01	0.13	-0.02	0.07	-0.10	-0.04	0.04	0.20	-0.03	0.10
VAR 64	0.20	-0.81	-0.25	0.07	0.01	0.16	-0.04	0.09	0.06	0.04	0.13	0.08	0.08	-0.01	-0.08	0.09	-0.07	0.07	0.09
VAR 65	0.25	-0.82	-0.35	-0.04	0.04	0.01	-0.11	0.05	0.11	0.01	0.16	0.05	0.06	-0.06	0.05	-0.04	-0.06	0.06	0.05
VAR 66	0.27	-0.41	0.08	0.01	0.10	-0.21	0.26	0.37	0.05	0.08	-0.23	-0.01	0.24	0.09	-0.24	0.02	0.22	-0.10	-0.00
VAR 67	-0.35	-0.02	-0.02	0.48	-0.22	0.23	0.10	-0.36	0.11	-0.11	0.18	-0.05	0.05	-0.03	-0.01	-0.16	0.03	0.32	0.15
VAR 68	0.64	-0.50	-0.31	0.20	0.04	-0.10	-0.05	0.16	0.08	0.03	0.03	-0.03	0.03	0.03	-0.01	0.06	-0.06	-0.05	0.07
VAR 69	-0.25	0.70	0.20	0.09	0.03	0.17	-0.24	-0.03	-0.14	-0.00	-0.23	0.02	-0.20	0.08	0.09	-0.14	0.02	0.02	0.15
VAR 70	-0.65	0.35	0.07	0.04	-0.05	0.01	-0.11	-0.20	-0.11	0.02	0.34	0.11	-0.10	0.25	0.07	-0.05	-0.04	0.09	-0.09
VAR 71	-0.45	0.52	0.01	0.12	-0.17	0.10	-0.41	-0.09	-0.16	0.09	0.17	0.12	-0.08	0.16	0.08	0.06	-0.19	0.05	-0.11
VAR 72	0.06	-0.63	-0.15	-0.06	0.03	0.43	-0.16	-0.03	0.17	0.11	-0.03	-0.07	0.08	-0.05	0.03	-0.05	-0.03	-0.12	0.17
VAR 73	-0.13	-0.13	0.15	-0.01	-0.13	-0.82	0.01	-0.08	0.09	-0.06	-0.05	-0.09	-0.02	0.04	0.01	-0.02	-0.01	0.06	0.06
VAR 74	0.45	-0.76	-0.11	0.07	0.13	0.05	0.05	0.05	0.16	0.01	-0.07	-0.08	-0.01	-0.00	0.05	-0.00	0.04	-0.00	0.15
VAR 75	0.40	0.04	0.08	0.16	0.17	-0.28	0.34	-0.02	0.08	0.05	0.04	-0.02	0.09	0.14	-0.08	-0.05	0.53	-0.00	-0.03
VAR 76	-0.23	0.20	-0.22	-0.13	-0.04	-0.03	0.00	-0.07	0.05	0.06	0.07	0.03	-0.06	-0.05	0.12	0.04	-0.79	-0.02	-0.00
VAR 77	0.00	0.63	0.02	0.10	0.05	0.51	0.07	0.00	-0.01	-0.04	-0.24	0.01	-0.14	0.05	0.13	0.18	0.03	0.06	-0.02
VAR 78	-0.45	0.31	0.14	0.10	-0.06	-0.45	0.02	-0.10	0.32	0.05	-0.08	-0.20	-0.03	0.09	-0.10	-0.11	-0.06	-0.03	-0.01
VAR 79	-0.09	-0.37	0.06	0.29	-0.17	0.19	0.16	-0.20	-0.00	-0.15	-0.15	0.25	-0.05	-0.37	-0.19	-0.23	0.26	0.24	0.04
VAR 80	0.38	-0.82	-0.17	-0.01	0.08	-0.02	0.02	0.10	0.07	0.04	0.12	0.01	0.01	-0.03	0.02	0.08	0.06	0.05	0.03
VAR 81	0.31	-0.80	-0.15	0.06	-0.02	-0.11	0.00	-0.00	0.12	0.13	0.19	-0.07	-0.05	0.01	0.03	-0.06	0.01	0.02	0.04
VAR 82	-0.12	0.43	0.29	-0.32	0.11	-0.29	-0.20	-0.05	-0.13	0.21	-0.20	-0.04	-0.17	0.16	0.10	-0.17	-0.15	-0.06	-0.08
VAR 83	-0.25	0.81	0.10	0.03	-0.05	-0.10	-0.05	0.02	0.14	0.03	-0.14	-0.10	-0.10	0.09	0.01	-0.02	-0.04	-0.11	-0.08
VAR 84	0.81	-0.29	-0.26	0.05	0.24	-0.10	-0.02	0.13	0.09	0.01	-0.02	0.05	-0.03	0.06	-0.04	0.11	0.21	-0.04	0.10
VAR 85	-0.04	0.65	-0.17	0.15	-0.04	0.39	-0.12	0.13	0.04	0.07	-0.20	0.11	-0.18	0.13	0.05	0.09	-0.21	-0.00	0.14
VAR 86	-0.24	0.80	-0.05	0.09	-0.15	0.13	-0.16	0.06	-0.14	0.12	0.03	0.04	-0.02	0.01	0.10	0.00	-0.15	-0.11	0.19
VAR 87	0.30	0.53	0.36	0.16	0.14	0.03	-0.06	0.04	-0.23	-0.01	-0.25	0.09	0.16	0.07	-0.16	0.15	0.25	0.09	0.10
VAR 88	0.46	0.01	-0.32	0.08	0.11	-0.06	-0.01	0.20	-0.07	0.03	-0.01	0.07	0.12	-0.23	0.04	0.56	0.10	-0.17	0.16
VAR 89	0.19	0.07	0.41	0.44	0.21	0.01	0.09	0.20	-0.09	-0.18	0.19	-0.25	0.06	0.47	-0.04	-0.05	0.17	-0.04	0.20
VAR 90	-0.09	0.07	0.36	0.39	0.22	0.08	0.12	0.12	-0.12	-0.23	-0.04	-0.08	-0.05	0.62	-0.01	-0.12	-0.00	0.12	-0.04
VAR 91	-0.34	-0.21	0.17	-0.76	-0.14	-0.05	0.10	-0.19	0.12	0.20	0.10	0.02	-0.00	-0.15	0.06	-0.05	0.01	0.00	0.00

TABLE 28

A COMPARISON OF FACTOR ANALYTIC STUDIES IN INTERNATIONAL RELATIONS

Reference	No. of Subjects	No. of Variables	Factor Method	Entries in Principal Diagonal	Factor Technique	Factors Interpreted	Rotation	Rotated Matrix Presented	Unrotated Matrix Presented	Missing Data	% Missing Data	Missing Data Handled	Data Transformed	Smallest N in Correlation Matrix
1. Adelman and Morris (Soc. Econ. Pol.), *Society, Politics, and Economic Development: A Quantitative Approach*	74	24	Prin. Comp.	Comm.	R	Causes	Orth.	Yes	No	No	0%	NA	Rank	74
2. Alker (Voting Data, Unrotated), "Dimensions of Conflict in the General Assembly"	101	70	Prin. Comp.	Un.	R	?	Orth.	Yes	Yes	?	?	?	Rank No.	?
3. Alker (Voting Data, Rotated), "Dimensions of Conflict in the General Assembly"	101	70	Prin. Comp.	Un.	R	?	Orth.	Yes	Yes	?	?	?	Rank No.	?
4. Alker and Russett (Voting Data, 1947, Unrotated), *World Politics in the General Assembly*	56	48	Prin. Comp.	Un.	R	?	Orth.	Yes	Yes	Yes	?	Est. (Mean-Other)	Rank No.	56

Table 28 continued

No.	Study														
5.	Alker and Russett (Voting Data, 1947, Rotated), *World Politics in the General Assembly*	56	48	Prin. Comp.	Un.	R	?	Yes		Orth.	Yes	?	Est. (Mean-Other)	Rank No.	56
6.	Alker and Russett (Voting Data, 1952, Unrotated), *World Politics in the General Assembly*	60	63	Prin. Comp.	Un.	R	?	Yes	Yes	Orth.		?	Est. (Mean-Other)	Rank No.	60
7.	Alker and Russett (Voting Data, 1952, Rotated), *World Politics in the General Assembly*	60	63	Prin. Comp.	Un.	R	?	Yes		Orth.	Yes	?	Est. (Mean-Other)	Rank No.	60
8.	Alker and Russett (Voting Data, 1957, Unrotated), *World Politics in the General Assembly*	81	50	Prin. Comp.	Un.	R	?	Yes	Yes	Orth.		?	Est. (Mean-Other)	Rank No.	81
9.	Alker and Russett (Voting Data, 1957, Rotated), *World Politics in the General Assembly*	81	50	Prin. Comp.	Un.	R	?	Yes		Orth.	Yes	?	Est. (Mean-Other)	Rank No.	81
10.	Alker and Russett (Voting Data, 1961, Unrotated), *World Politics in the General Assembly*	101	70	Prin. Comp.	Un.	R	?	Yes	Yes	Orth.		?	Est. (Mean-Other)	Rank No.	101

Table 28 continued

Is the reader warned that communalities & variance explained is inflated when subjects are dropped from the correlation matrix?	Factor scores calculated	Factors calculated to ? eigenvalue	Factors calculated	Factors presented	Factors discussed	% Variance explained by factor scores calculated	% Variance explained by factors calculated	% Variance explained by factors discussed	Is the correlation of factor scores relative to original variables given, if different from the loadings?	Variables heterogeneous by category	Variables homogeneous by category
NA	Comp.	?	4	4	4	53%	73.7%	73.7%	NA	Yes	
NA	Comp.	1.0	9	9	2	66%	83%	66%	NA		Yes
NA	Comp.	1.0	9	9	4	?	83%	?	NA		Yes
NA	Comp.	1.0	8	8	2	52.1%	81%	52.1%	NA		Yes
NA	Incomp.	1.0	8	8	8	80.8%	80.8%	80.8%	No		Yes
NA	Comp.	1.0	9	9	2	58.1%	82.3%	58.1%	NA		Yes
NA	Incomp.	1.0	9	9	9	80.3%	80.3%	80.3%	No		Yes
NA	Comp.	1.0	7	7	2	62.4%	80%	62.4%	NA		Yes
NA	Incomp.	1.0	6		6	80%	80%	80%	No		Yes
NA	Comp.	1.0	7	7	2	83%	64%	83%	NA		Yes

Table 28 continued

Reference	No. of Subjects	No. of Variables	Factor Method	Entries in Principal Diagonal	Factor Technique	Factors Interpreted	Rotation	Rotated Matrix Presented	Unrotated Matrix Presented	Missing Data	% Missing Data	Missing Data Handled	Data Transformed	Smallest N in Correlation Matrix
11. Alker and Russett (Voting Data, 1961, Rotated), *World Politics in the General Assembly*	101	70	Prin. Comp.	Un.	R	?	Orth.	Yes	Yes	Yes	?	Est. (Mean-Other)	Rank No.	101
12. Banks and Gregg (Political), "Grouping Political Systems: Q-Factor Analysis of *A Cross-Polity Survey*"	115	68	Prin. Comp.	Un.	Q	?	Orth.	Yes	No	Yes	?	Drop	No	?
13. Cattell (Soc. Cult. Pol.), "The Dimensions of Culture Patterns by Factorization of National Characteristics"	69	72	Centroid	Comm.	R	Causes	Obliq.	Yes	Yes	Yes	?	Drop	Rank	?
14. Cattell and Gorsuch (Soc. Cult. Pol.), "The Definition and Measurement of National Morale and Morality"	52	51	Prin. Comp.	Comm.	R	Causes	Obliq.	Yes	No	No	0%	NA	Rank	52

Table 28 continued

				Comm.	R	Causes	Obliq.	Yes	No	No	0%	NA	Rank	
15. Cattell, Breul, and Hartman (Soc. Cult. Pol.), "An Attempt at a More Refined Definition of the Cultural Dimensions of Syntality in Modern Nations"	40	72	Cent-roid											40
16. Gregg and Banks (Political), "Dimensions of Political Systems: Factor Analysis of *A Cross-Polity Survey*"	115	68	Prin. Comp.	Un.	R	?	Orth. & Obliq.	Yes	No	Yes	?	Drop	No	?
17. Rummel (Domestic Conflict), "Dimensions of Conflict Behavior Within Nations"	113	13	Prin. Comp.	Un.	R	?	Orth. & Obliq.	Yes	Yes	?	?	?	Logs	113
18. Rummel (Foreign Conflict), "Dimensions of Conflict Behavior Within and Between Nations"	77	13	Prin. Comp.	Un.	R	?	Orth.	Yes	Yes	?	?	?	Logs & Other	77
19. Rummel (Domestic Conflict), "Dimensions of Conflict Behavior Within and Between Nations"	77	9	Prin. Comp.	Un.	R	?	Orth.	Yes	Yes	?	?	?	Logs & Other	77
20. Rummel (Foreign and Domestic Conflict), "Dimensions of Conflict Behavior Within and Between Nations"	77	24	Prin. Comp.	Un.	R	?	Orth.	Yes	Yes	?	?	?	Logs & Other	77

Table 28 continued

Is the reader warned that communalities & variance explained is inflated when subjects are dropped from the correlation matrix?	Factor scores calculated	Factors calculated to ? eigenvalue	Factors calculated	Factors presented	Factors discussed	% Variance explained by factor scores calculated	% Variance explained by factors calculated	% Variance explained by factors discussed	Is the correlation of factor scores relative to original variables given, if different from the loadings?	Variables heterogeneous by category	Variables homogeneous by category
NA	Incomp.	1.0	7	7	7	82.9%	82.9%	82.9%	No		Yes
No	No	1.0	?	5	5	NA	?	88.7%	NA		Yes
No	No	?	12	12	12	NA	NA	NA	NA	Yes	
NA	Incomp.	1.0	14	14	14	NA	NA	NA	No	Yes	
NA	No	?	12	12	12	NA	NA	NA	NA	Yes	
Yes	No	?	?	7	7	NA	?	72%	NA		
NA	No	1.0	3	3	3	NA	64.3%	64.3%	NA		Yes
NA	Incomp.	1.0	3	3	3	66.4%	66.4%	66.4%	No		Yes
NA	Incomp.	1.0	3	3	3	70.8%	70.8%	70.8%	No		Yes
NA	No	1.0	6	6	5	NA	69.1%	69%	NA		Yes

Table 28 continued

Reference	No. of Subjects	No. of Variables	Factor Method	Entries in Principal Diagonal	Factor Tech- nique	Factors Inter- preted	Rota- tion	Rota- ted Matrix Pre- sented	Unro- tated Matrix Pre- sented	Miss- ing Data	% Miss- ing Data	Miss- ing Data Han- dled	Data Trans- formed	Small- est N in Corre- lation Matrix
21. Rummel (Attributes), "Indicators of Cross-National and International Patterns"	82	236	Prin. Comp.	Un.	R	?	Orth. & Obliq.	No	No	Yes	17%	Drop	?	?
22. Russett (Soc. Cult.), "Delineating International Regions"	82	54	Prin. Comp.	Un.	R	?	Orth.	Yes	No	Yes	15%	Est. (Regression-Mean)	Logs	82
23. Russett (Regions), "Delineating International Regions"	82	29	Prin. Comp.	Un.	Q	?	Orth.	Yes	No	No	0%	NA	?	29
24. Russett (Voting Data), "Discovering Voting Groups in the United Nations"	107	66	Prin. Comp.	Un.	Q Descrip.		Orth.	Yes	No	Yes	?	Est. (Other)	No	107
25. Sawyer (Soc. Econ. Pol.), "Dimensions of Nations: Size, Wealth and Politics"	82	236	Prin. Comp.	?	R	?	Orth.	No	No	Yes	18%	Drop	Logs & Other	25
26. Tanter (Foreign Conflict), "Dimensions of Conflict Behavior Within and Between Nations, 1958–1960"	83	13	Prin. Comp.	Un.	R	?	Orth. & Oblique	Yes	No	?	?	?	Yes?	?

Table 28 continued

27. Tanter (Domestic Conflict), "Dimensions of Conflict Behavior Within and Between Nations, 1958–1960"	83	9	Prin. Comp.	Un.	R	?	Orth. & Obliq.	Yes	No	?	?	Yes?	?
28. Tanter (Foreign and Domestic Conflict), "Dimensions of Conflict Behavior Within and Between Nations, 1958–1960"	83	25	Prin. Comp.	Un.	R	?	Orth. & Obliq.	Yes	No	?	?	Yes?	?
29. Vincent (Soc. Cult. Pol., Unrotated),	129	91	Prin. Comp.	Un.	R Descrip.		Orth.	Yes	Yes	15%	Drop	Rank No.	40
30. Vincent (Soc. Cult. Pol., Rotated),	129	91	Prin. Comp.	Un.	R Descrip.		Orth.	Yes	Yes	15%	Drop	Rank No.	40

Table 28 continued

Is the reader warned that communalities & variance explained is inflated when subjects are dropped from the correlation matrix?	Factor scores calculated	Factors calculated to ? eigenvalue	Factors calculated	Factors presented	Factors discussed	% Variance explained by factor scores calculated	% Variance explained by factors calculated	% Variance explained by factors discussed	Is the correlation of factor scores relative to original variables given, if different from the loadings?	Variables heterogeneous by category	Variables homogeneous by category
No	No	1.0	15	7	7	NA	77%	56%	NA	Yes	
NA	Yes?	1.0	10	5	5	53%	?	60%	NA	Yes	
NA	NA	1.0	9	4	4	NA	?	81%	NA	Yes	
NA	No	1.0	15	6	6	NA	?	73%	NA		Yes
No	Incomp.	?	15	3	3	?	77%	40%	Yes	Yes	
NA	No	1.0	9	3	3	NA	57.1%	57.1%	NA		Yes
NA	No	1.0	2	2	2	NA	64.4%	64.4%	NA		Yes
NA	Incomp.	1.0	7	7	7	?	?	?	No		Yes
Yes	Comp.	1.0	19	6	6	71.8%	71.8%	71.8%	No	Yes	
Yes	Comp. & Incomp.	1.0	19	19	19	89.2%	89.2%	89.2%	Yes	Yes	

TABLE 29

Factor Scores of the States on All 19 Dimensions, Using the Formula F = ZA,
Where Loadings in the Range of − .49 to .49 Are Set to Zero

Factor scores

AFGHANISTAN	16.96	-10.44	0.09	-2.68	0.77	-2.20	-0.92	-0.05	-2.51	-0.55
	-0.44	0.51	0.0	-2.86	-1.29	-1.23	-0.69	0.61	0.68	
ALBANIA	4.17	-15.23	-5.30	4.00	-0.96	-0.71	-0.38	0.08	1.89	2.40
	1.09	0.51	-2.20	-1.39	1.85	0.44	-1.28	-0.98	-0.37	
ALGERIA	7.07	-6.86	-2.22	-0.83	3.81	0.0	-0.63	0.0	-1.02	0.16
	1.09	0.0	-0.57	-0.30	-1.29	0.17	1.32	-1.50	0.0	
ARGENTINA	-16.64	1.59	2.55	-5.97	4.65	-0.39	0.25	2.23	0.19	0.02
	-0.44	-1.34	0.84	3.50	1.58	0.52	-0.92	0.08	0.69	-1.43
AUSTRALIA	-25.07	13.30	3.43	-4.44	6.11	0.52	-0.92	0.08	0.69	-0.10
	-0.44	0.51	0.50	0.20	2.42	1.29	-1.85	-0.58	0.68	
AUSTRIA	-21.70	12.74	-0.91	-2.61	-2.82	0.52	0.0	-1.50	0.0	-1.82
	-0.44	0.0	-1.78	1.54	1.30	0.30	-1.28	-0.16	0.0	
BELGIUM	-22.92	14.16	2.30	-3.27	-5.33	0.52	0.34	-0.05	-0.39	1.31
	-0.44	0.51	-0.39	2.01	-1.29	0.04	-1.85	-0.78	0.68	
BOLIVIA	3.94	6.68	2.75	2.18	4.14	1.68	0.68	1.53	0.69	-1.90
	0.65	-0.41	0.05	3.13	-1.29	-1.95	1.87	0.29	0.68	
BRAZIL	-5.61	3.72	4.21	-6.11	2.90	-1.10	-0.26	-1.50	0.16	-0.47
	-0.44	-1.16	1.52	3.05	-0.19	-2.02	2.23	0.0	0.68	
BULGARIA	-11.21	-15.23	-4.00	-2.74	-1.44	-0.71	-0.26	-0.77	1.09	2.24
	1.09	0.51	-2.51	-0.83	-1.29	0.52	-1.28	-0.98	0.68	
BURMA	13.20	-4.73	-2.41	-2.76	1.01	0.66	0.0	-0.92	-0.83	0.37
	1.09	-0.91	0.18	-3.54	1.80	0.66	0.70	0.42	0.68	
BURUNDI	12.81	-0.95	0.06	5.48	-1.52	1.68	0.0	-0.72	0.0	-0.98
	-0.44	0.0	-0.13	0.82	-1.29	1.27	0.0	0.0	0.0	
CAMBODIA	18.38	-9.90	-3.11	1.85	-1.04	1.16	-0.16	-0.77	-0.63	2.05
	0.65	0.51	0.0	-1.92	1.75	0.90	1.96	-0.51	0.68	
CAMEROUN	16.48	2.41	-1.10	2.21	1.18	1.68	0.0	1.53	0.0	-0.68
	-0.44	0.0	-0.30	0.44	-1.29	0.97	0.77	-0.36	0.0	
CANADA	-23.89	11.95	5.52	-6.33	4.38	0.52	-0.92	1.53	-0.46	1.26
	-0.44	0.51	0.63	1.51	2.52	-0.44	-1.85	0.05	-0.37	
CENTRAL AFRICAN RE	15.91	-7.11	-1.62	4.06	1.78	1.68	0.0	0.73	0.0	-0.46
	-0.44	0.0	0.78	-0.00	-1.29	1.04	0.77	-0.58	0.0	
CEYLON	5.70	4.97	-1.45	-0.64	-3.38	-1.68	0.20	1.53	0.63	-3.16
	-0.44	0.51	0.81	-2.49	2.47	1.14	2.23	0.51	0.68	
CHAD	17.70	-2.58	-3.15	3.16	2.11	1.68	0.0	1.53	0.0	-1.13
	-0.44	0.0	-0.36	-0.36	-1.29	-1.14	0.77	-0.25	0.0	
CHILE	-12.81	8.62	4.05	-1.21	0.63	-1.10	0.92	-1.50	-1.24	0.58
	-0.44	-0.55	0.88	3.62	0.72	-0.38	1.87	0.29	0.68	
CHINA, PR	3.83	-16.08	-5.30	-6.81	0.62	-0.71	-0.38	0.0	-1.11	-0.03
	1.09	-1.28	-0.95	-3.22	-1.29	-0.10	-1.28	-0.69	0.68	
COLOMBIA	-8.03	3.97	4.55	-2.20	1.22	-0.19	0.40	-1.50	0.10	-2.29
	-0.44	-0.75	0.65	4.26	-1.29	-2.12	1.11	0.29	0.68	
CONGO (BRA)	12.72	-4.78	-4.11	3.28	1.42	1.68	0.0	1.53	0.0	-0.69
	-0.44	0.0	0.0	0.73	-1.29	1.06	0.77	0.0	0.0	
CONGO (LEO)	15.61	-0.63	1.03	-0.57	2.87	0.52	0.0	1.53	0.0	-0.21
	-0.44	0.0	0.0	0.43	-1.29	1.09	2.23	-0.10	0.0	
COSTA RICA	-7.07	9.24	3.34	3.90	-0.46	0.52	-0.92	-1.50	0.35	-3.72
	-0.44	-0.41	0.07	3.92	-1.29	-0.25	1.33	0.36	-0.75	
CUBA	-10.37	-9.83	-3.06	-3.41	-0.81	0.0	0.0	-0.86	-0.18	-0.42
	1.09	0.51	-0.62	3.33	2.70	-2.26	-0.63	-0.86	-0.75	
CYPRUS	-2.75	6.50	-1.89	3.78	-1.04	0.0	0.82	1.53	0.65	0.94
	-0.44	0.0	-0.25	-0.96	-1.29	0.0	0.0	0.95	0.0	
CZECHOSLOVAKIA	-16.50	-13.40	-2.50	-3.26	-2.29	-0.71	0.64	1.53	0.69	1.99
	1.09	0.51	-2.45	0.88	1.70	0.28	-1.28	-0.98	-1.07	
DAHOMEY	16.95	-6.27	-4.11	4.38	-0.13	1.68	0.0	1.53	0.0	-0.79
	-0.44	0.0	0.47	0.17	-1.29	0.73	0.77	0.88	0.0	
DENMARK	-23.66	14.46	1.62	-1.43	-2.03	0.52	-0.92	-1.50	0.69	0.91
	-0.44	0.51	0.48	0.22	1.15	0.07	-1.85	0.77	-1.07	
DOMINICAN REPUBLIC	-4.26	7.71	4.16	2.84	-2.67	0.52	0.0	-1.50	1.14	0.84
	-0.44	0.0	-0.45	3.50	-1.29	-2.22	0.0	-0.63	0.0	
ECUADOR	-2.83	3.32	3.60	1.98	0.18	-1.62	0.52	1.53	0.69	-0.82
	-0.44	-0.75	0.86	3.15	-1.29	-2.06	2.85	0.36	-0.75	
EL SALVADOR	0.31	-6.65	2.62	3.83	-3.43	-2.20	-0.92	-1.50	0.63	-2.34
	-0.44	0.51	1.21	3.72	-1.29	-2.16	2.85	-0.23	0.68	
ETHIOPIA	15.96	-9.88	1.07	-1.61	1.00	-2.20	-0.92	1.53	0.0	-1.69
	-0.44	0.51	0.0	-1.01	0.81	0.0	0.74	0.34	-0.06	-0.37
FINLAND	-19.95	14.46	-0.41	-2.38	1.00	0.52	-0.16	-0.77	1.15	0.75
	-0.44	0.51	-1.07	-1.69	1.57	0.13	-1.85	0.24	0.68	
FRANCE	-21.73	6.89	2.52	-6.55	-0.93	1.68	1.36	-1.50	-1.38	3.21
	-0.44	0.51	-1.51	1.84	2.70	-0.05	0.87	-0.84	0.68	
GABON	10.19	-5.59	-0.49	3.91	1.41	1.68	0.0	1.53	0.0	-0.32
	-0.44	0.0	-1.00	1.06	-1.29	1.04	0.77	0.0	0.0	
EAST GERMANY	-17.04	-16.08	-1.74	-2.58	-3.25	-2.20	0.0	0.08	0.0	0.42
	1.09	0.0	-2.33	0.12	2.42	0.25	-1.28	0.0	0.0	
WEST GERMANY	-22.64	12.99	2.96	-6.22	-3.55	0.52	0.0	0.08	0.0	0.79
	-0.44	0.0	-1.26	0.96	2.90	0.17	-1.28	0.0	0.0	
GHANA	12.48	-11.15	0.02	-1.27	-0.15	0.97	0.0	1.53	0.0	-2.85
	1.09	0.0	0.38	0.42	1.27	0.61	0.87	0.09	0.0	
GREECE	-11.38	9.82	1.41	-2.66	-0.72	0.52	-0.92	-1.50	0.84	2.71
	-0.44	0.51	-0.79	-0.68	-1.29	0.56	2.49	-0.34	-0.75	
GUATEMALA	1.81	-0.90	3.57	2.66	-1.72	-1.62	1.18	1.53	0.44	-1.74
	-0.44	-1.22	1.21	3.09	-1.29	-2.20	2.85	0.05	-0.75	
GUINEA	15.70	-11.15	-0.35	4.01	0.64	0.97	0.0	1.53	0.0	-0.39
	1.09	0.0	-0.79	-0.06	-1.29	0.41	-1.28	0.09	0.0	
HAITI	12.30	-8.82	2.88	4.48	-2.38	-1.50	1.18	0.08	0.69	-1.07
	-0.44	-1.03	-1.00	1.91	-1.29	-2.23	0.0	-0.61	0.68	
HONDURAS	2.92	0.31	3.86	4.37	0.59	-1.62	0.20	-1.50	-1.55	-2.35
	-0.44	-0.91	0.61	2.91	-1.29	-2.18	2.85	0.0	-0.75	
HUNGARY	-15.63	-14.02	-3.72	-3.83	-2.11	-0.71	0.88	0.08	-2.29	0.68
	1.09	-1.08	-2.19	1.74	1.65	0.36	-1.16	-0.91	-1.34	
ICELAND	-19.62	13.76	0.86	3.41	4.24	0.52	0.0	-1.50	0.0	-1.13
	-0.44	0.0	-0.49	-0.15	-1.29	-0.13	-1.85	0.77	0.0	
INDIA	7.09	9.11	3.04	-7.49	-1.37	1.68	1.28	-0.05	-3.07	-2.39
	-0.44	-1.28	1.19	-3.30	-1.29	0.69	-1.28	0.15	-0.75	
INDONESIA	13.05	-5.52	2.16	-6.20	-0.31	1.68	0.96	-0.05	0.69	-0.57
	0.65	-0.91	-1.77	-2.83	-1.29	1.02	1.11	-0.21	-0.37	
IRAN	11.57	-7.12	2.00	-4.20	3.38	-0.91	-0.38	-0.05	0.69	2.32
	-0.44	-0.75	-0.25	-3.18	0.23	0.90	0.70	0.81	-0.37	
IRAQ	8.67	-3.56	-0.78	-2.08	1.83	0.0	1.12	-0.05	-0.84	2.68
	-0.44	0.51	1.21	-1.90	1.85	1.00	1.46	-0.67	-1.25	

Table 29 continued

Factor scores

IRELAND	-18.79	13.19	-0.24	1.20	-0.50	0.52	-0.92	-1.50	0.69	-0.94
	-0.44	0.51	-0.03	2.61	-0.23	-0.12	-1.85	-0.06	0.68	
ISRAEL	-12.70	12.03	1.80	0.49	-1.96	0.52	0.34	-0.05	-1.49	3.41
	0.65	0.51	-1.55	-2.55	-1.29	0.92	-1.28	0.58	0.68	
ITALY	-20.46	11.99	2.83	-6.20	-3.50	0.52	1.06	-1.50	0.22	1.23
	-0.44	0.51	-2.01	2.91	2.57	0.32	-0.53	0.40	0.68	
IVORY COAST	15.31	-6.00	-0.51	1.96	0.93	1.68	0.0	1.53	0.0	-0.35
	0.65	0.0	-1.21	0.28	-1.29	0.56	0.99	0.86	0.0	
JAMAICA	-2.24	12.00	2.58	4.27	-3.26	0.52	0.0	0.08	-0.25	-2.43
	-0.44	0.0	-0.43	0.17	-1.29	-2.24	-0.63	0.0	0.0	
JAPAN	-15.86	11.32	4.48	-6.29	-2.80	0.52	0.20	0.08	0.69	-1.54
	-0.44	0.51	-0.11	-3.64	3.10	-0.03	0.11	0.66	0.68	
JORDAN	10.59	-10.15	-1.94	3.10	0.27	-1.62	1.06	-1.50	-2.64	3.70
	-0.44	-0.75	-0.85	-2.28	0.95	0.97	1.46	-0.55	-1.07	
NORTH KOREA	3.74	-15.23	-5.30	-1.54	-0.91	-0.71	0.0	-0.72	0.0	2.64
	1.09	0.0	0.0	-2.58	2.75	0.02	-1.28	0.0	0.0	
SOUTH KOREA	3.19	-3.93	3.91	-1.33	-4.64	0.0	0.0	-0.72	0.0	3.55
	0.0	0.0	1.58	-2.26	-1.29	0.07	1.46	0.0	0.0	
LAOS	19.67	-7.82	0.33	3.02	1.29	-0.71	0.0	1.53	0.0	3.10
	-0.44	0.0	0.0	-2.32	-1.29	0.82	2.85	0.58	0.0	
LEBANON	-2.87	3.12	-0.45	3.26	-4.01	0.0	0.20	0.08	0.69	-0.99
	-0.44	0.51	0.10	0.06	-1.29	0.92	2.85	-0.16	-0.75	
LIBERIA	10.49	-8.01	2.21	4.63	0.21	-1.62	-0.92	1.53	0.69	-0.96
	-0.44	0.51	-0.55	-0.35	-1.29	0.56	-0.46	0.63	0.68	
LIBYA	10.22	2.74	-0.10	2.08	3.65	-1.10	0.0	-1.50	0.0	0.32
	-0.44	0.0	0.0	-0.88	0.16	0.50	-0.12	-0.19	0.0	
LUXEMBOURG	-20.51	14.46	2.22	2.72	-3.89	0.52	0.0	-1.50	0.0	0.84
	-0.44	0.0	-1.00	2.29	0.0	0.17	-1.85	0.44	0.0	
MALAGASY REP	14.43	8.08	-0.84	2.35	2.50	0.52	0.0	0.08	0.0	-0.85
	-0.44	0.0	-0.10	0.40	-1.29	0.92	-0.40	-0.48	0.0	
MALI	16.69	-3.78	-4.03	2.36	1.92	1.68	0.0	1.53	0.0	-0.74
	-0.44	0.0	1.00	-0.63	0.08	-1.53	0.77	-0.29	0.0	
MAURITANIA	16.98	-1.45	-1.44	4.08	2.14	0.97	0.0	-0.05	0.0	-0.85
	0.65	0.0	0.41	-1.50	-1.29	-1.70	0.77	-0.38	0.0	
MEXICO	-7.24	2.40	5.30	-3.40	1.60	0.97	-0.04	-1.50	0.53	0.49
	-0.44	0.51	1.40	3.53	-1.29	-2.31	-1.28	0.19	0.68	
MONGOLIA	4.51	-15.23	-2.55	2.38	3.85	-0.71	-0.92	-1.50	0.69	1.18
	1.09	0.51	0.0	-2.86	0.0	0.0	-1.28	-0.88	0.68	
MOROCCO	8.93	2.24	-1.11	-1.98	1.37	1.68	0.0	-0.05	0.0	-0.17
	0.65	0.0	0.67	-0.94	1.62	0.11	-0.75	0.24	0.0	
NEPAL	17.95	-9.16	-3.33	1.89	-0.79	-2.20	0.06	1.53	0.69	-0.98
	0.0	0.51	0.0	-2.86	-1.29	0.32	1.46	0.19	0.68	
NETHERLANDS	-22.06	14.16	2.04	-3.18	-5.13	0.52	-0.92	0.08	0.69	2.22
	-0.44	0.51	-1.22	0.45	2.17	0.07	-1.85	0.46	-0.17	
NEW ZEALAND	-23.71	13.30	2.27	-0.63	3.42	0.52	-0.92	0.08	0.69	0.21
	-0.44	0.51	-0.15	0.50	0.13	1.17	-1.85	0.0	0.68	
NICARAGUA	1.19	-4.64	3.00	4.03	0.90	-2.20	-0.26	-1.50	-1.18	-0.85
	-0.44	0.51	-1.68	3.18	-1.29	-2.18	1.96	0.55	-0.75	
NIGER	18.62	-2.95	-2.82	3.71	2.06	1.68	0.0	1.53	0.0	-0.98
	-0.44	0.0	1.03	-1.72	1.71	-1.28	0.77	-0.40	0.0	
NIGERIA	15.65	2.78	0.52	-1.33	0.82	-0.91	0.0	1.53	0.0	-2.30
	-0.44	0.0	0.49	-0.17	0.13	0.69	-0.63	0.92	0.0	
NORWAY	-22.37	14.46	1.44	-1.79	0.24	0.52	-0.92	-1.50	0.69	1.70
	-0.44	0.51	-0.37	-1.54	1.20	-0.03	-1.85	0.81	0.68	
PAKISTAN	11.63	-7.43	2.73	-5.14	-1.28	0.0	1.24	-0.05	-2.49	-0.60
	0.0	-0.75	2.44	-1.55	2.30	0.80	-0.05	0.53	0.68	
PANAMA	-5.78	4.94	3.96	4.52	-0.02	-1.10	-0.04	-0.77	1.55	-3.79
	-0.44	0.51	0.67	3.22	-1.29	-2.12	2.85	-0.76	0.68	
PARAGUAY	-0.19	-10.88	1.44	3.57	1.86	-2.20	-0.38	-1.50	0.69	0.98
	-0.44	-1.16	1.24	3.15	-1.29	-0.23	-0.40	0.72	0.68	
PERU	-3.84	0.01	4.39	-2.42	2.98	-0.19	0.20	1.53	-0.84	-1.34
	-0.44	-0.41	0.52	3.29	-1.29	-2.05	2.23	0.13	0.68	
PHILIPPINES	-2.64	10.65	4.43	-1.89	-2.82	0.97	-0.92	-0.05	1.71	-2.01
	-0.44	-0.55	1.10	0.25	-1.29	0.64	-1.28	-0.42	0.68	
POLAND	-13.60	-13.73	-1.15	-5.99	-2.25	-1.62	0.44	-1.50	-0.95	1.37
	1.09	0.51	-2.09	1.88	1.01	0.25	-1.28	-0.98	-0.37	
PORTUGAL	-9.93	-10.68	1.22	0.12	-2.80	-2.20	-0.92	-1.50	0.78	1.20
	-0.44	0.51	1.73	2.21	-1.29	0.00	-1.85	-0.80	0.68	
ROMANIA	-9.09	-15.23	-3.43	-5.33	-0.86	-0.71	-0.92	-0.77	1.64	1.50
	1.09	-0.55	-1.26	-0.25	1.92	0.41	-1.28	-0.98	-0.37	
RWANDA	9.34	-1.98	-4.60	5.50	-1.75	0.0	0.0	-0.72	0.0	-0.90
	-0.44	0.0	-1.14	0.80	-1.29	1.23	0.0	0.0	0.0	
SAUDI ARABIA	14.92	-8.36	1.38	-0.33	3.37	-2.20	-0.92	-1.50	-0.98	0.02
	-0.44	0.51	1.24	-2.57	-1.29	0.52	-0.69	-0.74	0.68	
SENEGAL	15.14	-5.39	-3.13	2.76	0.35	1.68	0.0	-0.05	0.0	-1.93
	0.65	0.0	0.04	-1.06	-0.17	0.30	-0.40	0.84	0.0	
SIERRA LEONE	19.70	6.59	-1.65	4.61	0.03	-0.19	0.0	1.53	0.0	-0.74
	-0.44	0.0	1.09	-0.86	-0.29	0.44	0.13	0.99	0.0	
SOMALIA	16.33	-1.89	-2.52	3.63	2.99	-1.10	0.0	-1.50	0.0	-0.67
	-0.44	0.0	0.0	-2.34	-1.29	1.23	0.13	0.36	0.0	
SOUTH AFRICA	-8.65	-0.55	2.23	-1.86	4.22	-0.71	0.0	1.53	0.0	-3.11
	-0.44	0.0	1.52	0.22	-1.29	-0.53	-1.85	-0.82	0.0	
SPAIN	-11.98	-11.35	2.50	-4.13	-1.38	-1.62	0.58	-0.77	0.90	1.59
	0.0	0.51	1.91	2.60	-0.19	-0.06	-1.85	-0.44	0.68	
SUDAN	16.17	-8.34	-2.01	-1.83	3.37	0.0	0.0	1.53	0.0	-2.56
	0.65	0.0	1.24	-1.25	1.77	-0.90	0.70	-0.53	0.0	
SWEDEN	-24.33	14.46	0.95	-4.11	0.56	0.52	-0.16	-1.50	0.69	1.89
	-0.44	0.51	-0.17	-0.97	1.97	0.13	-1.85	0.66	-1.34	
SWITZERLAND	-21.94	14.86	1.29	-1.36	-3.17	-0.39	-0.92	1.53	0.69	-1.75
	-0.44	0.51	1.98	0.80	2.02	0.21	-1.85	0.0	-0.37	
SYRIA	6.28	-1.98	-2.14	-0.28	1.83	0.0	0.78	-0.77	-2.82	1.99
	-0.44	0.0	1.49	-1.31	1.95	0.87	2.85	-0.46	-1.07	
THAILAND	9.04	-10.51	1.83	-2.14	-0.11	-2.20	0.48	-0.77	1.55	0.46
	-0.44	-0.41	0.67	-3.51	-1.29	0.85	2.49	0.72	0.68	
TOGO	16.34	-3.43	-3.44	4.81	-0.02	0.52	0.0	1.53	0.0	-0.74
	-0.44	0.0	0.0	0.50	-0.32	0.69	0.0	-0.27	0.0	
TRINIDAD	-2.81	12.00	2.72	4.57	-3.48	0.52	0.0	0.81	0.65	-1.50
	-0.44	0.0	-0.57	-0.07	-1.29	-2.07	-0.63	0.0	0.0	
TUNISIA	6.11	-2.17	-2.52	0.54	0.47	1.68	0.0	-1.50	0.0	-1.16
	0.65	0.0	0.08	-1.10	0.79	0.39	-1.28	0.97	0.0	

Table 29 continued

Factor Scores

TURKEY	-1.06	5.66	3.21	-5.27	0.99	0.52	0.78	-0.77	1.21	1.99
	-0.44	0.51	-0.20	-3.14	-1.29	0.61	-0.40	0.48	-1.07	
UGANDA	13.65	5.95	-0.79	2.70	-0.09	-0.91	0.0	1.53	0.0	-0.98
	-0.44	0.0	0.75	0.49	3.02	1.17	-0.63	0.0	0.0	
U.S.S.R.	-15.54	-14.27	-2.50	-5.13	4.36	-0.71	0.58	0.73	1.77	3.35
	1.09	0.51	-2.32	-1.62	3.02	-0.44	-1.85	-0.98	-1.49	
U.A.R.	0.22	-8.86	0.17	-5.43	-2.08	1.16	0.06	-1.50	-3.85	0.49
	1.09	-0.99	2.36	-1.28	2.37	0.77	1.27	-0.65	-1.07	
UNITED KINGDOM	-24.95	12.39	3.10	-6.29	-2.40	-0.19	0.0	0.08	0.0	2.46
	-0.44	0.0	0.48	0.06	2.90	-0.09	-1.85	-0.32	0.0	
UNITED STATES	-25.94	10.99	-3.61	-7.56	2.56	0.97	0.0	0.08	-1.20	2.96
	-0.44	0.0	-0.54	1.06	-1.29	-2.31	-1.85	-0.06	0.0	
UPPER VOLTA	17.32	-4.81	-4.60	4.40	0.49	1.68	0.0	1.53	0.0	-0.98
	-0.44	0.0	0.44	-1.20	-1.29	0.56	0.77	0.92	0.0	
URUGUAY	-14.87	13.28	1.16	0.72	2.63	0.52	-0.92	-1.50	0.69	-1.68
	-0.44	0.51	0.04	3.21	-1.29	0.50	1.27	0.72	0.68	
VENEZUELA	-10.47	6.14	-4.96	-1.24	3.74	1.68	0.72	-1.50	1.34	-0.77
	0.65	0.51	-1.37	3.51	-1.29	-2.12	0.83	1.01	-1.43	
NORTH VIETNAM	5.28	-15.23	-5.30	-1.61	-2.21	-0.71	0.0	0.01	0.0	2.11
	1.09	0.0	0.0	-2.05	2.75	0.75	-1.28	0.0	0.0	
SOUTH VIETNAM	8.63	-10.89	1.43	-0.19	-1.81	-1.68	0.0	0.01	0.0	2.77
	1.09	0.0	0.0	-1.08	-1.29	0.87	1.33	0.0	0.0	
YEMEN	12.36	-1.98	-4.60	3.86	1.17	0.0	-0.92	-1.50	-1.23	-0.25
	0.0	-0.63	1.24	-2.34	-1.29	0.72	0.0	-0.72	0.68	
YUGOSLAVIA	-5.28	-13.90	0.55	-5.30	-0.86	-1.62	-0.92	1.53	0.96	3.31
	1.09	0.51	-2.04	-0.43	0.74	0.36	-1.28	-0.13	-1.25	
BOTSWANA	8.52	2.42	-4.60	3.33	1.99	-0.19	0.0	0.81	0.0	-1.13
	0.65	0.0	0.0	0.15	0.0	1.30	-0.53	0.0	0.0	
GAMBIA	10.93	5.66	-4.16	5.94	-0.99	-0.19	0.0	-0.77	0.0	-1.13
	0.65	0.0	0.0	-2.74	-1.29	0.36	0.10	0.0	0.0	
KUWAIT	-6.26	1.18	0.21	2.66	0.72	0.97	0.0	-0.77	0.0	0.40
	0.65	0.0	1.24	-1.42	0.26	1.07	-0.75	0.0	0.0	
LESOTHO	8.82	6.04	-4.60	4.56	-0.64	-0.19	0.0	0.81	0.0	-0.65
	0.65	0.0	0.0	0.97	0.0	1.36	-0.06	0.0	0.0	
MALAWI	9.22	4.19	-3.38	4.98	-0.26	0.97	0.0	1.53	0.0	-1.13
	0.65	0.0	0.0	0.20	-1.29	-0.55	0.10	0.0	0.0	
MALTA	-11.31	11.27	-1.08	5.44	-5.61	0.52	0.0	-0.77	0.0	0.0
	1.09	0.0	0.67	2.31	-1.29	0.47	-0.19	0.0	0.0	
TANZANIA	12.09	6.06	-1.80	0.63	2.40	1.68	0.0	0.81	0.0	-2.25
	0.65	0.0	0.69	-0.13	1.60	1.23	1.33	0.0	0.0	
ZAMBIA	3.42	3.83	-3.40	1.73	1.72	-0.19	0.0	1.53	0.0	-0.59
	0.65	0.0	0.0	0.25	-1.29	1.33	-0.53	0.0	0.0	
KENYA	11.28	4.79	-0.98	0.60	0.02	1.68	0.0	0.81	0.0	-0.98
	0.65	0.0	0.86	-0.17	1.12	1.20	-0.06	0.0	0.0	
W. SAMOA	6.55	3.12	-2.63	4.72	-0.99	0.0	0.0	0.81	0.0	0.0
	-0.44	0.0	0.0	0.85	-1.29	0.80	0.0	0.0	0.0	
SINGAPORE	-8.01	9.79	-0.25	3.22	-4.10	0.52	0.0	1.53	-0.71	0.0
	1.09	0.0	-0.52	-1.98	1.72	-0.77	-0.53	0.0	0.0	
GUYANA	7.98	7.05	0.73	3.93	3.65	0.52	0.0	0.08	0.77	-0.39
	0.65	0.0	0.02	0.74	-1.29	-2.02	0.30	0.0	0.0	
MALAYSIA	1.67	6.92	0.55	0.59	-1.52	-0.46	0.0	1.53	0.34	0.80
	-0.44	0.0	0.59	-2.94	0.18	-0.75	-0.53	1.03	0.0	
MALDIVE ISLANDS	8.65	1.21	-4.60	4.80	0.27	0.52	0.0	-1.50	0.0	0.0
	-0.44	0.0	0.0	-2.31	-1.29	1.26	-0.63	0.0	0.0	
BARBADOS	-5.97	10.05	-0.61	6.00	-3.88	-0.46	0.0	0.08	0.0	-1.13
	0.65	0.0	0.0	1.41	-1.29	-2.09	-1.85	0.0	0.0	
TAIWAN	-0.06	-3.94	3.46	-1.00	-5.22	0.45	-0.04	-0.12	-2.64	3.96
	0.65	0.51	1.01	-2.70	-1.29	0.47	0.74	0.0	0.68	

TABLE 30
Factor Scores of the States on All 19 Dimensions, Using the Formula F = ZA, Where Loadings in the Range of − .49 to .49 Are Set to Zero and All Other Loadings Set to ± 1.0

Factor scores

AFGHANISTAN	20.67	-13.40	-0.27	-4.24	1.08	-2.85	-1.17	0.12	-3.83	-0.68
	-0.68	0.74	0.0	-4.60	-1.59	-1.42	-1.34	1.01	0.85	
ALBANIA	6.30	-19.71	-7.83	6.01	-1.11	-1.30	-0.48	-0.07	2.66	3.05
	1.67	0.74	-2.95	-2.41	2.28	0.55	-1.95	-1.62	-0.46	
ALGERIA	9.44	-8.06	-3.06	-0.79	4.99	0.0	0.0	0.12	-1.41	0.22
	1.67	0.0	-0.82	-0.60	-1.59	0.05	-0.77	0.0	0.0	
ARGENTINA	-22.07	2.25	3.97	-8.69	6.29	0.02	1.68	-2.06	0.99	0.08
	-0.68	-1.95	1.20	5.82	1.97	0.19	3.63	0.31	-1.78	
AUSTRALIA	-33.44	17.70	5.27	-6.55	8.28	0.96	-1.17	-0.07	0.99	-0.07
	-0.68	0.74	0.81	0.31	3.00	2.11	-3.05	0.90	0.85	
AUSTRIA	-28.82	16.76	-1.14	-3.97	-3.95	0.96	0.0	-2.06	0.0	-2.28
	-0.68	0.0	-2.39	2.48	1.61	0.28	-1.95	-0.26	0.0	
BELGIUM	-30.70	18.84	3.89	-4.79	-7.23	0.96	0.43	0.12	-0.51	1.71
	-0.68	0.74	-0.43	3.17	-1.59	-0.20	-3.05	-1.29	0.85	
BOLIVIA	5.00	8.35	3.82	2.95	5.63	2.15	0.86	2.11	0.99	-2.47
	0.99	-0.60	0.02	5.11	-1.59	-2.75	2.93	0.49	0.85	
BRAZIL	-8.36	4.28	6.27	-8.90	3.86	-1.28	1.07	-2.06	0.26	-0.56
	-0.68	-1.69	2.06	5.03	-0.23	-2.89	3.63	0.0	0.85	
BULGARIA	-13.87	-19.71	-5.94	-4.17	-1.70	-1.30	-0.33	-0.96	1.54	2.85
	1.67	0.74	-3.37	-1.10	-1.59	0.69	-1.95	-1.62	0.85	
BURMA	17.58	-5.91	-3.46	-3.78	1.83	0.0	-1.17	0.12	-0.95	0.57
	1.67	-1.32	0.22	-5.98	2.23	0.95	1.36	0.70	0.85	
BURUNDI	16.88	-0.99	-0.30	8.04	-2.08	2.15	0.0	-1.08	0.0	-1.38
	-0.68	0.0	-0.16	1.21	-1.59	2.09	0.0	0.0	0.0	
CAMBODIA	24.17	-12.38	-4.49	2.38	-1.24	1.19	-0.20	-0.96	-0.69	2.66
	0.99	0.74	0.0	-3.20	2.16	1.39	2.76	-0.84	0.85	
CAMEROUN	21.85	2.84	-1.68	2.76	1.48	2.15	0.0	2.11	0.0	-0.97
	-0.68	0.0	-0.38	0.64	-1.59	1.52	0.94	-0.59	0.0	

Table 30 continued

Factor Scores

CANADA	-31.94	15.81	8.02	-9.29	5.71	0.96	-1.17	2.11	-0.59	1.65
	-0.68	0.74	0.96	2.52	3.13	-1.09	-3.05	0.09	-0.46	
CENTRAL AFRICAN RE	21.40	-8.61	-2.52	5.80	2.19	2.15	0.0	1.09	0.0	-0.65
	-0.68	0.0	0.99	-0.01	-1.59	1.65	0.94	-0.96	0.0	
CEYLON	8.60	5.66	-2.04	-1.02	-4.14	-1.89	0.25	2.11	0.00	-4.00
	-0.68	0.74	1.14	-4.38	3.06	1.85	3.63	0.84	0.85	
CHAD	23.74	-3.31	-4.68	4.30	2.73	2.15	0.0	2.11	0.0	-1.59
	-0.68	0.0	0.0	-0.71	-1.59	-1.26	0.94	-0.42	0.0	
CHILE	-17.19	11.49	5.93	-1.26	0.67	-1.28	1.17	-2.06	-2.26	0.77
	-0.68	-0.80	1.16	6.06	0.90	-0.70	2.93	0.49	0.85	
CHINA, PR	3.90	-21.32	-7.83	-9.94	1.74	-1.30	-0.48	0.0	-1.30	0.04
	1.67	-1.86	-1.35	-5.35	-1.59	-0.47	-1.95	-1.15	0.85	
COLOMBIA	-11.50	4.89	6.58	-2.66	1.67	-0.34	0.51	-2.06	0.18	-2.89
	-0.68	-1.09	0.84	7.01	-1.59	-3.08	1.60	0.49	0.85	
CONGO (BRA)	17.26	-6.04	-5.79	4.30	1.70	2.15	0.0	2.11	0.0	-0.97
	-0.68	0.0	0.0	1.10	-1.59	1.69	0.94	0.0	0.0	
CONGO (LEO)	20.36	-1.35	1.18	-0.44	3.63	0.96	0.0	2.11	0.0	-0.29
	-0.68	0.0	0.0	0.58	-1.59	1.76	3.63	-0.17	0.0	
COSTA RICA	-8.92	11.86	4.70	5.47	-0.84	0.96	-1.17	-2.06	0.52	-4.76
	-0.68	-0.60	0.08	6.51	-1.59	-0.74	1.99	0.59	-0.93	
CUBA	-13.65	-12.68	-4.04	-5.27	-1.26	0.0	1.27	-2.06	-0.21	-0.50
	1.67	0.74	-0.86	5.61	3.34	-3.33	-0.77	-1.43	-0.93	
CYPRUS	-1.63	7.86	-2.90	5.10	-1.63	0.0	0.0	2.11	0.90	1.32
	-0.68	0.0	-0.35	-1.41	-1.59	1.26	0.0	1.57	0.0	
CZECHOSLOVAKIA	-21.14	-17.08	-3.75	-4.64	-2.81	-1.30	0.81	2.11	0.99	2.53
	1.67	0.74	-3.29	1.39	2.11	0.24	-1.95	-1.62	-1.34	
DAHOMEY	22.66	-7.74	-5.78	6.21	-0.32	2.15	0.0	2.11	0.0	-1.11
	-0.68	0.0	0.59	0.21	-1.59	1.07	0.94	1.46	0.0	
DENMARK	-31.72	19.27	2.87	-2.48	-2.74	0.96	0.94	-2.06	0.99	1.18
	-0.68	0.74	0.78	0.85	1.42	-0.14	-3.05	1.27	-1.34	
DOMINICAN REPUBLIC	-6.52	9.60	5.91	4.32	-3.68	0.96	0.0	-2.06	1.58	1.11
	-0.68	0.0	-0.63	5.77	-1.59	-3.25	0.0	-1.04	0.0	
ECUADOR	-4.01	3.62	5.16	2.98	0.20	-2.24	0.66	2.11	0.99	-1.02
	-0.62	-1.09	1.09	5.14	-1.59	-2.96	4.40	0.59	-0.93	
EL SALVADOR	0.54	-8.17	3.77	5.59	-4.66	-2.85	-1.17	-2.06	0.90	-2.94
	-0.68	0.74	1.60	6.07	-1.59	-3.14	4.40	-0.38	0.85	
ETHIOPIA	19.39	-12.70	1.17	-2.38	1.45	-2.85	-1.17	2.11	0.99	-2.11
	-0.68	0.74	0.0	-1.43	1.01	-0.51	0.66	-0.10	-0.46	
FINLAND	-26.36	19.27	-0.42	-4.25	1.42	0.96	-0.20	-0.96	1.63	0.97
	-0.68	0.74	-1.46	-2.39	1.95	-0.03	-3.05	0.40	0.85	
FRANCE	-29.06	9.15	4.18	-9.52	-1.19	2.15	1.73	-2.06	-1.66	4.11
	-0.68	0.74	-2.02	3.05	3.34	-0.36	1.10	-1.39	0.85	
GABON	13.86	-6.95	-1.01	5.42	1.60	2.15	0.0	2.11	0.0	-0.44
	-0.68	0.0	-1.27	1.60	-1.59	1.65	0.94	0.0	0.0	
EAST GERMANY	-21.58	-21.32	-2.39	-3.75	-4.10	-2.85	0.0	-0.07	0.0	0.57
	1.67	0.0	-3.13	0.24	3.00	0.19	-1.95	0.0	0.0	
WEST GERMANY	-29.57	17.27	4.77	-9.03	-4.56	0.96	0.0	-0.07	0.0	1.07
	-0.68	0.0	-1.68	1.55	3.59	0.05	-1.95	0.0	0.0	
GHANA	16.30	-14.13	-0.06	-2.08	-0.19	0.85	0.0	2.11	0.0	-3.61
	1.67	0.0	0.48	0.63	1.58	0.86	1.10	0.16	0.0	
GREECE	-14.56	12.94	2.37	-4.00	-1.15	0.96	1.50	-2.06	1.20	3.49
	-0.68	0.74	-1.06	-0.83	-1.59	0.76	3.70	-0.56	-0.93	
GUATEMALA	1.70	-1.32	5.11	4.03	-2.31	-2.24	1.50	2.11	0.65	-2.18
	-0.68	-1.78	1.57	5.13	-1.59	-3.22	4.40	0.09	-0.93	
GUINEA	21.20	-14.13	-0.81	5.83	0.73	0.85	0.0	2.11	0.0	-0.56
	1.67	0.0	-1.00	-0.26	-1.59	0.49	-1.95	0.16	0.0	
HAITI	15.35	-11.22	3.85	6.68	-3.11	-1.54	1.50	-0.07	0.99	-1.29
	-0.68	-1.49	-1.36	3.16	-1.59	-3.27	0.0	-1.01	0.85	
HONDURAS	3.97	-0.23	5.45	6.43	0.47	-2.24	0.25	-2.06	-1.95	-2.97
	-0.68	-1.32	0.77	4.65	-1.59	-3.18	4.40	0.0	-0.93	
HUNGARY	-20.41	-17.76	-5.51	-5.87	-2.64	-1.30	1.12	-0.07	-2.92	0.91
	1.67	-1.57	-2.94	2.97	2.04	0.40	-1.71	-1.50	-1.67	
ICELAND	-24.58	18.18	1.33	4.36	5.41	0.96	0.0	-2.06	0.0	-1.59
	-0.68	0.0	-0.58	0.16	-1.59	-0.52	-3.05	1.27	0.0	
INDIA	8.47	11.25	4.22	-10.91	-1.07	2.15	1.63	0.12	-4.64	-2.99
	-0.68	-1.86	1.65	-5.66	-1.59	1.01	-1.95	0.24	-0.93	
INDONESIA	16.70	-6.56	2.83	-9.10	0.31	2.15	1.22	0.12	0.99	-0.61
	0.99	-1.32	-2.37	4.83	-1.59	1.61	1.60	-0.35	-0.46	
IRAN	14.60	-9.04	3.24	-6.09	4.86	-0.94	-0.48	0.12	0.99	2.99
	-0.68	-1.09	-0.35	-5.18	0.30	1.39	1.36	1.34	-0.46	
IRAQ	12.10	-5.64	-1.24	-3.12	2.65	0.0	1.42	0.12	-1.70	3.43
	-0.68	0.74	1.54	-3.27	2.28	1.59	2.69	-1.11	-1.56	
IRELAND	-24.75	17.34	-0.21	1.77	-1.10	0.96	-1.17	-2.06	0.99	-1.18
	-0.68	0.74	-0.01	4.36	-0.29	-0.49	-3.05	-0.10	0.85	
ISRAEL	-15.05	15.63	2.38	0.65	-2.56	0.96	0.43	0.12	-1.79	4.37
	0.99	0.74	-2.07	-4.40	-1.59	1.44	-1.95	0.96	0.85	
ITALY	-27.76	15.47	4.57	-9.02	-4.56	0.96	1.35	-2.06	0.35	1.61
	-0.68	0.74	-2.69	4.76	3.19	0.32	-0.61	0.66	0.85	
IVORY COAST	20.17	-7.35	-0.76	2.59	1.13	2.15	0.0	2.11	0.0	-0.49
	0.99	0.0	1.54	0.41	-1.59	0.76	1.33	1.43	0.0	
JAMAICA	-2.43	15.33	3.42	6.35	-0.18	0.96	0.0	-0.07	-0.34	-3.29
	-0.68	0.0	-0.59	0.24	-1.59	-1.59	0.0	0.0	0.0	
JAPAN	-21.19	14.38	6.71	-9.23	-3.23	0.96	0.25	-0.07	0.99	-1.91
	-0.68	0.74	-0.13	-6.10	3.84	-0.32	-0.24	1.10	0.85	
JORDAN	15.31	-13.51	-2.94	4.40	0.14	-2.24	1.35	-2.06	4.11	4.74
	-0.68	-1.09	-1.21	-3.91	1.17	1.52	2.69	-0.90	-1.34	
NORTH KOREA	4.84	-19.71	-7.83	-2.19	-1.17	-1.30	0.0	1.08	0.0	3.35
	1.67	0.0	0.0	-4.22	3.40	-0.23	-1.95	0.0	0.0	
SOUTH KOREA	4.09	-5.20	5.65	-0.99	-5.94	0.0	0.0	-1.08	0.0	4.55
	0.0	0.0	2.13	-3.74	-1.59	-0.14	2.69	0.0	0.0	
LAOS	26.08	-10.48	-0.11	4.31	1.40	-1.30	0.0	2.11	0.0	3.99
	-0.68	0.0	0.0	-3.88	-1.59	1.26	4.40	0.96	0.0	
LEBANON	-2.79	3.84	-0.81	4.96	-5.25	0.0	0.25	-0.07	0.99	-1.05
	-0.68	0.74	0.09	0.01	-1.59	1.44	4.40	-0.26	-0.93	
LIBERIA	13.31	-9.86	2.64	6.79	0.03	-2.24	-1.17	2.11	0.99	-1.24
	-0.68	0.74	-0.70	-0.43	-1.59	0.76	-1.34	1.04	0.85	
LIBYA	14.31	3.37	-0.23	2.47	4.48	-1.28	0.0	-2.06	0.0	0.48
	-0.68	0.0	0.0	-1.51	0.21	0.65	-0.24	-0.31	0.0	
LUXEMBOURG	-27.21	19.27	3.77	3.77	-5.29	0.96	0.0	-2.06	0.0	1.11
	-0.68	0.0	-1.40	3.68	0.0	0.05	-3.05	0.73	0.0	

Table 30 continued

Factor Scores

MALAGASY REP	19.27	10.02	-1.36	3.03	2.98	0.96	0.0	-0.07	0.0	-1.20
	-0.68	0.0	-0.13	0.58	-1.59	1.71	-0.37	-0.80	0.0	
MALI	22.10	-4.69	-5.82	3.08	2.50	2.15	0.0	2.11	0.0	-1.04
	-0.68	0.0	1.27	-0.99	0.09	-1.98	0.94	-0.49	0.0	
MAURITANIA	23.15	-2.04	-2.42	5.78	2.70	0.85	0.0	0.12	0.0	-1.20
	0.99	0.0	0.52	-2.34	-1.59	-2.29	0.94	-0.63	0.0	
MEXICO	-11.12	3.07	7.68	-4.44	2.05	0.85	-0.05	-2.06	0.77	0.60
	-0.68	0.74	1.89	5.88	-1.59	-3.42	-1.95	0.31	0.85	
MONGOLIA	7.38	-19.71	-4.11	2.93	4.68	-1.30	-1.17	-2.06	0.99	1.66
	1.67	0.74	0.0	-4.60	0.0	-0.27	-1.95	-1.46	0.85	
MOROCCO	11.34	2.96	-1.67	-3.01	1.96	2.15	0.0	0.12	0.0	-0.13
	0.99	0.0	0.95	-1.72	2.26	-0.07	-1.01	0.40	0.0	
NEPAL	22.74	-11.46	-4.83	3.26	-0.99	-2.85	0.08	2.11	0.99	-1.25
	0.0	0.74	0.0	-4.60	-1.59	0.59	2.69	0.31	0.85	
NETHERLANDS	-29.86	18.84	3.52	-4.63	-6.93	0.96	-1.17	-0.07	0.99	2.85
	-0.68	0.74	-1.59	0.56	2.69	-0.14	-3.05	0.77	-0.22	
NEW ZEALAND	-31.18	17.70	3.63	-1.04	4.39	0.96	-1.17	-0.07	0.99	0.29
	-0.68	0.74	-0.11	0.80	0.18	1.90	-3.05	0.0	0.85	
NICARAGUA	1.32	-5.34	4.22	5.92	0.99	-2.85	-0.33	-2.06	-1.40	-1.02
	-0.68	0.74	-2.24	5.27	-1.59	-3.18	2.76	0.90	-0.93	
NIGER	24.84	-3.75	-4.24	5.27	2.65	2.15	0.0	2.11	0.0	-1.38
	-0.68	0.0	1.31	-2.71	2.11	-1.51	0.94	-0.66	0.0	
NIGERIA	19.82	3.13	0.66	-1.44	1.19	-0.94	0.0	2.11	0.0	-3.12
	-0.68	0.0	0.63	-0.26	0.15	1.01	-0.77	1.51	0.0	
NORWAY	-30.03	19.27	2.56	-3.11	0.44	0.96	-1.17	-2.06	0.99	2.20
	-0.68	0.74	-0.45	-2.20	1.49	-0.32	-3.05	1.34	0.85	
PAKISTAN	14.71	-10.13	4.06	-7.43	-1.10	0.0	1.57	0.12	-3.92	-0.69
	0.0	-1.09	3.28	-2.53	2.85	1.21	0.02	0.87	0.85	
PANAMA	-6.93	5.54	5.59	6.67	-0.38	-1.28	-0.05	-0.95	2.19	-4.84
	-0.68	0.74	0.93	5.47	-1.59	-3.08	4.40	-1.25	0.85	
PARAGUAY	-0.17	-14.15	2.00	5.18	2.62	-2.85	-0.48	-2.06	0.99	1.29
	-0.68	-1.69	1.63	5.07	-1.59	-0.43	-0.37	1.18	0.85	
PERU	-5.35	-0.06	6.40	-3.66	4.14	-0.34	0.25	2.11	-1.70	-1.64
	-0.68	-0.60	0.66	5.35	-1.59	-2.93	3.63	0.21	0.85	
PHILIPPINES	-3.94	13.76	6.50	-1.96	-3.53	0.85	-1.17	0.12	2.40	-2.53
	-0.68	-0.80	1.54	0.29	-1.59	0.92	-1.95	-0.70	0.85	
POLAND	-17.95	-17.55	-1.82	-8.84	-2.90	-2.24	0.56	-2.06	-1.10	1.77
	1.67	0.74	-2.80	2.98	1.25	0.19	-1.95	-1.62	-0.46	
PORTUGAL	-14.39	-12.67	2.13	0.85	-4.10	-2.85	-1.17	-2.06	1.12	1.57
	-0.68	0.74	2.30	3.62	-1.59	-0.27	-3.05	-1.32	0.85	
ROMANIA	-11.86	-19.71	-5.05	-7.86	-0.98	-1.30	-1.17	-0.96	2.31	1.93
	1.67	-0.80	-1.60	-0.31	2.38	0.49	-1.95	-1.62	-0.46	
RWANDA	12.38	-3.19	-6.55	8.07	-2.35	0.0	0.0	-1.08	0.0	-1.27
	-0.68	0.0	-1.45	1.18	-1.59	2.01	0.0	0.0	0.0	
SAUDI ARABIA	18.69	-10.65	1.88	-0.09	4.16	-2.85	-1.17	-2.06	-1.14	0.08
	-0.68	0.74	1.58	-4.11	-1.59	0.96	-1.34	-1.22	0.85	
SENEGAL	19.93	-7.03	-4.52	4.08	0.34	2.15	0.0	0.12	0.0	-2.40
	0.99	0.0	0.05	-1.54	-0.22	0.28	-0.37	1.39	0.0	
SIERRA LEONE	26.13	7.45	-2.51	6.80	-0.25	-0.34	0.0	2.11	0.0	-1.04
	-0.68	0.0	1.38	-1.42	-0.37	0.55	0.57	1.64	0.0	
SOMALIA	22.37	-2.74	-3.82	5.21	3.55	-1.28	0.0	-2.06	0.0	-0.89
	-0.68	0.0	0.0	-3.74	-1.59	2.01	0.57	0.59	0.0	
SOUTH AFRICA	-12.71	-1.16	3.15	-2.36	5.86	-1.30	0.0	2.11	0.0	-3.94
	-0.68	0.0	1.99	0.40	-1.59	-0.11	-3.05	-1.36	0.0	
SPAIN	-16.99	-13.96	3.47	-5.75	-1.95	-2.24	0.74	-0.96	1.29	2.02
	0.0	0.74	2.53	4.11	-0.25	-0.39	-3.05	-0.73	0.85	
SUDAN	20.93	-11.15	-2.94	-2.92	4.20	0.0	0.0	2.11	0.0	-3.23
	0.99	0.0	1.58	-2.08	2.20	-0.81	1.36	-0.87	0.0	
SWEDEN	-32.78	19.27	1.54	-6.23	0.84	0.96	-0.20	-2.06	0.99	2.45
	-0.68	0.74	-0.16	-1.20	2.45	-0.03	-3.05	1.10	-1.67	
SWITZERLAND	-29.22	19.69	2.03	-2.31	-4.26	0.02	-1.17	2.11	0.99	-2.15
	-0.68	0.74	2.67	1.33	2.51	0.13	-3.05	0.0	-0.46	
SYRIA	8.65	-3.05	-3.10	-0.91	2.53	0.0	0.99	-0.96	-4.22	2.57
	-0.68	-1.57	1.98	-2.20	2.42	1.34	4.40	-0.77	-1.34	
THAILAND	11.15	-13.58	2.89	-2.58	0.25	-2.85	0.61	-0.96	2.19	0.63
	-0.68	-0.60	0.95	-5.90	-1.59	1.30	3.70	1.18	0.85	
TOGO	21.57	-4.86	-4.97	6.80	-0.33	0.96	0.0	2.11	0.0	-1.04
	-0.68	0.0	0.0	0.73	-0.40	1.01	0.0	-0.45	0.0	
TRINIDAD	-2.55	15.33	4.09	6.77	-4.49	0.96	0.0	1.03	0.90	-1.99
	-0.68	0.0	-0.82	-0.20	-1.59	-2.98	-0.77	0.0	0.0	
TUNISIA	8.14	-2.03	-3.52	0.40	0.65	2.15	0.0	-2.06	0.0	-1.45
	0.99	0.0	0.06	-1.94	0.98	0.45	-1.95	1.60	0.0	
TURKEY	-2.11	6.71	4.78	-7.55	1.37	0.96	0.99	-0.96	1.72	2.52
	-0.68	0.74	-0.24	-5.10	-1.59	0.86	-0.37	0.80	-1.34	
UGANDA	17.57	6.98	-1.24	4.32	-0.11	-0.94	0.0	2.11	0.0	-1.38
	-0.68	0.0	0.95	0.73	3.74	1.90	-0.77	0.0	0.0	
U.S.S.R.	-20.73	-18.18	-3.65	-6.56	6.22	-1.30	0.74	1.09	2.49	4.31
	1.67	0.74	-3.10	-2.73	3.74	-1.09	-3.05	-1.62	-1.86	
U.A.R.	0.41	-10.56	0.11	-8.07	-2.60	1.19	0.08	-2.06	-5.72	0.71
	1.67	-1.43	3.17	-1.99	2.94	1.15	1.47	-1.08	-1.34	
UNITED KINGDOM	-33.70	16.39	4.97	-9.14	-2.88	-0.34	0.0	-0.07	0.0	3.19
	-0.68	0.0	0.76	0.17	3.59	-0.44	-3.05	-0.52	0.0	
UNITED STATES	-34.75	13.99	-4.76	-10.89	3.46	0.85	0.0	-0.07	-1.67	3.82
	-0.68	0.0	-0.75	1.83	-1.59	-3.42	-3.05	-0.10	0.0	
UPPER VOLTA	22.94	-6.02	-6.55	6.36	0.60	2.15	0.0	2.11	0.0	-1.38
	-0.68	0.0	0.56	-1.77	-1.59	0.76	0.94	1.51	0.0	
URUGUAY	-19.49	17.14	1.80	0.78	3.38	0.96	-1.17	-2.06	0.99	-2.10
	-0.68	0.74	0.05	5.30	-1.59	0.65	1.47	1.18	0.85	
VENEZUELA	-14.29	7.95	7.22	-1.32	5.10	2.15	0.99	-2.06	1.89	-0.94
	0.99	0.74	-1.85	5.79	-1.59	-3.08	1.92	1.67	-1.78	
NORTH VIETNAM	7.31	-19.71	-7.83	-2.31	-2.56	-1.30	0.0	0.01	0.0	2.69
	1.67	0.0	0.0	-3.51	3.40	1.11	-1.95	0.0	0.0	
SOUTH VIETNAM	11.19	-14.37	2.03	0.32	-2.10	-1.89	0.0	0.01	0.0	3.57
	1.67	0.0	0.0	-1.85	-1.59	1.34	1.99	0.0	0.0	
YEMEN	17.09	-3.00	-6.55	5.66	1.24	0.0	-1.17	-2.06	-1.46	-0.27
	0.0	-0.92	1.58	-3.74	-1.59	1.34	0.0	-1.18	0.85	
YUGOSLAVIA	-6.63	-17.90	0.77	-7.75	-0.91	-2.24	-1.17	2.11	1.37	4.25
	1.67	0.74	-2.76	-0.88	0.91	0.40	-1.95	-0.21	-1.56	
BOTSWANA	11.39	3.33	-6.55	4.31	2.46	-0.34	0.0	1.03	0.0	-1.59
	0.99	0.0	0.0	0.14	0.0	2.14	-0.61	0.0	0.0	

Table 30 continued

Factor Scores

GAMBIA	14.87	7.09	-5.94	8.69	-1.55	-0.34	0.0	-0.96	0.0	-1.59
	0.99	0.0	0.0	-4.40	-1.59	0.40	0.16	0.0	0.0	
KUWAIT	-6.45	1.83	0.15	3.77	0.37	0.85	0.0	-0.96	0.0	0.57
	0.99	0.0	1.58	-2.30	0.30	1.72	-1.01	0.0	0.0	
LESOTHO	12.18	7.83	-6.55	6.29	-1.08	-0.34	0.0	1.03	0.0	-0.92
	0.99	0.0	0.0	1.46	0.0	2.25	0.29	0.0	0.0	
MALAWI	12.81	5.23	-4.80	7.24	-0.45	0.85	0.0	2.11	0.0	-1.59
	0.99	0.0	0.0	0.29	-1.59	-0.16	0.16	0.0	0.0	
MALTA	-14.10	14.77	-1.92	7.89	-7.56	0.96	0.0	-0.96	0.0	0.0
	1.67	0.0	0.95	3.69	-1.59	0.60	-0.60	0.0	0.0	
TANZANIA	15.21	7.91	-2.62	0.88	2.96	2.15	0.0	1.03	0.0	-3.04
	0.99	0.0	0.88	-0.29	1.98	2.01	1.99	0.0	0.0	
ZAMBIA	3.57	5.43	-4.68	2.27	2.18	-0.34	0.0	2.11	0.0	-0.83
	0.99	0.0	0.0	0.41	-1.59	2.19	-0.61	0.0	0.0	
KENYA	14.87	6.22	-1.48	0.81	0.22	2.15	0.0	1.03	0.0	-1.28
	0.99	0.0	1.09	-0.30	1.39	1.96	0.29	0.0	0.0	
W. SAMOA	8.86	4.04	-4.09	7.06	-1.58	0.0	0.0	1.03	0.0	0.0
	-0.68	0.0	0.0	1.49	-1.59	1.21	0.0	0.0	0.0	
SINGAPORE	-9.46	13.28	-0.57	4.26	-5.05	0.96	0.0	2.11	-0.98	0.0
	1.67	0.0	-0.66	-2.97	2.12	-0.57	-0.61	0.0	0.0	
GUYANA	11.66	9.56	0.66	5.49	4.65	0.96	0.0	-0.07	1.07	-0.56
	0.99	0.0	-0.01	1.15	-1.59	-2.89	0.98	0.0	0.0	
MALAYSIA	2.56	8.43	0.80	1.43	-1.79	-0.05	0.0	2.11	0.47	1.04
	-0.68	0.0	0.81	-5.12	0.21	-0.53	-0.61	1.71	0.0	
MALDIVE ISLANDS	11.78	1.73	-6.55	7.19	-0.21	0.96	0.0	-2.06	0.0	0.0
	-0.68	0.0	0.0	-3.63	-1.59	2.06	-0.77	0.0	0.0	
BARBADOS	-6.96	13.08	-1.21	8.80	-4.82	-0.05	0.0	-0.07	0.0	-1.59
	0.99	0.0	0.0	2.37	-1.59	-3.01	-3.05	0.0	0.0	
TAIWAN	-0.53	-4.02	5.10	-0.74	-6.86	-0.11	-0.05	0.01	-3.99	5.07
	0.99	0.74	1.44	-4.53	-1.59	0.60	0.53	0.0	0.85	

TABLE 31
Correlation of Factor Scores Using the Formula $F = ZA$, Where Loadings in the Range of $-.49$ to $.49$ Are Set to Zero

X	Y	R	X	Y	R	X	Y	R	X	Y	R
1,	2	-0.48230	4,	5	-0.01738	7,	8	0.00108	11,	12	-0.01112
1,	3	-0.40171	4,	6	0.04761	7,	9	-0.26837	11,	13	-0.30460
1,	4	0.49391	4,	7	-0.15960	7,	10	0.04793	11,	14	-0.20640
1,	5	0.23141	4,	8	0.17456	7,	11	0.03161	11,	15	0.20933
1,	6	0.03434	4,	9	0.11745	7,	12	-0.37268	11,	16	0.09861
1,	7	0.05313	4,	10	-0.35743	7,	13	-0.04496	11,	17	-0.18751
1,	8	0.38087	4,	11	-0.00829	7,	14	0.12290	11,	18	-0.26023
1,	9	-0.16220	4,	12	0.03328	7,	15	0.08412	11,	19	-0.12776
1,	10	-0.27436	4,	13	0.09773	7,	16	-0.11658	12,	13	-0.23719
1,	11	0.08114	4,	14	0.08027	7,	17	0.29139	12,	14	-0.05131
1,	12	-0.23701	4,	15	-0.49660	7,	18	-0.05729	12,	15	0.11151
1,	13	0.25493	4,	16	-0.03404	7,	19	-0.25349	12,	16	0.07484
1,	14	-0.39089	4,	17	0.24753	8,	9	0.08599	12,	17	-0.30719
1,	15	-0.32328	4,	18	0.13332	8,	10	-0.21573	12,	18	-0.02013
1,	16	0.17708	4,	19	0.10242	8,	11	0.05990	12,	19	0.12904
1,	17	0.44360	5,	6	0.09319	8,	12	0.02183	13,	14	-0.01424
1,	18	0.08035	5,	7	-0.04209	8,	13	0.03572	13,	15	-0.08824
1,	19	0.06795	5,	8	0.06879	8,	14	-0.17395	13,	16	-0.04726
2,	3	0.41494	5,	9	0.04880	8,	15	-0.01782	13,	17	0.26827
2,	4	0.01373	5,	10	-0.13309	8,	16	0.08491	13,	18	0.11538
2,	5	-0.09343	5,	11	-0.07935	8,	17	0.07367	13,	19	0.18904
2,	6	0.38402	5,	12	-0.11130	8,	18	0.09375	14,	15	-0.14527
2,	7	-0.06869	5,	13	0.13967	8,	19	-0.03185	14,	16	-0.47200
2,	8	-0.10004	5,	14	-0.02007	9,	10	-0.09885	14,	17	0.11203
2,	9	0.10478	5,	15	-0.09122	9,	11	-0.07149	14,	18	-0.07237
2,	10	-0.19472	5,	16	-0.02851	9,	12	0.33544	14,	19	-0.01827
2,	11	-0.47087	5,	17	0.11924	9,	13	-0.19110	15,	16	0.18511
2,	12	0.15097	5,	18	0.04129	9,	14	0.16979	15,	17	-0.25339
2,	13	0.10973	5,	19	-0.04631	9,	15	-0.06625	15,	18	-0.14282
2,	14	0.24487	6,	7	0.08612	9,	16	-0.08979	15,	19	-0.19864
2,	15	-0.00631	6,	8	0.12603	9,	17	-0.11266	16,	17	-0.12787
2,	16	-0.09886	6,	9	-0.05971	9,	18	0.06967	16,	18	0.04117
2,	17	-0.15089	6,	10	-0.10734	9,	19	0.06742	16,	19	-0.01874
2,	18	0.32808	6,	11	0.03307	10,	11	0.19159	17,	18	0.10009
2,	19	0.07940	6,	12	0.03550	10,	12	0.18354	17,	19	-0.07609
3,	4	-0.29473	6,	13	-0.00252	10,	13	-0.29132	18,	19	0.06379
3,	5	-0.06089	6,	14	-0.02774	10,	14	-0.27610			
3,	6	-0.09662	6,	15	-0.00098	10,	15	0.25453			
3,	7	0.08434	6,	16	0.13677	10,	16	0.23556			
3,	8	-0.20858	6,	17	-0.07798	10,	17	-0.21723			
3,	9	0.08312	6,	18	0.11205	10,	18	-0.17094			
3,	10	-0.07100	6,	19	-0.04832	10,	19	-0.15985			
3,	11	-0.42750									
3,	12	-0.02333									
3,	13	0.21161									
3,	14	0.39348									
3,	15	-0.09771									
3,	16	-0.37894									
3,	17	0.14141									
3,	18	0.20104									
3,	19	0.08625									

TABLE 32
Correlation of Factor Scores Using the Formula F = ZA, Where Loadings in the Range of −.49 to .49 Are Set to Zero and All Other Loadings Set to ±1.0

X	Y	R		X	Y	R
1,	2	−0.49048		7,	8	0.00659
1,	3	−0.44263		7,	9	−0.28343
1,	4	0.51414		7,	10	0.04724
1,	5	0.22124		7,	11	0.03161
1,	6	−0.00106		7,	12	−0.37267
1,	7	0.05254		7,	13	−0.05094
1,	8	0.39504		7,	14	0.11349
1,	9	−0.16739		7,	15	0.08408
1,	10	−0.28059		7,	16	−0.11289
1,	11	0.09053		7,	17	0.31553
1,	12	−0.23431		7,	18	−0.05728
1,	13	0.23408		7,	19	−0.25349
1,	14	−0.40608		8,	9	0.06615
1,	15	−0.32412		8,	10	−0.22130
1,	16	0.25158		8,	11	0.06643
1,	17	0.46016		8,	12	0.01703
1,	18	0.07801		8,	13	0.04474
1,	19	0.06247		8,	14	−0.19963
2,	3	0.43651		8,	15	−0.02584
2,	4	0.00086		8,	16	0.14400
2,	5	−0.11072		8,	17	0.08121
2,	6	0.48185		8,	18	0.09826
2,	7	−0.07772		8,	19	−0.02909
2,	8	−0.11676		9,	10	−0.09988
2,	9	0.10143		9,	11	−0.05597
2,	10	−0.18864		9,	12	0.33822
2,	11	−0.46397		9,	13	−0.21110
2,	12	0.15990		9,	14	0.17557
2,	13	0.11427		9,	15	−0.06973
2,	14	0.25360		9,	16	−0.10673
2,	15	−0.00247		9,	17	−0.11470
2,	16	−0.10469		9,	18	0.05998
2,	17	−0.16609		9,	19	0.07413
2,	18	0.32684		10,	11	0.19106
2,	19	0.08174		10,	12	0.17914
3,	4	−0.29990		10,	13	−0.28502
3,	5	−0.05897		10,	14	−0.27355
3,	6	−0.03208		10,	15	0.25940
3,	7	0.08265		10,	16	0.18534
3,	8	−0.22040		10,	17	−0.21370
3,	9	0.07351		10,	18	−0.17214
3,	10	−0.04774		10,	19	−0.15911
3,	11	−0.43680		11,	12	−0.01112
3,	12	−0.01517		11,	13	−0.30832
3,	13	0.21668		11,	14	−0.21082
3,	14	0.39806		11,	15	0.20911
3,	15	−0.07858		11,	16	0.08783
3,	16	−0.39666		11,	17	−0.17051
3,	17	0.11540		11,	18	−0.26024
3,	18	0.20561		11,	19	−0.12776
3,	19	0.08807		12,	13	−0.23172
4,	5	−0.07210		12,	14	−0.04506
4,	6	0.03923		12,	15	0.11117
4,	7	−0.15507		12,	16	0.06283
4,	8	0.16672		12,	17	−0.32307
4,	9	0.12616		12,	18	−0.02013
4,	10	−0.36520		12,	19	0.12904
4,	11	−0.09027		13,	14	−0.01407
4,	12	0.03215		13,	15	−0.08412
4,	13	0.09920		13,	16	−0.01087
4,	14	0.08151		13,	17	0.25341
4,	15	−0.50970		13,	18	0.11997
4,	16	0.00489		13,	19	0.19073
4,	17	0.25394		14,	15	−0.14364
4,	18	0.13209		14,	16	−0.50011
4,	19	0.10721		14,	17	0.09898
5,	6	0.07511		14,	18	−0.06936
5,	7	−0.02416		14,	19	−0.02022
5,	8	0.08671		15,	16	0.15308
5,	9	0.03855		15,	17	−0.24655
5,	10	−0.12527		15,	18	−0.14290
5,	11	−0.06051		15,	19	−0.19871
5,	12	−0.13391		16,	17	−0.08573
5,	13	0.12861		16,	18	0.03562
5,	14	−0.03869		16,	19	−0.01035
5,	15	−0.06820		17,	18	0.10557
5,	16	−0.00063		17,	19	−0.09339
5,	17	0.12450		18,	19	0.06379
5,	18	0.04306				
5,	19	−0.05426				
6,	7	0.07713				
6,	8	0.10581				
6,	9	−0.03141				
6,	10	−0.14284				
6,	11	−0.04499				
6,	12	0.05047				
6,	13	0.00545				
6,	14	−0.00267				
6,	15	0.00517				
6,	16	0.13288				
6,	17	−0.09322				
6,	18	0.16402				
6,	19	−0.04196				

TABLE 33
Correlation of Raw Factor Scores with Original Variables Using Factor Score Formula F = ZA, Where Loadings in the Range of −.49 to .49 Are Set to Zero

Factor Scores

	1	2	3	4	5	6	7	8	9	10	11	12	13	14	15	16	17	18	19
VAR 1	-0.44	-0.18	0.09	-0.48	-0.18	-0.25	0.00	-0.34	-0.00	0.87	0.16	0.22	-0.26	-0.17	0.29	0.14	-0.22	-0.17	-0.10
VAR 2	-0.19	0.09	0.05	-0.10	-0.85	0.03	0.07	0.08	-0.07	0.15	0.12	0.04	-0.13	-0.12	0.18	0.04	-0.10	0.03	0.03
VAR 3	0.92	-0.34	-0.36	0.48	0.22	0.12	0.00	0.41	-0.10	-0.34	0.00	-0.16	0.25	-0.33	-0.30	0.11	0.37	0.14	0.12
VAR 4	-0.62	0.12	0.39	-0.91	-0.12	-0.04	0.12	-0.18	-0.01	0.30	-0.00	0.03	-0.11	0.05	0.42	-0.03	-0.29	-0.10	-0.09
VAR 5	0.83	-0.43	-0.14	0.24	0.13	-0.02	0.11	0.22	-0.26	-0.19	0.05	-0.29	0.36	-0.35	-0.30	-0.02	0.43	0.11	0.13
VAR 6	0.70	-0.53	-0.19	0.22	0.20	-0.04	0.08	0.17	-0.06	-0.09	0.07	-0.23	0.14	-0.26	-0.32	0.05	0.41	0.01	0.03
VAR 7	0.43	-0.14	0.08	0.32	0.11	0.03	0.10	0.11	-0.10	-0.12	0.02	-0.15	0.32	-0.07	-0.24	-0.17	0.42	0.04	-0.03
VAR 8	0.59	-0.27	-0.33	0.11	0.24	0.21	0.03	0.17	-0.26	-0.01	0.12	-0.06	0.13	-0.53	-0.04	0.30	0.11	-0.02	-0.11
VAR 9	-0.38	0.19	0.39	-0.03	-0.17	-0.06	0.19	-0.12	0.12	-0.15	-0.13	-0.07	-0.08	0.89	-0.07	-0.33	0.10	-0.16	-0.02
VAR 10	-0.65	0.37	0.46	-0.09	-0.23	-0.09	0.04	-0.27	0.22	-0.11	-0.23	0.03	-0.15	0.83	0.01	-0.41	-0.09	-0.12	-0.06
VAR 11	0.52	-0.14	-0.02	0.13	-0.08	-0.01	0.07	0.14	-0.14	-0.06	-0.05	-0.25	0.31	-0.58	-0.15	0.15	0.25	0.19	0.03
VAR 12	0.14	-0.02	0.25	0.21	0.16	-0.06	0.12	-0.22	0.01	-0.31	-0.20	-0.22	0.37	0.59	-0.35	-0.41	0.47	0.05	-0.00
VAR 13	-0.34	0.38	0.09	-0.01	-0.61	0.10	-0.16	-0.13	-0.03	0.16	-0.20	0.40	0.01	0.13	0.12	0.06	-0.20	-0.01	0.09
VAR 14	-0.01	-0.25	-0.33	0.05	-0.05	-0.05	0.14	0.01	0.09	0.18	0.32	0.10	-0.82	0.01	-0.00	-0.06	-0.09	-0.18	-0.19
VAR 15	-0.21	0.34	-0.08	-0.01	0.55	0.21	-0.19	0.05	0.13	0.06	-0.15	0.20	-0.03	-0.12	0.13	0.06	-0.17	0.21	-0.11
VAR 16	-0.33	0.08	0.12	-0.22	-0.90	-0.15	0.09	-0.11	-0.05	0.18	0.12	0.06	-0.17	-0.04	0.20	0.02	-0.20	-0.06	0.00
VAR 17	0.13	-0.16	0.12	-0.52	0.70	0.09	0.10	0.09	-0.09	-0.01	-0.01	-0.19	0.18	-0.10	0.12	-0.04	0.06	-0.05	0.01
VAR 18	-0.90	0.42	0.28	-0.30	-0.20	-0.04	-0.09	-0.42	0.14	0.28	0.02	0.27	-0.30	0.37	0.25	-0.15	-0.43	-0.06	-0.08
VAR 19	-0.06	-0.17	-0.09	-0.25	-0.03	0.02	0.09	-0.10	-0.26	0.78	0.14	-0.06	-0.14	-0.29	0.18	0.24	-0.03	-0.13	-0.18
VAR 20	-0.39	0.03	-0.07	-0.19	-0.18	0.05	-0.04	-0.06	0.22	0.30	0.22	0.28	-0.89	0.02	0.14	0.12	-0.34	-0.04	-0.14
VAR 21	-0.18	-0.14	-0.19	-0.17	-0.12	-0.03	0.03	-0.10	-0.01	0.89	0.18	0.28	-0.32	-0.24	0.17	0.22	-0.28	-0.14	-0.13
VAR 22	0.24	-0.40	-0.52	-0.06	-0.04	0.01	-0.00	0.15	-0.25	0.13	0.44	0.09	-0.14	-0.43	0.28	0.27	-0.19	-0.28	-0.24
VAR 23	0.70	-0.48	-0.46	0.32	0.17	0.11	0.09	0.37	-0.13	-0.13	0.24	-0.15	-0.01	-0.25	-0.12	0.15	0.25	-0.15	-0.13
VAR 24	0.08	0.33	0.20	0.13	0.04	0.11	-0.06	0.09	0.07	-0.17	-0.26	-0.02	0.12	-0.07	-0.14	0.04	0.10	0.99	0.06
VAR 25	-0.72	0.31	0.20	-0.39	-0.14	0.12	-0.05	-0.23	0.10	0.31	-0.10	0.19	-0.18	0.15	0.23	0.01	-0.31	-0.11	-0.09
VAR 26	-0.86	0.43	0.21	-0.33	-0.32	-0.04	-0.06	-0.31	0.17	0.34	0.05	0.26	-0.30	0.22	0.25	-0.04	-0.40	-0.13	-0.03
VAR 27	-0.81	0.40	0.15	-0.28	-0.24	0.09	-0.12	-0.24	0.27	0.24	-0.02	0.25	-0.34	0.28	0.25	-0.08	-0.43	-0.03	-0.03
VAR 28	0.10	-0.12	-0.04	0.18	0.00	-0.08	-0.16	0.01	0.57	0.08	0.04	0.19	-0.15	-0.16	-0.06	0.07	0.00	-0.04	0.02
VAR 29	0.05	-0.07	0.08	-0.16	-0.04	0.09	0.99	0.00	-0.27	0.05	0.03	-0.37	-0.04	0.12	0.08	-0.12	0.29	-0.06	-0.25
VAR 30	0.24	-0.15	0.02	-0.03	0.11	-0.04	0.37	-0.02	-0.34	-0.18	0.01	-1.00	0.24	0.05	-0.11	-0.07	0.31	0.02	-0.13
VAR 31	0.06	0.08	0.07	-0.02	-0.03	0.17	0.12	-0.07	0.07	-0.33	0.01	-0.27	0.07	0.09	-0.09	-0.04	0.16	-0.03	-0.09
VAR 32	0.21	-0.20	-0.19	-0.02	-0.07	0.02	0.13	-0.14	-0.82	0.15	0.18	-0.25	0.01	-0.27	0.02	0.13	0.07	-0.15	-0.03
VAR 33	-0.07	-0.08	-0.09	-0.10	0.05	0.05	0.25	0.03	-0.07	0.16	0.13	-0.13	-0.19	0.02	0.20	0.02	0.08	-0.06	-1.00
VAR 34	0.20	-0.09	0.02	-0.08	-0.02	0.05	0.30	-0.01	-0.71	0.10	-0.04	-0.27	0.30	-0.19	0.07	0.10	0.18	0.00	-0.10
VAR 35	-0.16	0.25	0.83	-0.01	0.00	-0.11	-0.01	-0.05	0.16	-0.22	-0.36	-0.01	0.18	0.37	-0.26	-0.40	0.16	0.15	0.16
VAR 36	-0.12	0.23	0.88	-0.10	0.01	-0.15	0.15	-0.14	-0.03	-0.12	-0.32	-0.16	0.27	0.30	-0.24	-0.39	0.32	0.18	0.03
VAR 37	-0.32	0.16	0.41	-0.49	0.02	0.06	0.23	-0.17	-0.06	0.11	-0.02	0.00	-0.07	0.16	0.19	-0.17	-0.05	-0.12	-0.15
VAR 38	0.39	-0.11	-0.56	0.38	-0.11	0.18	-0.07	0.32	0.02	-0.11	0.30	0.02	-0.03	-0.33	-0.01	0.28	-0.14	0.06	0.12
VAR 39	0.35	-0.50	-0.77	0.13	0.02	-0.01	-0.04	0.27	-0.09	0.04	0.53	0.07	-0.26	-0.42	0.18	0.35	-0.24	-0.30	-0.17
VAR 40	-0.56	0.40	0.86	-0.55	-0.13	-0.05	0.09	-0.25	0.01	0.12	-0.29	0.05	0.13	0.24	0.15	-0.17	-0.05	0.12	-0.02
VAR 41	-0.60	0.43	0.84	-0.52	-0.13	-0.06	0.08	-0.21	0.13	0.01	-0.32	0.12	0.03	0.33	0.20	-0.24	-0.14	0.10	0.02
VAR 42	-0.50	0.07	0.07	-0.75	-0.09	0.06	0.04	-0.15	-0.15	0.24	0.14	0.03	-0.26	-0.06	0.55	0.13	-0.29	-0.11	-0.12
VAR 43	-0.40	0.06	-0.04	-0.70	-0.09	0.09	0.12	-0.13	-0.18	0.27	0.14	-0.01	-0.20	-0.15	0.54	0.16	-0.26	-0.09	-0.16
VAR 44	-0.35	0.01	-0.04	-0.54	-0.03	-0.04	0.08	-0.07	-0.05	0.27	0.18	0.07	-0.11	-0.12	0.96	0.19	-0.22	-0.15	-0.20
VAR 45	-0.27	-0.02	-0.15	-0.42	-0.15	0.03	0.09	0.03	-0.08	0.22	0.22	0.14	-0.06	-0.16	0.96	0.17	-0.27	-0.13	-0.18
VAR 46	0.27	0.15	-0.26	0.88	0.03	0.14	-0.18	0.09	0.09	-0.28	-0.13	0.09	0.01	0.13	-0.38	-0.04	0.15	0.13	0.06
VAR 47	0.77	-0.31	-0.39	0.51	0.11	0.06	-0.09	0.44	-0.05	-0.32	0.13	-0.12	0.19	-0.28	-0.33	0.22	0.24	0.09	0.18
VAR 48	-0.85	0.54	0.38	-0.34	-0.24	-0.03	-0.02	-0.27	0.06	0.13	-0.19	0.19	-0.09	0.35	0.27	-0.13	-0.37	0.01	0.05
VAR 49	0.91	-0.47	-0.38	0.41	0.20	0.00	0.05	0.30	-0.14	-0.24	0.09	-0.29	0.25	-0.32	-0.32	0.20	0.40	0.08	0.15
VAR 50	0.67	-0.20	-0.46	0.87	0.13	0.00	-0.13	0.23	0.05	-0.31	0.04	-0.07	0.06	-0.05	-0.42	0.05	0.32	0.07	0.11
VAR 51	0.85	-0.47	-0.30	0.53	0.16	-0.05	0.05	0.19	-0.09	-0.27	0.02	-0.22	0.22	-0.20	-0.35	0.07	0.50	0.07	0.05
VAR 52	0.28	-0.82	-0.32	-0.17	0.03	-0.35	0.10	-0.06	-0.17	0.28	0.55	-0.10	-0.14	-0.26	0.11	0.21	0.06	-0.28	-0.10
VAR 53	0.90	-0.37	-0.35	0.48	0.23	0.07	0.02	0.37	-0.14	-0.29	-0.01	-0.16	0.29	-0.24	-0.36	0.18	0.35	0.08	0.08
VAR 54	0.19	0.01	-0.18	0.21	0.00	0.07	-0.05	0.85	0.16	-0.21	-0.01	0.06	-0.07	-0.01	0.05	0.00	-0.01	0.04	-0.05
VAR 55	0.08	0.06	0.25	0.18	0.12	-0.11	0.11	0.10	0.03	-0.36	-0.12	-0.10	0.15	0.34	-0.27	-0.89	0.24	-0.05	0.04
VAR 56	0.46	-0.19	-0.16	0.07	0.12	0.14	0.06	0.79	-0.03	-0.14	0.12	-0.03	0.14	-0.30	-0.09	0.15	0.14	0.12	0.00
VAR 57	0.60	-0.09	-0.42	0.47	0.03	0.37	0.10	0.35	-0.24	-0.21	0.26	-0.16	0.12	-0.27	-0.13	0.32	0.14	0.07	-0.01
VAR 58	0.90	-0.35	-0.40	0.49	0.25	0.12	0.11	0.37	-0.16	-0.22	0.08	-0.20	0.25	-0.38	-0.22	0.24	0.39	0.12	0.06
VAR 59	0.71	-0.14	-0.31	0.60	0.11	0.11	-0.09	0.42	-0.03	-0.33	-0.00	-0.05	0.15	-0.21	-0.35	0.14	0.28	0.13	0.20
VAR 60	-0.08	0.47	0.43	0.09	0.08	-0.03	-0.03	-0.06	0.07	-0.19	-1.00	0.01	0.30	0.21	-0.21	-0.10	0.19	0.26	0.13
VAR 61	0.21	-0.90	-0.32	-0.19	-0.02	-0.42	-0.04	0.01	-0.08	0.32	0.46	-0.04	-0.20	-0.22	0.13	0.10	-0.03	-0.39	-0.07
VAR 62	0.47	-0.25	0.06	0.13	0.12	-0.12	0.37	0.08	-0.08	-0.17	-0.00	-0.40	0.16	0.03	-0.13	-0.10	0.83	0.06	-0.15
VAR 63	0.52	-0.87	-0.37	0.04	0.14	-0.36	0.03	0.15	-0.15	0.13	0.36	-0.19	-0.02	-0.29	-0.03	0.07	0.23	-0.23	-0.01
VAR 64	0.41	-0.83	-0.47	0.11	0.03	-0.15	-0.04	0.15	-0.04	0.17	0.46	0.00	-0.16	-0.18	-0.06	0.18	-0.01	-0.28	0.00
VAR 65	0.42	-0.89	-0.55	0.02	0.10	-0.26	-0.05	0.15	-0.01	0.14	0.50	-0.01	-0.17	-0.27	0.06	0.13	-0.04	-0.35	-0.04
VAR 66	0.41	-0.41	-0.00	0.11	0.12	-0.23	0.21	0.30	-0.10	0.02	-0.13	-0.21	0.21	0.02	-0.21	-0.07	0.46	-0.11	-0.01
VAR 67	-0.23	0.09	0.00	0.27	-0.26	0.04	-0.10	-0.31	0.12	-0.05	0.08	0.11	-0.11	0.19	-0.15	-0.16	-0.08	0.07	0.04
VAR 68	0.77	-0.68	-0.54	0.41	0.10	-0.17	0.01	0.31	-0.08	-0.08	0.22	-0.11	0.02	-0.31	-0.17	0.17	0.18	-0.11	0.04
VAR 69	-0.41	0.78	0.33	0.03	-0.02	0.32	-0.19	-0.11	0.10	-0.19	-0.46	0.24	-0.01	0.23	-0.00	-0.16	-0.19	0.21	0.16
VAR 70	-0.74	0.47	0.24	-0.15	-0.17	0.03	-0.07	-0.35	0.13	0.12	0.08	0.20	-0.21	0.42	0.17	-0.17	-0.30	0.02	-0.04
VAR 71	-0.56	0.59	0.15	0.01	-0.25	0.21	-0.31	-0.20	0.12	0.06	-0.03	0.30	-0.13	0.19	0.11	0.02	-0.48	0.09	0.05
VAR 72	0.21	-0.64	-0.31	-0.09	0.06	0.05	-0.11	0.03	-0.02	0.19	0.28	-0.00	-0.14	-0.15	0.02	0.08	-0.12	-0.29	-0.03
VAR 73	-0.12	-0.13	0.21	-0.02	-0.12	-0.93	-0.10	-0.16	0.12	0.03	-0.21	0.00	0.01	0.08	0.02	-0.14	0.06	0.02	0.06
VAR 74	0.62	-0.82	-0.36	0.16	0.19	-0.19	0.05	0.17	-0.03	-0.00	0.20	-0.15	-0.04	-0.19	-0.09	0.08	0.26	-0.17	0.03
VAR 75	0.41	-0.01	0.09	0.32	0.17	-0.12	0.31	0.05	-0.03	-0.21	-0.21	-0.24	0.26	0.12	-0.28	-0.11	0.86	0.16	-0.12
VAR 76	-0.31	0.13	-0.18	-0.20	-0.05	0.00	-0.14	-0.06	0.15	0.18	0.23	0.20	-0.25	-0.12	0.24	0.11	-0.88	-0.05	-0.02
VAR 77	-0.13	0.67	0.12	0.07	0.02	0.77	0.04	0.02	0.07	-0.19	-0.27	0.09	0.01	0.07	0.02	0.09	-0.08	0.29	-0.02
VAR 78	-0.48	0.36	0.33	-0.01	-0.10	-0.37	-0.12	-0.21	0.36	0.02	-0.29	0.11	-0.07	0.32	-0.07	-0.21	-0.09	0.07	0.06
VAR 79	0.09	-0.25	-0.03	0.22	-0.20	-0.05	-0.03	-0.12	0.02	0.05	0.03	0.06	-0.08	-0.06	-0.18	-0.10	0.17	-0.05	0.10
VAR 80	0.56	-0.89	-0.42	0.09	0.12	-0.24	0.09	0.19	-0.11	0.14	0.43	-0.11	-0.08	-0.26	-0.01	0.12	0.17	-0.20	-0.04
VAR 81	0.48	-0.88	-0.38	0.12	0.03	-0.37	0.06	0.07	-0.05	0.20	0.47	-0.12	-0.14	-0.19	-0.05	0.04	0.17	-0.27	-0.07
VAR 82	-0.34	0.40	0.39	-0.31	0.08	-0.10	-0.00	-0.13	0.02	0.05	-0.35	0.08	0.03	0.13	0.12	-0.18	-0.14	0.14	-0.00
VAR 83	-0.42	0.81	0.33	0.02	-0.07	0.18	-0.11	-0.04	0.25	-0.14	-0.38	0.13	0.04	0.21	-0.01	-0.11	-0.11	0.27	0.05
VAR 84	0.85	-0.47	-0.41	0.37	0.29	-0.05	0.03	0.32	-0.07	-0.21	0.06	-0.16	0.12	-0.28	-0.25	0.18	0.44	0.04	-0.00
VAR 85	-0.18	0.63	-0.05	0.18	-0.06	0.52	-0.17	0.12	0.16	-0.15	-0.20	0.29	-0.10	0.06	-0.01	0.10	-0.33	0.26	0.12
VAR 86	-0.41	0.79	0.14	0.03	-0.20	0.32	-0.15	-0.03	0.05	-0.05	-0.21	0.25	0.02	0.05	0.09	0.03	-0.34	0.16	0.17
VAR 87	0.17	0.56	0.39	0.28	0.09	0.28	0.02	0.05	-0.11	-0.34	-0.59	-0.06	0.50	0.10	-0.32	-0.03	0.32	0.44	0.21
VAR 88	0.48	-0.11	-0.41	0.20	0.12	0.12	-0.07	0.32	-0.14	-0.05	0.02	-0.00	0.12	-0.47	-0.02	0.73	0.10	0.00	0.03
VAR 89	0.24	0.14	0.26	0.47	0.20	0.04	0.11	0.14	0.10	-0.55	-0.18	-0.23	0.30	0.52	-0.37	-0.34	0.40	0.20	0.19
VAR 90	-0.03	0.21	0.28	0.34	0.16	0.10	0.12	0.05	0.14	-0.49	-0.24	-0.13	0.13	0.78	-0.27	-0.46	0.18	0.14	0.04
VAR 91	-0.44	-0.17	0.29	-0.88	-0.10	-0.23	0.16	-0.27	-0.12	0.56	0.15	-0.02	-0.13	-0.12	0.38	0.02	-0.15	-0.17	-0.07

TABLE 34

Correlation of Raw Factor Scores with Original Variables Using Factor Score Formula F = ZA, Where Loadings in the Range of −.49 to .49 Are Set to Zero and All Other Loadings Set to ±1.0

Factor scores

			3	4	5	6	7	8	9	10	11	12	13	14	15	16	17	18	19
VAR 1	-0.44	-0.18	0.11	-0.48	-0.17	-0.27	0.00	-0.34	-0.00	0.88	0.16	0.22	-0.26	-0.17	0.29	0.08	-0.22	-0.17	-0.10
VAR 2	-0.19	0.09	0.06	-0.10	-0.83	0.04	0.07	0.07	-0.07	0.15	0.12	0.04	-0.12	-0.12	0.18	0.02	-0.11	0.03	0.03
VAR 3	0.91	-0.34	-0.38	0.48	0.22	0.09	0.00	0.42	-0.11	-0.35	0.00	-0.16	0.24	-0.33	-0.30	0.18	0.38	0.14	0.12
VAR 4	-0.64	0.13	0.41	-0.90	-0.08	-0.03	0.12	-0.18	-0.01	0.31	-0.00	0.03	-0.10	0.05	0.42	-0.08	-0.29	-0.10	-0.09
VAR 5	0.82	-0.44	-0.16	0.25	0.14	-0.07	0.11	0.24	-0.27	-0.20	0.05	-0.29	0.35	-0.36	-0.30	0.03	0.44	0.11	0.13
VAR 6	0.70	-0.54	-0.22	0.22	0.20	-0.10	0.08	0.19	-0.07	-0.10	0.07	-0.23	0.13	-0.27	-0.32	0.08	0.42	0.01	0.03
VAR 7	0.43	-0.15	0.06	0.33	0.11	0.00	0.10	0.12	-0.11	-0.12	0.02	-0.15	0.32	-0.08	-0.24	-0.12	0.44	0.04	-0.03
VAR 8	0.60	-0.27	-0.34	0.10	0.24	0.18	0.03	0.19	-0.26	-0.02	0.12	-0.06	0.12	-0.53	-0.04	0.34	0.12	-0.02	-0.11
VAR 9	-0.39	0.19	0.40	-0.02	-0.18	-0.04	0.19	-0.14	0.13	-0.15	-0.13	-0.07	-0.08	0.87	-0.07	-0.35	0.09	-0.16	-0.02
VAR 10	-0.66	0.38	0.48	-0.10	-0.24	-0.05	0.04	-0.29	0.22	-0.10	-0.23	0.03	-0.14	0.84	0.01	-0.45	-0.10	-0.12	-0.06
VAR 11	0.52	-0.14	-0.03	0.14	-0.05	-0.01	0.07	0.15	-0.16	-0.06	-0.05	-0.25	0.31	-0.59	-0.15	0.18	0.26	0.19	0.03
VAR 12	0.13	-0.03	0.24	0.21	0.14	-0.07	0.12	-0.22	0.00	-0.30	-0.20	-0.22	0.36	0.60	-0.35	-0.40	0.47	0.05	-0.00
VAR 13	-0.34	0.38	0.11	-0.01	-0.64	0.13	-0.16	-0.14	-0.02	0.15	-0.20	0.40	0.02	0.13	0.12	0.03	-0.23	-0.01	0.09
VAR 14	-0.00	-0.25	-0.33	0.04	-0.05	-0.08	0.14	0.00	0.11	0.18	0.32	0.10	-0.84	0.01	-0.00	-0.07	-0.07	-0.18	-0.19
VAR 15	-0.21	0.35	-0.07	-0.03	0.55	0.25	-0.19	0.04	0.13	0.06	-0.15	0.20	-0.03	-0.10	0.13	0.06	-0.17	0.21	-0.11
VAR 16	-0.33	0.08	0.13	-0.21	-0.87	-0.13	0.09	-0.11	-0.04	0.19	0.12	0.06	-0.17	-0.05	0.20	-0.02	-0.20	-0.06	0.00
VAR 17	0.12	-0.16	0.12	-0.52	0.73	0.06	0.10	0.10	-0.10	-0.01	-0.01	-0.19	0.18	-0.10	0.12	-0.03	0.06	-0.05	0.01
VAR 18	-0.90	0.43	0.30	-0.31	-0.20	-0.01	-0.09	-0.43	0.15	0.28	0.02	0.27	-0.29	0.38	0.25	-0.21	-0.44	-0.06	-0.08
VAR 19	-0.05	-0.16	-0.08	-0.25	-0.01	-0.00	0.09	-0.09	-0.27	0.78	0.14	-0.06	-0.14	-0.30	0.18	0.21	-0.04	-0.13	-0.18
VAR 20	-0.38	0.04	-0.06	-0.20	-0.17	0.06	-0.04	-0.07	0.24	0.30	0.22	0.28	-0.87	0.02	0.14	0.08	-0.35	-0.04	-0.14
VAR 21	-0.17	-0.14	-0.18	-0.16	-0.12	-0.06	0.03	-0.10	-0.01	0.87	0.18	0.28	-0.32	-0.24	0.17	0.19	-0.27	-0.14	-0.13
VAR 22	0.25	-0.39	-0.53	-0.08	-0.03	-0.04	-0.00	0.16	-0.25	0.13	0.44	0.09	-0.14	-0.43	0.28	0.28	-0.19	-0.28	-0.24
VAR 23	0.71	-0.49	-0.48	0.32	0.16	0.06	0.09	0.38	-0.14	-0.14	0.24	-0.15	-0.02	-0.26	-0.12	0.20	0.25	-0.15	-0.13
VAR 24	0.08	0.33	0.21	0.13	0.04	0.16	-0.06	0.10	0.06	-0.17	-0.26	-0.02	0.12	-0.07	-0.14	0.04	0.11	0.99	0.06
VAR 25	-0.72	0.31	0.22	-0.39	-0.13	0.13	-0.05	-0.24	0.10	0.31	-0.10	0.19	-0.16	0.16	0.23	-0.04	-0.33	-0.11	-0.09
VAR 26	-0.85	0.43	0.24	-0.33	-0.32	0.00	-0.06	-0.33	0.18	0.35	0.05	0.26	-0.29	0.22	0.25	-0.10	-0.41	-0.13	-0.03
VAR 27	-0.81	0.41	0.18	-0.29	-0.24	0.13	-0.12	-0.26	0.28	0.24	-0.02	0.25	-0.33	0.29	0.25	-0.14	-0.45	-0.03	-0.03
VAR 28	0.11	-0.12	-0.05	0.19	0.00	-0.08	-0.16	0.02	0.55	0.08	0.04	0.19	-0.15	-0.16	-0.06	0.07	0.01	-0.04	0.02
VAR 29	0.05	-0.08	0.08	-0.16	-0.02	0.08	0.99	0.01	-0.28	0.05	0.03	-0.37	-0.05	0.11	0.08	-0.11	0.32	-0.06	-0.25
VAR 30	0.23	-0.16	0.02	-0.03	0.13	-0.05	0.37	-0.02	-0.34	-0.18	0.01	-1.00	0.23	0.05	-0.11	-0.06	0.32	0.02	-0.13
VAR 31	0.06	0.08	0.07	-0.02	-0.02	0.17	0.12	-0.06	0.07	-0.32	0.01	-0.27	0.08	0.09	-0.09	-0.03	0.17	-0.03	-0.09
VAR 32	0.21	-0.20	-0.19	-0.03	-0.07	-0.02	0.13	-0.13	-0.78	0.15	0.18	-0.25	0.00	-0.27	0.02	0.15	0.06	-0.15	-0.03
VAR 33	-0.06	-0.08	-0.09	-0.11	0.05	0.04	0.25	0.03	-0.07	0.16	0.13	-0.13	-0.19	0.02	0.20	0.01	0.09	-0.06	-1.00
VAR 34	0.20	-0.09	0.01	-0.08	-0.01	0.02	0.30	0.01	-0.77	0.10	-0.04	-0.27	0.30	-0.20	0.07	0.12	0.18	0.00	-0.10
VAR 35	-0.16	0.25	0.81	0.00	0.00	-0.08	-0.01	-0.05	0.15	-0.21	-0.36	-0.01	0.18	0.37	-0.26	-0.40	0.15	0.15	0.16
VAR 36	-0.13	0.23	0.86	-0.08	0.02	-0.12	0.15	-0.13	-0.05	-0.11	-0.32	-0.16	0.27	0.30	-0.24	-0.40	0.32	0.18	0.03
VAR 37	-0.33	0.16	0.41	-0.49	0.04	0.05	0.23	-0.16	-0.07	0.12	-0.02	0.00	-0.07	0.16	0.19	-0.20	-0.06	-0.12	-0.15
VAR 38	0.41	-0.11	-0.57	0.37	-0.12	0.18	-0.07	0.31	0.02	-0.13	0.30	0.02	-0.03	-0.33	-0.01	0.30	-0.11	0.06	0.12
VAR 39	0.36	-0.50	-0.79	0.12	0.03	-0.07	-0.04	0.27	-0.08	0.04	0.53	0.07	-0.27	-0.42	0.18	0.37	-0.22	-0.30	-0.17
VAR 40	-0.58	0.41	0.87	-0.54	-0.12	-0.01	0.09	-0.24	0.00	0.13	-0.29	0.05	0.14	0.24	0.15	-0.21	-0.06	0.12	-0.02
VAR 41	-0.62	0.43	0.85	-0.51	-0.12	-0.00	0.08	-0.22	0.12	0.02	-0.32	0.12	0.04	0.33	0.20	-0.28	-0.15	0.10	0.02
VAR 42	-0.50	0.08	0.09	-0.78	-0.06	0.07	0.04	-0.15	-0.15	0.25	0.14	0.03	-0.25	-0.05	0.56	0.09	-0.30	-0.11	-0.12
VAR 43	-0.40	0.06	-0.02	-0.74	-0.06	0.09	0.12	-0.13	-0.17	0.28	0.14	-0.01	-0.20	-0.14	0.54	0.12	-0.26	-0.09	-0.16
VAR 44	-0.35	0.01	-0.02	-0.55	-0.00	-0.02	0.08	-0.08	-0.05	0.27	0.18	0.07	-0.10	-0.11	0.96	0.15	-0.21	-0.15	-0.20
VAR 45	-0.27	-0.02	-0.14	-0.43	-0.13	0.03	0.09	0.03	-0.08	0.23	0.22	0.14	-0.06	-0.16	0.96	0.14	-0.26	-0.13	-0.18
VAR 46	0.29	0.15	-0.27	0.86	-0.01	0.16	-0.18	0.08	0.10	-0.29	-0.13	0.09	0.01	0.13	-0.38	-0.01	0.14	0.13	0.06
VAR 47	0.77	-0.32	-0.41	0.51	0.09	0.03	-0.09	0.45	-0.04	-0.33	0.13	-0.12	0.18	-0.29	-0.33	0.27	0.24	0.09	0.18
VAR 48	-0.84	0.55	0.41	-0.34	-0.24	0.03	-0.02	-0.29	0.06	0.14	-0.19	0.19	-0.08	0.36	0.27	-0.17	-0.37	0.01	0.05
VAR 49	0.91	-0.48	-0.40	0.42	0.20	-0.04	0.05	0.31	-0.15	-0.25	0.09	-0.29	0.25	-0.33	-0.32	0.25	0.41	0.08	0.15
VAR 50	0.68	-0.21	-0.48	0.86	0.09	0.00	-0.13	0.22	0.05	-0.33	0.04	-0.07	0.05	-0.05	-0.42	0.10	0.33	0.07	0.11
VAR 51	0.85	-0.47	-0.33	0.53	0.14	-0.08	0.05	0.21	-0.10	-0.28	0.02	-0.22	0.21	-0.21	-0.35	0.12	0.50	0.07	0.05
VAR 52	0.28	-0.81	-0.33	-0.16	0.05	-0.43	0.10	-0.05	-0.16	0.28	0.55	-0.10	-0.15	-0.27	0.11	0.20	0.07	-0.28	-0.10
VAR 53	0.90	-0.38	-0.37	0.49	0.22	0.03	0.02	0.38	-0.14	-0.30	-0.01	-0.16	0.28	-0.25	-0.36	0.25	0.36	0.08	0.08
VAR 54	0.19	0.01	-0.19	0.21	0.01	0.07	-0.05	0.82	0.16	-0.22	-0.01	0.06	-0.07	-0.02	0.05	0.05	-0.00	0.04	-0.05
VAR 55	0.07	0.05	0.24	0.19	0.12	-0.10	0.11	0.09	0.03	-0.36	-0.12	-0.10	0.14	0.34	-0.27	-0.82	0.25	-0.05	0.04
VAR 56	0.45	-0.20	-0.17	0.07	0.13	0.10	0.06	0.83	-0.05	-0.15	0.12	-0.03	0.14	-0.31	-0.09	0.19	0.13	0.12	0.00
VAR 57	0.62	-0.10	-0.44	0.47	0.02	0.35	0.10	0.35	-0.24	-0.23	0.26	-0.16	0.11	-0.28	-0.14	0.37	0.15	0.07	-0.01
VAR 58	0.90	-0.36	-0.42	0.49	0.24	0.09	0.11	0.38	-0.17	-0.23	0.08	-0.20	0.24	-0.38	-0.22	0.31	0.41	0.12	0.06
VAR 59	0.71	-0.14	-0.33	0.61	0.09	0.10	-0.09	0.42	-0.02	-0.34	-0.00	-0.05	0.14	-0.21	-0.35	0.21	0.28	0.13	0.20
VAR 60	-0.09	0.46	0.44	0.09	0.06	0.04	-0.03	-0.07	0.06	-0.19	-1.00	0.01	0.31	0.21	-0.21	-0.09	0.17	0.26	0.13
VAR 61	0.21	-0.89	-0.33	-0.19	-0.00	-0.50	-0.04	0.02	-0.08	0.33	0.46	-0.04	-0.21	-0.23	0.13	0.08	-0.04	-0.39	-0.07
VAR 62	0.47	-0.26	0.04	0.14	0.13	-0.13	0.37	0.08	-0.10	-0.17	-0.00	-0.40	0.15	0.02	-0.13	-0.06	0.87	0.06	-0.15
VAR 63	0.52	-0.87	-0.38	0.05	0.15	-0.43	0.03	0.17	-0.14	0.13	0.36	-0.19	-0.03	-0.30	-0.03	0.10	0.22	-0.23	-0.01
VAR 64	0.41	-0.82	-0.49	0.11	0.03	-0.24	-0.04	0.16	-0.03	0.17	0.46	0.00	-0.16	-0.19	-0.06	0.18	-0.02	-0.28	0.00
VAR 65	0.42	-0.89	-0.56	0.02	0.10	-0.35	-0.05	0.16	-0.00	0.14	0.50	-0.01	-0.17	-0.27	0.06	0.14	-0.05	-0.35	-0.04
VAR 66	0.41	-0.42	-0.02	0.13	0.12	-0.26	0.21	0.31	-0.12	0.02	-0.13	-0.21	0.20	0.01	-0.21	-0.04	0.46	-0.11	-0.01
VAR 67	-0.22	0.10	0.00	0.27	-0.27	0.06	-0.10	-0.33	0.13	-0.05	0.08	0.11	-0.11	0.20	-0.15	-0.17	-0.09	0.07	0.04
VAR 68	0.77	-0.69	-0.57	0.41	0.10	-0.22	0.01	0.32	-0.09	-0.09	0.22	-0.11	0.01	-0.32	-0.17	0.22	0.19	-0.11	0.04
VAR 69	-0.41	0.79	0.35	0.02	-0.03	0.39	-0.19	-0.12	0.11	-0.19	-0.46	0.24	0.00	0.24	-0.00	-0.18	-0.21	0.21	0.16
VAR 70	-0.74	0.49	0.27	-0.16	-0.18	0.06	-0.07	-0.37	0.13	0.12	0.08	0.20	-0.20	0.43	0.17	-0.20	-0.31	0.02	-0.04
VAR 71	-0.55	0.61	0.17	0.00	-0.26	0.25	-0.31	-0.22	0.12	0.06	-0.03	0.30	-0.12	0.20	0.11	-0.02	-0.49	0.09	0.05
VAR 72	0.21	-0.63	-0.32	-0.09	0.06	-0.03	-0.11	0.04	-0.01	0.19	0.28	-0.00	-0.14	-0.15	0.02	0.07	-0.14	-0.29	-0.03
VAR 73	-0.13	-0.14	0.20	-0.01	-0.12	-0.85	-0.10	-0.17	0.12	0.04	-0.21	0.00	0.01	0.08	0.02	-0.14	0.08	0.02	0.06
VAR 74	0.61	-0.83	-0.39	0.17	0.19	-0.26	0.05	0.17	-0.02	-0.00	0.20	-0.15	-0.05	-0.20	-0.09	0.11	0.26	-0.17	0.03
VAR 75	0.41	-0.02	0.08	0.32	0.16	-0.10	0.31	0.06	-0.04	-0.21	-0.21	-0.24	0.25	0.12	-0.28	-0.07	0.89	0.16	-0.12
VAR 76	-0.31	0.14	-0.16	-0.20	-0.05	0.02	-0.14	-0.07	0.15	0.18	0.23	0.20	-0.25	-0.12	0.24	0.09	-0.82	-0.05	-0.02
VAR 77	-0.12	0.67	0.14	0.05	0.01	0.87	0.04	0.02	0.06	-0.20	-0.27	0.09	0.02	0.07	0.02	0.09	-0.08	0.29	-0.02
VAR 78	-0.48	0.36	0.34	0.00	-0.11	-0.29	-0.12	-0.23	0.36	0.02	-0.29	0.11	-0.06	0.33	-0.07	-0.23	-0.07	0.07	0.06
VAR 79	0.10	-0.25	-0.04	0.22	-0.21	-0.07	-0.03	-0.11	0.02	0.05	0.03	0.06	-0.09	-0.05	-0.18	-0.10	0.14	-0.05	0.10
VAR 80	0.56	-0.89	-0.44	0.09	0.13	-0.31	0.09	0.20	-0.10	0.14	0.43	-0.11	-0.09	-0.26	-0.01	0.14	0.18	-0.20	-0.04
VAR 81	0.48	-0.88	-0.41	0.12	0.04	-0.44	0.06	0.08	-0.05	0.20	0.47	-0.12	-0.15	-0.19	-0.05	0.04	0.18	-0.27	-0.07
VAR 82	-0.35	0.40	0.41	-0.31	0.09	-0.05	-0.00	-0.13	0.01	0.06	-0.35	0.08	0.03	0.13	0.12	-0.19	-0.13	0.14	-0.00
VAR 83	-0.42	0.81	0.34	0.01	-0.08	0.27	-0.11	-0.06	0.25	-0.14	-0.38	0.13	0.04	0.21	-0.01	-0.12	-0.10	0.27	0.05
VAR 84	0.84	-0.48	-0.43	0.37	0.29	-0.08	0.03	0.33	-0.08	-0.22	0.06	-0.16	0.11	-0.29	-0.25	0.24	0.46	0.04	-0.00
VAR 85	-0.17	0.64	-0.04	0.17	-0.08	0.56	-0.17	0.11	0.16	-0.16	-0.20	0.29	-0.10	0.06	-0.01	0.10	-0.34	0.26	0.12
VAR 86	-0.41	0.80	0.16	0.02	-0.21	0.37	-0.15	-0.04	0.05	-0.06	-0.21	0.25	0.03	0.05	0.09	0.02	-0.34	0.16	0.17
VAR 87	0.17	0.56	0.40	0.28	0.07	0.35	0.02	0.05	-0.12	-0.34	-0.59	-0.06	0.50	0.10	-0.32	0.01	0.31	0.44	0.21
VAR 88	0.49	-0.12	-0.41	0.20	0.12	0.11	-0.07	0.33	-0.15	-0.06	0.02	-0.00	0.12	-0.48	-0.02	0.82	0.11	0.00	0.03
VAR 89	0.24	0.14	0.25	0.48	0.19	0.06	0.11	0.13	0.09	-0.56	-0.18	-0.23	0.30	0.52	-0.37	-0.29	0.41	0.20	0.19
VAR 90	-0.04	0.21	0.28	0.33	0.14	0.13	0.12	0.04	0.13	-0.49	-0.24	-0.13	0.13	0.78	-0.27	-0.45	0.18	0.14	0.04
VAR 91	-0.45	-0.16	0.31	-0.86	-0.06	-0.25	0.16	-0.26	-0.12	0.57	0.15	-0.02	-0.13	-0.12	0.38	-0.03	-0.15	-0.17	-0.07

TABLE 35
Correlation of Ranked Factor Scores with Original Variables Using Factor Score Formula F = ZA, Where Loadings in the Range of −.49 to .49 Are Set to Zero

Factor Scores Ranked

	1	2	3	4	5	6	7	8	9	10	11	12	13	14	15	16	17	18	19
VAR 1	-0.45	-0.17	0.10	-0.50	-0.19	-0.25	-0.03	-0.33	0.12	0.87	0.17	0.32	-0.25	-0.19	0.29	0.01	-0.26	-0.14	-0.05
VAR 2	-0.20	0.09	0.07	-0.09	-0.84	0.03	0.08	0.09	-0.03	0.12	0.12	0.05	-0.15	-0.10	0.15	-0.03	-0.10	0.03	0.01
VAR 3	0.93	-0.35	-0.36	0.48	0.25	0.13	-0.00	0.39	-0.21	-0.36	0.01	-0.26	0.22	-0.32	-0.28	0.23	0.43	0.10	0.05
VAR 4	-0.61	0.13	0.39	-0.90	-0.15	-0.04	0.12	-0.17	0.10	0.32	0.00	0.15	-0.08	0.05	0.40	-0.16	-0.33	-0.08	-0.04
VAR 5	0.82	-0.45	-0.13	0.23	0.16	-0.01	0.08	0.19	-0.25	-0.23	0.08	-0.31	0.37	-0.38	-0.30	0.09	0.48	0.10	0.11
VAR 6	0.68	-0.54	-0.19	0.22	0.20	-0.03	0.07	0.14	-0.04	-0.11	0.08	-0.24	0.14	-0.28	-0.30	0.12	0.43	0.01	0.02
VAR 7	0.37	-0.15	0.09	0.32	0.12	0.04	0.08	0.09	-0.12	-0.11	0.01	-0.17	0.33	-0.08	-0.24	-0.01	0.42	0.04	-0.05
VAR 8	0.60	-0.28	-0.34	0.10	0.26	0.22	0.03	0.13	-0.23	-0.03	0.12	-0.09	0.13	-0.54	-0.02	0.34	0.16	-0.06	-0.13
VAR 9	-0.41	0.18	0.39	-0.03	-0.17	-0.05	0.20	-0.11	0.10	-0.13	-0.15	-0.04	-0.03	0.90	-0.08	-0.27	0.06	-0.14	-0.01
VAR 10	-0.66	0.38	0.46	-0.09	-0.24	-0.08	0.01	-0.25	0.25	-0.07	-0.25	0.09	-0.14	0.83	-0.00	-0.41	-0.14	-0.09	-0.04
VAR 11	0.46	-0.15	-0.01	0.13	-0.06	-0.01	0.05	0.11	-0.06	-0.09	-0.04	-0.26	0.25	-0.58	-0.14	0.21	0.25	0.17	0.01
VAR 12	0.09	-0.04	0.25	0.19	0.18	-0.05	0.10	-0.25	0.04	-0.28	-0.20	-0.22	0.37	0.54	-0.35	-0.31	0.46	0.06	0.01
VAR 13	-0.30	0.37	0.08	-0.01	-0.57	0.12	-0.11	-0.12	-0.05	0.16	-0.19	0.36	-0.05	0.16	0.12	-0.05	-0.21	-0.04	0.06
VAR 14	-0.00	-0.24	-0.32	0.04	-0.06	-0.05	0.12	0.03	0.01	0.15	0.29	0.09	-0.76	0.01	-0.01	-0.06	-0.06	-0.18	-0.19
VAR 15	-0.18	0.35	-0.09	-0.01	0.52	0.19	-0.18	0.06	0.07	0.06	-0.17	0.18	-0.08	-0.10	0.14	0.03	-0.18	0.18	-0.08
VAR 16	-0.35	0.09	0.13	-0.20	-0.91	-0.15	0.11	-0.08	0.01	0.15	0.13	0.11	-0.17	-0.02	0.18	-0.07	-0.22	-0.05	-0.01
VAR 17	0.16	-0.17	0.11	-0.52	0.71	0.09	0.09	0.06	-0.06	0.01	-0.01	-0.14	0.20	-0.12	0.12	-0.04	0.07	-0.05	0.04
VAR 18	-0.91	0.43	0.27	-0.30	-0.21	-0.04	-0.09	-0.39	0.18	0.30	0.01	0.32	-0.29	0.38	0.25	-0.26	-0.48	-0.03	-0.04
VAR 19	-0.05	-0.16	-0.09	-0.25	-0.01	-0.00	0.05	-0.10	-0.21	0.75	0.14	-0.04	-0.16	-0.28	0.18	0.18	-0.03	-0.15	-0.16
VAR 20	-0.36	0.05	-0.07	-0.17	-0.20	0.05	-0.07	-0.04	0.20	0.30	0.18	0.30	-0.88	0.06	0.11	0.04	-0.33	-0.02	-0.15
VAR 21	-0.17	-0.13	-0.19	-0.17	-0.11	-0.05	-0.03	-0.09	-0.01	0.85	0.18	0.28	-0.33	-0.23	0.17	0.15	-0.27	-0.14	-0.11
VAR 22	0.26	-0.39	-0.52	-0.06	-0.04	0.00	0.01	0.14	-0.23	0.12	0.45	0.10	-0.10	-0.42	0.28	0.23	-0.17	-0.26	-0.25
VAR 23	0.71	-0.49	-0.47	0.33	0.20	0.09	0.10	0.36	-0.20	-0.15	0.23	-0.21	0.02	-0.24	-0.10	0.26	0.32	-0.18	-0.16
VAR 24	0.09	0.32	0.19	0.13	0.06	0.12	-0.04	0.08	0.07	-0.18	-0.26	-0.07	0.07	-0.09	-0.12	0.04	0.11	0.99	0.05
VAR 25	-0.71	0.32	0.21	-0.37	-0.16	0.12	-0.04	-0.22	0.20	0.31	-0.10	0.26	-0.17	0.16	0.21	-0.11	-0.33	-0.09	-0.03
VAR 26	-0.86	0.44	0.22	-0.32	-0.35	-0.05	-0.05	-0.27	0.22	0.35	0.05	0.32	-0.30	0.24	0.23	-0.15	-0.45	-0.12	0.01
VAR 27	-0.78	0.41	0.16	-0.27	-0.26	0.09	-0.11	-0.21	0.27	0.26	-0.04	0.28	-0.33	0.31	0.23	-0.21	-0.47	-0.01	-0.01
VAR 28	0.09	-0.10	-0.03	0.17	-0.02	-0.08	-0.20	0.02	0.60	0.07	0.04	0.15	-0.11	-0.16	-0.05	0.09	0.01	-0.05	0.03
VAR 29	0.03	-0.08	0.10	-0.14	-0.05	0.06	0.96	-0.00	-0.24	0.02	0.04	-0.34	-0.02	0.10	0.07	-0.04	0.31	-0.02	-0.20
VAR 30	0.20	-0.16	0.04	-0.02	0.11	-0.06	0.35	-0.03	-0.25	-0.16	0.01	-0.96	0.24	0.03	-0.11	-0.01	0.31	0.04	-0.12
VAR 31	0.04	0.07	0.08	-0.01	-0.03	0.14	0.09	-0.08	0.04	-0.29	0.00	-0.26	0.11	0.09	-0.10	0.01	0.16	-0.02	-0.10
VAR 32	0.20	-0.20	-0.18	-0.01	-0.06	0.02	0.11	-0.15	-0.75	0.14	0.19	-0.22	0.01	-0.27	0.01	0.15	0.09	-0.15	-0.01
VAR 33	-0.08	-0.07	-0.09	-0.09	0.04	0.03	0.25	0.03	-0.00	0.14	0.10	-0.09	-0.16	0.02	0.19	0.02	0.07	-0.07	-0.97
VAR 34	0.18	-0.10	0.02	-0.07	0.00	0.03	0.29	-0.03	-0.54	0.07	-0.03	-0.27	0.27	-0.20	0.07	0.14	0.19	0.01	-0.07
VAR 35	-0.18	0.25	0.83	-0.01	-0.00	-0.08	-0.01	-0.05	0.25	-0.21	-0.36	0.04	0.16	0.35	-0.28	-0.35	0.14	0.17	0.17
VAR 36	-0.15	0.22	0.88	-0.09	0.01	-0.12	0.14	-0.15	0.10	-0.12	-0.32	-0.10	0.27	0.26	-0.26	-0.32	0.29	0.21	0.06
VAR 37	-0.32	0.16	0.40	-0.48	0.01	0.05	0.20	-0.18	0.05	0.11	-0.04	0.09	-0.04	0.15	0.18	-0.17	-0.07	-0.10	-0.09
VAR 38	0.39	-0.11	-0.56	0.38	-0.10	0.15	-0.06	0.34	-0.17	-0.12	0.30	-0.14	-0.07	-0.28	-0.00	0.30	-0.08	0.03	0.04
VAR 39	0.35	-0.48	-0.78	0.12	0.02	-0.04	-0.02	0.27	-0.16	0.05	0.53	0.03	-0.22	-0.40	0.19	0.31	-0.20	-0.31	-0.20
VAR 40	-0.57	0.40	0.85	-0.55	-0.15	-0.04	0.09	-0.25	0.16	0.13	-0.28	0.14	0.13	0.22	0.15	-0.24	-0.10	0.15	0.02
VAR 41	-0.59	0.43	0.83	-0.52	-0.15	-0.04	0.08	-0.21	0.23	0.04	-0.32	0.20	0.03	0.32	0.20	-0.31	-0.19	0.13	0.05
VAR 42	-0.47	0.08	0.07	-0.75	-0.12	0.06	0.02	-0.14	-0.06	0.27	0.13	0.13	-0.26	-0.04	0.54	0.01	-0.31	-0.09	-0.11
VAR 43	-0.37	0.07	-0.04	-0.71	-0.10	0.08	0.11	-0.13	-0.07	0.30	0.13	0.07	-0.22	-0.14	0.54	0.04	-0.28	-0.09	-0.15
VAR 44	-0.32	0.02	-0.04	-0.54	-0.06	-0.06	0.06	-0.05	-0.02	0.28	0.17	0.13	-0.08	-0.11	0.94	0.08	-0.23	-0.12	-0.18
VAR 45	-0.24	-0.01	-0.15	-0.43	-0.18	0.01	0.08	0.05	-0.07	0.23	0.22	0.17	-0.02	-0.16	0.96	0.04	-0.27	-0.12	-0.17
VAR 46	0.27	0.15	-0.26	0.86	0.05	0.15	-0.18	0.09	0.01	-0.29	-0.14	0.00	-0.03	0.13	-0.36	0.06	0.17	0.11	0.02
VAR 47	0.77	-0.32	-0.39	0.52	0.12	0.08	-0.09	0.42	-0.13	-0.33	0.13	-0.20	0.19	-0.26	-0.32	0.33	0.29	0.07	0.13
VAR 48	-0.82	0.55	0.38	-0.34	-0.26	-0.02	0.00	-0.25	0.09	0.18	-0.19	0.24	-0.10	0.36	0.26	-0.22	-0.42	0.02	0.08
VAR 49	0.89	-0.49	-0.37	0.41	0.23	-0.01	0.03	0.27	-0.18	-0.26	0.11	-0.36	0.24	-0.32	-0.31	0.31	0.47	0.07	0.11
VAR 50	0.65	-0.21	-0.46	0.85	0.15	0.01	-0.13	0.22	-0.11	-0.32	0.04	-0.19	0.04	-0.05	-0.41	0.18	0.36	0.04	0.05
VAR 51	0.82	-0.48	-0.30	0.53	0.17	-0.04	0.06	0.17	-0.14	-0.29	0.03	-0.29	0.20	-0.21	-0.34	0.20	0.54	0.04	0.01
VAR 52	0.24	-0.81	-0.31	-0.18	0.01	-0.35	0.08	-0.07	-0.10	0.26	0.57	-0.04	-0.10	-0.28	0.11	0.19	0.06	-0.26	-0.07
VAR 53	0.90	-0.39	-0.35	0.48	0.27	0.07	0.03	0.34	-0.23	-0.31	0.01	-0.25	0.27	-0.24	-0.34	0.34	0.40	0.05	0.04
VAR 54	0.21	0.02	-0.18	0.22	-0.01	0.06	0.02	0.89	0.05	-0.25	-0.02	-0.03	-0.07	0.05	0.04	0.06	0.01	0.02	-0.11
VAR 55	0.06	0.06	0.26	0.18	0.13	-0.08	0.11	0.10	0.01	-0.38	-0.13	-0.11	0.15	0.30	-0.27	-0.74	0.23	-0.06	0.04
VAR 56	0.48	-0.20	-0.17	0.05	0.14	0.13	0.08	0.73	-0.09	-0.16	0.12	-0.08	0.16	-0.29	-0.08	0.17	0.15	0.12	-0.03
VAR 57	0.59	-0.10	-0.44	0.48	0.04	0.35	0.11	0.33	-0.37	-0.22	0.25	-0.30	0.11	-0.24	-0.13	0.44	0.20	0.04	-0.06
VAR 58	0.89	-0.37	-0.40	0.48	0.27	0.10	0.10	0.34	-0.28	-0.25	0.09	-0.30	0.23	-0.37	-0.20	0.37	0.45	0.09	0.01
VAR 59	0.72	-0.15	-0.31	0.60	0.13	0.11	-0.07	0.41	-0.19	-0.33	-0.01	-0.18	0.14	-0.19	-0.33	0.29	0.34	0.10	0.14
VAR 60	-0.04	0.46	0.43	0.10	0.09	-0.01	-0.04	-0.06	0.11	-0.19	-0.99	0.03	0.24	0.21	-0.20	-0.06	0.16	0.23	0.13
VAR 61	0.20	-0.89	-0.31	-0.20	-0.04	-0.41	-0.06	0.00	-0.03	0.33	0.48	0.03	-0.16	-0.24	0.12	0.02	-0.03	-0.36	-0.06
VAR 62	0.42	-0.26	0.07	0.12	0.12	-0.13	0.37	0.06	-0.06	-0.19	0.01	-0.39	0.21	-0.01	-0.13	0.07	0.83	0.06	-0.13
VAR 63	0.52	-0.87	-0.36	0.02	0.14	-0.35	0.02	0.13	-0.14	0.13	0.37	-0.17	0.02	-0.31	-0.02	0.07	0.24	-0.23	-0.02
VAR 64	0.38	-0.83	-0.47	0.10	0.03	-0.14	-0.07	0.14	-0.08	0.16	0.46	-0.01	-0.11	-0.19	-0.06	0.17	0.02	-0.28	-0.00
VAR 65	0.42	-0.89	-0.54	0.01	0.08	-0.24	-0.05	0.15	-0.05	0.15	0.51	-0.01	-0.12	-0.27	0.06	0.10	-0.01	-0.34	-0.05
VAR 66	0.39	-0.42	0.00	0.11	0.14	-0.25	0.18	0.28	-0.06	-0.01	-0.10	-0.18	0.25	-0.01	-0.21	0.02	0.45	-0.12	-0.00
VAR 67	-0.26	0.09	0.01	0.28	-0.28	0.08	-0.10	-0.29	0.11	-0.02	0.07	0.10	-0.15	0.18	-0.15	-0.18	-0.08	0.07	0.05
VAR 68	0.76	-0.69	-0.54	0.40	0.10	-0.18	0.00	0.30	-0.18	-0.10	0.24	-0.18	0.03	-0.31	-0.15	0.23	0.24	-0.13	0.00
VAR 69	-0.37	0.79	0.33	0.03	-0.02	0.33	-0.19	-0.09	0.05	-0.20	-0.47	0.20	-0.06	0.25	-0.00	-0.18	-0.21	0.20	0.15
VAR 70	-0.75	0.48	0.25	-0.15	-0.17	0.03	-0.10	-0.34	0.17	0.13	0.06	0.23	-0.19	0.41	0.17	-0.25	-0.34	0.04	-0.01
VAR 71	-0.54	0.60	0.15	0.01	-0.25	0.21	-0.31	-0.19	0.11	0.07	-0.05	0.29	-0.17	0.22	0.12	-0.08	-0.49	0.09	0.04
VAR 72	0.21	-0.64	-0.31	-0.09	0.04	0.08	-0.13	0.03	-0.04	0.21	0.28	-0.02	-0.10	-0.15	0.02	0.04	-0.07	-0.29	-0.04
VAR 73	-0.15	-0.13	0.21	-0.03	-0.14	-0.91	-0.07	-0.15	0.19	0.06	-0.19	0.08	0.02	0.06	0.03	-0.16	0.01	0.03	0.07
VAR 74	0.61	-0.83	-0.36	0.15	0.18	-0.17	0.03	0.16	-0.06	0.00	0.21	-0.17	0.01	-0.21	-0.10	0.14	0.29	-0.17	0.03
VAR 75	0.38	-0.02	0.10	0.30	0.19	-0.13	0.34	0.03	-0.05	-0.21	-0.20	-0.26	0.29	0.09	-0.27	0.05	0.85	0.16	-0.09
VAR 76	-0.30	0.15	-0.18	-0.19	-0.07	-0.02	-0.15	-0.04	0.08	0.21	0.23	0.17	-0.27	-0.08	0.26	-0.01	-0.85	-0.04	-0.03
VAR 77	-0.08	0.66	0.11	0.08	0.04	0.77	0.04	0.03	0.00	-0.18	-0.28	0.02	-0.06	0.10	0.02	0.11	-0.04	0.26	-0.04
VAR 78	-0.49	0.37	0.33	-0.00	-0.12	-0.38	-0.07	-0.18	0.40	0.04	-0.29	0.14	-0.04	0.32	-0.07	-0.23	-0.14	0.10	0.09
VAR 79	0.08	-0.24	-0.03	0.21	-0.18	-0.01	-0.01	-0.12	0.04	0.05	0.03	0.06	-0.08	-0.09	-0.19	-0.14	0.17	-0.04	0.10
VAR 80	0.54	-0.89	-0.42	0.07	0.12	-0.23	0.07	0.18	-0.08	0.14	0.44	-0.11	-0.03	-0.27	-0.00	0.14	0.20	-0.20	-0.04
VAR 81	0.45	-0.88	-0.38	0.10	0.02	-0.35	0.05	0.07	-0.01	0.19	0.48	-0.11	-0.08	-0.20	-0.05	0.06	0.20	-0.26	-0.06
VAR 82	-0.34	0.40	0.39	-0.32	0.08	-0.10	0.00	-0.13	0.06	0.03	-0.36	0.11	-0.01	0.12	0.13	-0.19	-0.17	0.15	0.03
VAR 83	-0.40	0.82	0.33	0.01	-0.07	0.15	-0.06	-0.02	0.23	-0.13	-0.40	0.10	-0.02	0.22	-0.01	-0.09	-0.13	0.25	0.05
VAR 84	0.85	-0.48	-0.41	0.35	0.31	-0.05	0.04	0.30	-0.14	-0.23	0.07	-0.23	0.12	-0.28	-0.24	0.30	0.50	0.01	-0.03
VAR 85	-0.13	0.63	-0.07	0.17	-0.05	0.52	-0.14	0.13	0.05	-0.15	-0.22	0.17	-0.15	0.10	-0.00	0.09	-0.30	0.24	0.07
VAR 86	-0.39	0.79	0.13	0.05	-0.19	0.31	-0.16	-0.01	0.01	-0.06	-0.22	0.21	-0.02	0.08	0.09	0.01	-0.34	0.14	0.15
VAR 87	0.19	0.53	0.38	0.28	0.13	0.31	0.03	0.03	-0.11	-0.34	-0.58	-0.11	0.42	0.10	-0.31	0.07	0.33	0.41	0.19
VAR 88	0.49	-0.12	-0.42	0.19	0.12	0.11	-0.05	0.31	-0.18	-0.04	0.03	-0.07	0.11	-0.43	-0.00	0.81	0.14	-0.03	-0.01
VAR 89	0.21	0.13	0.26	0.48	0.20	0.04	0.10	0.14	0.02	-0.53	-0.18	-0.28	0.29	0.51	-0.37	-0.16	0.42	0.21	0.18
VAR 90	-0.04	0.20	0.28	0.35	0.16	0.11	0.10	0.05	0.08	-0.46	-0.25	-0.17	0.13	0.78	-0.26	-0.35	0.19	0.15	0.03
VAR 91	-0.44	-0.16	0.30	-0.87	-0.12	-0.24	0.14	-0.27	0.04	0.55	0.16	0.13	-0.10	-0.14	0.35	-0.10	-0.19	-0.14	-0.01

TABLE 36

Correlation of Ranked Factor Scores with Original Variables Using Factor Score Formula F = ZA, Where Loadings in the Range of −.49 to .49 Are Set to Zero and All Other Loadings Set to ±1.0

Factor Scores Ranked

	1	2	3	4	5	6	7	8	9	10	11	12	13	14	15	16	17	18	19
VAR 1	-0.45	-0.17	0.11	-0.49	-0.19	-0.27	-0.03	-0.33	0.12	0.89	0.17	0.32	-0.24	-0.19	0.29	-0.00	-0.26	-0.14	-0.05
VAR 2	-0.21	0.09	0.07	-0.09	-0.81	0.05	0.08	0.08	-0.03	0.13	0.12	0.05	-0.13	-0.11	0.15	-0.04	-0.12	0.03	0.01
VAR 3	0.93	-0.36	-0.38	0.48	0.25	0.09	-0.00	0.43	-0.21	-0.37	0.01	-0.26	0.21	-0.33	-0.28	0.25	0.44	0.10	0.05
VAR 4	-0.63	0.14	0.41	-0.89	-0.11	-0.02	0.12	-0.16	0.11	0.33	0.00	0.15	-0.07	0.05	0.40	-0.17	-0.32	-0.08	-0.04
VAR 5	0.81	-0.45	-0.16	0.25	0.18	-0.09	0.08	0.28	-0.25	-0.24	0.08	-0.31	0.35	-0.39	-0.30	0.11	0.49	0.10	0.11
VAR 6	0.67	-0.54	-0.21	0.23	0.21	-0.11	0.07	0.21	-0.04	-0.11	0.08	-0.24	0.14	-0.30	-0.30	0.13	0.43	0.01	0.02
VAR 7	0.37	-0.15	0.07	0.33	0.11	-0.03	0.08	0.13	-0.12	-0.12	0.01	-0.17	0.32	-0.10	-0.24	-0.01	0.43	0.04	-0.05
VAR 8	0.60	-0.28	-0.34	0.10	0.27	0.17	0.03	0.23	-0.23	-0.03	0.12	-0.09	0.12	-0.55	-0.02	0.35	0.17	-0.06	-0.13
VAR 9	-0.41	0.19	0.40	-0.02	-0.18	-0.03	0.20	-0.16	0.10	-0.13	-0.15	-0.04	-0.03	0.88	-0.08	-0.29	0.06	-0.14	-0.01
VAR 10	-0.66	0.38	0.47	-0.09	-0.26	-0.02	0.01	-0.32	0.25	-0.07	-0.25	0.09	-0.14	0.84	-0.00	-0.43	-0.14	-0.09	-0.04
VAR 11	0.45	-0.15	-0.02	0.14	-0.03	-0.01	0.05	0.19	-0.06	-0.09	-0.04	-0.26	0.25	-0.59	-0.14	0.22	0.26	0.17	0.01
VAR 12	0.09	-0.04	0.24	0.20	0.15	-0.07	0.10	-0.22	0.03	-0.27	-0.20	-0.22	0.37	0.54	-0.35	-0.31	0.46	0.06	0.01
VAR 13	-0.30	0.38	0.10	-0.01	-0.61	0.15	-0.11	-0.16	-0.05	0.15	-0.19	0.36	-0.05	0.16	0.12	-0.06	-0.24	-0.04	0.06
VAR 14	0.01	-0.24	-0.33	0.04	-0.05	-0.08	0.12	-0.01	0.02	0.16	0.29	0.09	-0.78	0.00	-0.01	-0.06	-0.06	-0.18	-0.19
VAR 15	-0.17	0.35	-0.08	-0.04	0.51	0.24	-0.18	0.03	0.06	0.06	-0.17	0.18	-0.09	-0.08	0.15	0.04	-0.17	0.18	-0.08
VAR 16	-0.36	0.09	0.14	-0.18	-0.89	-0.14	0.11	-0.11	0.02	0.16	0.13	0.11	-0.16	-0.02	0.17	-0.08	-0.22	-0.05	-0.01
VAR 17	0.14	-0.17	0.11	-0.53	0.75	0.06	0.09	0.12	-0.06	0.01	-0.01	-0.14	0.20	-0.13	0.12	-0.04	0.08	-0.05	0.04
VAR 18	-0.90	0.44	0.29	-0.31	-0.23	-0.01	-0.09	-0.45	0.18	0.30	0.01	0.32	-0.29	0.39	0.25	-0.27	-0.48	-0.03	-0.04
VAR 19	-0.04	-0.16	-0.08	-0.25	0.00	-0.01	0.05	-0.06	-0.21	0.75	0.14	-0.04	-0.15	-0.29	0.18	0.16	-0.03	-0.15	-0.16
VAR 20	-0.36	0.05	-0.06	-0.18	-0.20	0.07	-0.07	-0.08	0.20	0.31	0.18	0.30	-0.87	0.06	0.11	0.02	-0.35	-0.02	-0.15
VAR 21	-0.16	-0.13	-0.18	-0.17	-0.12	-0.07	-0.03	-0.11	-0.01	0.83	0.18	0.28	-0.33	-0.23	0.17	0.13	-0.28	-0.14	-0.11
VAR 22	0.26	-0.39	-0.53	-0.08	-0.03	-0.04	0.01	0.17	-0.23	0.12	0.45	0.10	-0.10	-0.42	0.28	0.23	-0.17	-0.26	-0.25
VAR 23	0.72	-0.50	-0.49	0.32	0.19	0.04	0.10	0.39	-0.20	-0.16	0.23	-0.21	0.02	-0.26	-0.10	0.26	0.30	-0.18	-0.16
VAR 24	0.08	0.32	0.20	0.12	0.07	0.18	-0.04	0.10	0.07	-0.18	-0.26	-0.07	0.08	-0.08	-0.12	0.03	0.12	0.99	0.05
VAR 25	-0.70	0.32	0.23	-0.37	-0.16	0.12	-0.04	-0.25	0.20	0.31	-0.10	0.26	-0.15	0.17	0.21	-0.13	-0.35	-0.09	-0.03
VAR 26	-0.86	0.44	0.24	-0.32	-0.35	-0.00	-0.05	-0.36	0.22	0.36	0.05	0.32	-0.30	0.25	0.23	-0.17	-0.45	-0.12	0.01
VAR 27	-0.77	0.42	0.18	-0.28	-0.27	0.13	-0.11	-0.29	0.27	0.26	-0.04	0.28	-0.32	0.32	0.23	-0.23	-0.48	-0.01	-0.01
VAR 28	0.08	-0.11	-0.04	0.18	-0.02	-0.08	-0.20	0.01	0.60	0.07	0.04	0.15	-0.13	-0.17	-0.05	0.09	0.01	-0.05	0.03
VAR 29	0.03	-0.08	0.09	-0.13	-0.04	0.06	0.96	0.02	-0.24	0.03	0.04	-0.34	-0.03	0.09	0.07	-0.04	0.31	-0.02	-0.20
VAR 30	0.19	-0.16	0.03	-0.00	0.14	-0.07	0.35	-0.00	-0.25	-0.16	0.01	-0.96	0.24	0.02	-0.11	-0.01	0.33	0.04	-0.12
VAR 31	0.04	0.07	0.08	-0.00	-0.02	0.14	0.09	-0.04	0.04	-0.28	0.00	-0.26	0.11	0.08	-0.10	0.00	0.17	-0.02	-0.10
VAR 32	0.20	-0.20	-0.18	-0.01	-0.06	-0.01	0.11	-0.11	-0.73	0.14	0.19	-0.22	0.01	-0.27	0.01	0.16	0.08	-0.15	-0.01
VAR 33	-0.08	-0.08	-0.09	-0.09	0.04	0.04	0.25	0.02	-0.01	0.15	0.10	-0.09	-0.16	0.02	0.19	0.01	0.08	-0.07	-0.97
VAR 34	0.18	-0.10	0.02	-0.07	0.02	0.01	0.29	0.04	-0.56	0.07	-0.03	-0.27	0.27	-0.21	0.07	0.14	0.19	0.01	-0.07
VAR 35	-0.19	0.25	0.81	0.00	-0.00	-0.05	-0.01	-0.06	0.24	-0.20	-0.36	0.04	0.16	0.35	-0.28	-0.35	0.12	0.17	0.17
VAR 36	-0.16	0.22	0.86	-0.07	0.02	-0.08	0.14	-0.12	0.09	-0.11	-0.32	-0.10	0.26	0.26	-0.25	-0.32	0.29	0.21	0.06
VAR 37	-0.33	0.16	0.41	-0.47	0.02	0.04	0.20	-0.14	0.05	0.12	-0.04	0.09	-0.03	0.14	0.18	-0.18	-0.06	-0.10	-0.09
VAR 38	0.41	-0.11	-0.56	0.38	-0.11	0.14	-0.06	0.30	-0.17	-0.14	0.30	-0.14	-0.08	-0.28	-0.00	0.30	-0.07	0.03	0.04
VAR 39	0.36	-0.48	-0.79	0.11	0.02	-0.10	-0.02	0.28	-0.16	0.04	0.53	0.03	-0.22	-0.40	0.19	0.32	-0.20	-0.31	-0.20
VAR 40	-0.59	0.40	0.87	-0.55	-0.13	0.03	0.09	-0.23	0.16	0.14	-0.28	0.14	0.13	0.22	0.15	-0.25	-0.10	0.15	0.02
VAR 41	-0.61	0.44	0.85	-0.52	-0.14	0.03	0.08	-0.22	0.22	0.04	-0.32	0.20	0.03	0.32	0.20	-0.31	-0.19	0.13	0.05
VAR 42	-0.48	0.09	0.09	-0.78	-0.09	0.09	0.02	-0.14	-0.06	0.28	0.13	0.13	-0.25	-0.03	0.54	-0.00	-0.30	-0.09	-0.11
VAR 43	-0.38	0.07	-0.02	-0.74	-0.08	0.11	0.11	-0.12	-0.07	0.30	0.13	0.07	-0.21	-0.13	0.54	0.02	-0.27	-0.09	-0.15
VAR 44	-0.33	0.03	-0.01	-0.55	-0.04	-0.01	0.06	-0.08	-0.02	0.28	0.17	0.13	-0.07	-0.11	0.94	0.06	-0.22	-0.12	-0.18
VAR 45	-0.23	-0.01	-0.13	-0.44	-0.17	0.04	0.08	0.02	-0.07	0.23	0.22	0.17	-0.02	-0.16	0.96	0.02	-0.27	-0.12	-0.17
VAR 46	0.30	0.15	-0.28	0.84	0.00	0.17	-0.18	0.05	0.00	-0.29	-0.14	0.00	-0.03	0.13	-0.36	0.07	0.15	0.11	0.02
VAR 47	0.77	-0.32	-0.41	0.52	0.12	0.01	-0.09	0.45	-0.13	-0.34	0.13	-0.20	0.18	-0.27	-0.32	0.34	0.28	0.07	0.13
VAR 48	-0.82	0.55	0.40	-0.34	-0.27	0.04	0.00	-0.31	0.09	0.18	-0.19	0.24	-0.09	0.37	0.26	-0.23	-0.42	0.02	0.08
VAR 49	0.88	-0.49	-0.40	0.42	0.23	-0.06	0.03	0.33	-0.18	-0.27	0.11	-0.36	0.24	-0.33	-0.31	0.33	0.47	0.07	0.11
VAR 50	0.67	-0.21	-0.48	0.85	0.12	0.00	-0.13	0.20	-0.11	-0.33	0.04	-0.19	0.02	-0.05	-0.41	0.19	0.36	0.04	0.05
VAR 51	0.82	-0.49	-0.33	0.54	0.16	-0.06	0.06	0.22	-0.14	-0.29	0.03	-0.29	0.20	-0.22	-0.34	0.21	0.53	0.04	0.01
VAR 52	0.23	-0.81	-0.32	-0.17	0.03	-0.43	0.08	-0.04	-0.09	0.27	0.57	-0.04	-0.09	-0.30	0.11	0.19	0.08	-0.26	-0.07
VAR 53	0.90	-0.39	-0.37	0.49	0.26	0.02	0.03	0.40	-0.23	-0.33	0.01	-0.25	0.27	-0.25	-0.34	0.36	0.41	0.05	0.04
VAR 54	0.22	0.01	-0.19	0.21	-0.01	0.04	0.02	0.75	0.05	-0.26	-0.02	-0.03	-0.08	0.05	0.04	0.08	0.01	0.02	-0.11
VAR 55	0.05	0.05	0.24	0.19	0.13	-0.11	0.11	0.07	0.01	-0.37	-0.13	-0.11	0.13	0.30	-0.28	-0.71	0.24	-0.06	0.04
VAR 56	0.47	-0.20	-0.18	0.05	0.16	0.07	0.08	0.88	-0.10	-0.18	0.12	-0.08	0.16	-0.30	-0.09	0.18	0.15	0.12	-0.03
VAR 57	0.61	-0.11	-0.45	0.47	0.03	0.31	0.11	0.37	-0.37	-0.23	0.25	-0.30	0.10	-0.25	-0.13	0.45	0.20	0.04	-0.06
VAR 58	0.89	-0.37	-0.42	0.48	0.27	0.06	0.10	0.40	-0.28	-0.26	0.09	-0.30	0.23	-0.39	-0.20	0.40	0.46	0.09	0.01
VAR 59	0.72	-0.15	-0.33	0.61	0.12	0.08	-0.07	0.42	-0.19	-0.35	-0.01	-0.18	0.12	-0.19	-0.34	0.32	0.33	0.10	0.14
VAR 60	-0.05	0.45	0.43	0.10	0.07	0.09	-0.04	-0.09	0.10	-0.20	-0.99	0.03	0.25	0.22	-0.20	-0.05	0.14	0.23	0.13
VAR 61	0.19	-0.89	-0.32	-0.19	-0.02	-0.49	-0.06	0.03	-0.02	0.33	0.48	0.03	-0.15	-0.24	0.12	0.02	-0.03	-0.36	-0.06
VAR 62	0.42	-0.27	0.05	0.14	0.13	-0.13	0.37	0.09	-0.07	-0.18	0.01	-0.39	0.19	-0.02	-0.13	0.08	0.87	0.06	-0.13
VAR 63	0.51	-0.87	-0.38	0.03	0.15	-0.43	0.02	0.19	-0.13	0.14	0.37	-0.17	0.02	-0.32	-0.02	0.08	0.25	-0.23	-0.02
VAR 64	0.39	-0.82	-0.48	0.11	0.03	-0.26	-0.07	0.16	-0.07	0.16	0.46	-0.01	-0.10	-0.20	-0.06	0.16	-0.00	-0.28	-0.00
VAR 65	0.41	-0.88	-0.56	0.01	0.09	-0.35	-0.05	0.16	-0.05	0.14	0.51	-0.01	-0.12	-0.28	0.06	0.10	-0.02	-0.34	-0.05
VAR 66	0.39	-0.42	-0.02	0.13	0.14	-0.26	0.18	0.31	-0.06	-0.00	-0.10	-0.18	0.23	-0.02	-0.21	0.04	0.45	-0.12	-0.00
VAR 67	-0.25	0.10	0.01	0.27	-0.30	0.09	-0.10	-0.36	0.11	-0.02	0.07	0.10	-0.15	0.19	-0.15	-0.18	-0.10	0.07	0.05
VAR 68	0.77	-0.70	-0.56	0.40	0.10	-0.24	0.00	0.32	-0.18	-0.11	0.24	-0.18	0.03	-0.32	-0.15	0.25	0.23	-0.13	0.00
VAR 69	-0.36	0.79	0.35	0.01	-0.03	0.41	-0.19	-0.14	0.05	-0.21	-0.47	0.20	-0.06	0.27	-0.00	-0.18	-0.22	0.20	0.15
VAR 70	-0.75	0.49	0.27	-0.16	-0.19	0.06	-0.10	-0.39	0.17	0.13	0.06	0.23	-0.19	0.42	0.17	-0.26	-0.34	0.04	-0.01
VAR 71	-0.54	0.62	0.17	-0.00	-0.27	0.26	-0.31	-0.24	0.11	0.07	-0.05	0.29	-0.16	0.23	0.12	-0.09	-0.50	0.09	0.04
VAR 72	0.21	-0.64	-0.31	-0.08	0.04	-0.03	-0.13	0.04	-0.04	0.20	0.28	-0.02	-0.09	-0.15	0.01	0.03	-0.10	-0.29	-0.04
VAR 73	-0.16	-0.14	0.20	-0.01	-0.14	-0.80	-0.07	-0.18	0.19	0.07	-0.19	0.08	0.03	0.07	0.03	-0.16	0.02	0.03	0.07
VAR 74	0.61	-0.83	-0.38	0.16	0.19	-0.25	0.03	0.17	-0.06	-0.00	0.21	-0.17	0.01	-0.22	-0.10	0.14	0.28	-0.17	0.03
VAR 75	0.38	-0.03	0.08	0.31	0.18	-0.08	0.34	0.05	-0.05	-0.21	-0.20	-0.26	0.28	0.08	-0.27	0.06	0.87	0.16	-0.09
VAR 76	-0.30	0.15	-0.17	-0.19	-0.07	-0.02	-0.15	-0.07	0.08	0.20	0.23	0.17	-0.27	-0.08	0.26	-0.01	-0.82	-0.04	-0.03
VAR 77	-0.07	0.67	0.13	0.05	0.03	0.89	0.04	0.02	-0.00	-0.19	-0.28	0.02	-0.06	0.11	0.02	0.10	-0.04	0.26	-0.04
VAR 78	-0.49	0.37	0.33	0.00	-0.14	-0.29	-0.07	-0.25	0.40	0.04	-0.29	0.14	-0.05	0.33	-0.07	-0.24	-0.13	0.10	0.09
VAR 79	0.09	-0.25	-0.04	0.21	-0.20	-0.04	-0.01	-0.11	0.04	0.05	0.03	0.06	-0.09	-0.08	-0.20	-0.14	0.14	-0.04	0.10
VAR 80	0.54	-0.89	-0.43	0.08	0.13	-0.31	0.07	0.21	-0.08	0.14	0.44	-0.11	-0.04	-0.28	-0.01	0.13	0.21	-0.20	-0.04
VAR 81	0.45	-0.88	-0.40	0.12	0.03	-0.43	0.05	0.08	-0.01	0.19	0.48	-0.11	-0.09	-0.21	-0.05	0.05	0.20	-0.26	-0.06
VAR 82	-0.35	0.40	0.41	-0.31	0.08	-0.04	0.00	-0.14	0.06	0.03	-0.36	0.11	-0.00	0.12	0.13	-0.19	-0.16	0.15	0.03
VAR 83	-0.40	0.81	0.34	0.00	-0.08	0.26	-0.06	-0.09	0.23	-0.13	-0.40	0.10	-0.02	0.23	-0.01	-0.10	-0.12	0.25	0.05
VAR 84	0.85	-0.49	-0.43	0.36	0.31	-0.07	0.04	0.35	-0.14	-0.24	0.07	-0.23	0.10	-0.29	-0.24	0.32	0.51	0.01	-0.03
VAR 85	-0.12	0.64	-0.05	0.16	-0.07	0.55	-0.14	0.09	0.05	-0.17	-0.22	0.17	-0.15	0.10	-0.00	0.08	-0.31	0.24	0.07
VAR 86	-0.38	0.80	0.15	0.03	-0.20	0.34	-0.16	-0.05	0.01	-0.08	-0.22	0.21	-0.02	0.09	0.09	0.01	-0.34	0.14	0.15
VAR 87	0.19	0.53	0.38	0.28	0.11	0.37	0.03	0.06	-0.11	-0.35	-0.58	-0.11	0.42	0.11	-0.31	0.08	0.33	0.41	0.19
VAR 88	0.49	-0.13	-0.42	0.19	0.13	0.11	-0.05	0.34	-0.18	-0.05	0.03	-0.07	0.11	-0.44	-0.00	0.84	0.15	-0.03	-0.01
VAR 89	0.21	0.12	0.25	0.48	0.18	0.06	0.10	0.12	0.01	-0.54	-0.18	-0.28	0.28	0.50	-0.37	-0.14	0.42	0.21	0.18
VAR 90	-0.04	0.20	0.28	0.35	0.14	0.13	0.10	0.01	0.07	-0.47	-0.25	-0.17	0.12	0.77	-0.26	-0.35	0.18	0.15	0.03
VAR 91	-0.46	-0.16	0.32	-0.86	-0.09	-0.24	0.14	-0.24	0.05	0.56	0.16	0.13	-0.09	-0.14	0.35	-0.12	-0.18	-0.14	-0.01

TABLE 37
Factor Scores, $F = ZA(A'A)^{-1}$, Regression Estimates for Missing Data

1	AFGHANISTAN	1.44	-0.80	0.35	-1.17	-0.09	-1.18	-2.18	-0.43	-1.41	-0.22
		-0.53	1.13	-0.01	-1.67	-0.46	-1.86	-0.99	1.18	0.19	
2	ALBANIA	-0.20	-1.68	-0.51	1.45	-0.41	-0.50	-1.09	0.22	1.48	1.25
		0.52	-0.32	-0.74	-0.25	0.97	-0.34	-0.78	-1.21	-1.48	
3	ALGERIA	0.03	-0.95	-0.51	-0.62	1.18	0.89	1.04	-0.63	-0.93	-0.01
		1.02	1.01	0.83	-0.07	-1.89	0.27	-1.29	0.92	-0.72	
4	ARGENTINA	-1.14	0.30	0.15	-1.17	2.06	-0.50	1.47	-0.30	0.59	-0.12
		0.58	-1.47	0.73	1.68	0.64	2.00	1.72	0.19	-0.57	
5	AUSTRALIA	-1.54	0.85	1.15	-0.39	2.48	0.09	-1.29	0.73	0.06	0.09
		0.67	-0.56	-0.06	-0.18	1.77	1.09	-0.10	-1.03	1.52	
6	AUSTRIA	-0.97	1.15	-1.34	-0.83	-0.91	-0.15	-0.42	-1.19	-0.07	-1.02
		-0.54	0.04	-0.66	0.98	0.34	0.31	-0.20	-0.30	-0.34	
7	BELGIUM	-1.36	0.93	-0.09	-1.31	-1.51	-0.04	0.02	1.17	-0.63	0.49
		-1.21	1.02	-0.24	0.88	-2.08	0.62	-0.21	-0.25	0.23	
8	BOLIVIA	0.17	0.48	0.52	0.21	1.25	0.21	1.10	0.73	0.69	-0.19
		0.43	-0.73	-0.81	0.80	-1.15	-1.37	0.48	2.11	-0.12	
9	BRAZIL	.0.10	0.47	0.44	-1.81	1.13	-1.00	1.12	-0.13	0.19	0.31
		0.30	-0.90	0.99	1.66	0.09	-0.77	0.60	-0.23	0.64	
10	BULGARIA	-1.14	-1.75	-1.04	-0.40	-0.15	-0.39	-0.40	-0.29	0.80	0.26
		0.77	0.34	-1.39	-0.59	-1.22	0.32	-0.62	-0.60	-0.11	
11	BURMA	1.09	-0.16	-0.56	-1.65	-0.22	-0.48	-1.24	-0.17	-0.14	-0.44
		1.59	-1.69	-0.67	-1.00	0.25	0.86	1.46	-0.69	0.34	
12	BURUNDI	0.81	-0.17	0.49	1.40	-1.11	1.24	-0.58	-0.71	0.53	-0.60
		-0.60	-0.36	-0.41	0.68	0.15	1.04	-0.54	-0.74	0.51	
13	CAMBODIA	1.52	-0.53	-0.40	0.02	-0.63	0.40	-1.25	-1.00	-0.63	1.30
		0.52	0.99	-0.71	-0.08	1.83	-0.60	1.80	-0.70	1.11	
14	CAMEROUN	1.20	0.57	-0.38	-0.04	0.29	0.60	-0.48	0.54	-0.28	-0.38
		-0.87	-0.89	-0.94	0.61	-0.46	1.19	0.14	-0.35	-0.57	
15	CANADA	-1.32	0.78	1.14	-1.22	1.57	0.84	-1.72	1.63	0.38	0.98
		-0.19	-0.08	0.67	0.78	1.04	-0.04	-0.16	-0.17	-0.30	
16	CENTRAL AFRICAN REP	0.60	-0.88	-0.16	0.94	0.94	1.52	0.00	-0.06	0.18	-0.28
		-1.19	0.02	0.85	0.18	-0.51	1.15	0.23	-0.15	0.30	
17	CEYLON	0.61	0.78	-0.22	-0.15	-1.55	-1.34	0.73	1.11	0.66	-1.40
		-0.67	-0.03	-0.24	-1.12	2.26	0.99	1.02	0.30	1.16	
18	CHAD	0.99	-0.04	-0.90	0.27	0.76	1.15	-0.41	0.73	-0.37	-0.99
		-1.79	-0.38	-0.59	-0.66	-0.73	-0.66	0.55	-0.68	0.09	
19	CHILE	-0.70	0.58	0.99	0.14	0.28	-0.37	1.31	-0.21	-0.57	0.97
		-0.10	0.19	1.03	1.60	0.55	-0.71	0.75	0.46	0.89	
20	CHINA, PR	-0.13	-1.56	-2.36	-2.23	-0.47	-0.31	-0.07	0.08	-0.44	-0.37
		-0.29	-2.29	-0.23	-1.40	-1.67	-1.86	-0.70	0.89	1.99	
21	COLOMBIA	-0.23	0.37	0.90	-1.31	0.58	0.10	0.58	-0.49	0.29	-0.94
		0.12	-0.44	-0.25	1.62	-1.31	-0.80	0.57	-0.69	0.97	
22	CONGO (BRA)	0.35	-0.50	-1.07	0.65	0.76	1.22	0.02	0.75	-0.25	-0.67
		-1.59	-0.47	-0.41	0.51	-0.59	1.51	0.99	0.16	0.08	
23	CONGO (LEO)	1.11	-0.00	0.62	-0.38	0.98	0.18	1.77	0.99	0.41	0.26
		-0.48	0.18	-0.65	0.51	-0.40	1.53	0.14	-0.47	-0.03	
24	COSTA RICA	-0.53	0.62	0.61	1.02	-0.28	0.91	-1.19	-1.47	0.79	-1.66
		0.34	-1.19	-0.09	1.12	-0.14	0.08	1.38	0.08	-1.11	
25	CUBA	-0.88	-1.41	-1.26	-0.64	-0.27	0.56	0.16	-1.18	0.27	-1.17
		1.59	1.29	-0.09	1.69	1.67	-1.80	0.20	-0.25	-0.88	
26	CYPRUS	-0.61	0.77	-0.76	1.12	-0.45	-0.01	1.50	0.85	-0.12	0.54
		-1.39	0.38	0.18	-1.44	-1.66	0.64	-0.61	0.70	-1.28	
27	CZECHOSLOVAKIA	-1.60	-1.73	-0.89	-0.60	-0.76	-0.40	0.24	2.00	0.28	0.31
		0.86	1.08	-0.87	0.67	-0.26	0.32	-0.11	-0.55	-1.54	
28	DAHOMEY	1.03	-0.49	-0.69	0.69	-0.28	1.34	-0.39	0.55	0.34	-0.39
		-1.09	-0.23	0.58	0.19	-0.39	0.82	0.71	0.70	-0.25	
29	DENMARK	-1.46	0.93	-0.17	0.09	-0.39	0.45	-1.29	-0.81	0.47	-0.05
		0.18	0.94	0.65	-0.53	0.27	0.23	0.56	1.98	-1.21	
30	DOMINICAN REPUBLIC	0.13	0.89	1.51	0.81	-0.95	0.54	0.20	-1.41	0.57	1.27
		-0.27	0.13	-0.82	1.29	-0.17	-1.81	-0.16	-1.57	-0.63	
31	ECUADOR	-0.26	0.23	1.06	0.41	-0.05	-0.95	0.47	1.81	0.56	0.39
		-0.41	-0.61	0.47	0.77	-0.64	-1.68	1.10	0.04	-1.04	
32	EL SALVADOR	-0.51	-1.11	0.88	1.28	-0.89	-1.11	-0.63	-0.33	0.20	-1.03
		0.10	0.56	1.12	0.98	-0.00	-0.73	1.85	0.09	1.10	
33	ETHIOPIA	1.35	-0.89	0.35	-1.02	0.17	-1.40	-1.54	1.22	0.18	-1.01
		-0.59	2.00	-0.25	-0.32	1.01	-1.15	0.86	0.66	-0.57	
34	FINLAND	-1.32	0.96	-0.66	-0.14	0.23	0.38	0.35	-0.34	0.83	-0.30
		-0.30	0.98	-0.43	-1.77	0.14	-0.01	-0.14	1.44	0.18	
35	FRANCE	-1.08	0.49	0.27	-1.31	0.21	0.98	1.33	-0.98	-1.51	1.20
		-0.61	2.00	-1.67	0.51	1.21	-0.22	1.65	-0.25	1.51	
36	GABON	0.07	-0.80	0.52	1.25	0.97	1.44	0.44	1.29	-0.42	-0.08
		-0.91	-0.37	-1.24	0.35	-0.45	1.73	0.80	-0.12	0.15	
37	EAST GERMANY	-1.86	-2.11	-1.18	-0.21	-1.05	-0.74	0.88	0.71	-0.06	-1.06
		-0.51	-0.63	-0.48	0.12	1.24	0.53	-0.38	0.34	1.23	
38	WEST GERMANY	-0.95	1.18	-0.43	-1.26	-1.17	-0.13	-0.32	0.52	-0.03	0.89
		-0.88	-0.89	-0.42	-1.16	1.40	0.38	-0.26	0.21	1.15	
39	GHANA	0.63	-1.16	0.57	-0.28	-0.56	1.33	0.09	0.41	-0.57	-1.87
		1.04	0.76	0.26	0.19	1.38	0.39	0.50	0.50	-0.46	
40	GREECE	-0.65	-0.98	0.91	-0.40	-0.16	0.16	-0.11	-1.32	0.88	1.11
		-0.43	1.15	-0.43	-0.56	-0.86	0.15	2.32	0.81	-0.56	
41	GUATEMALA	-0.31	-0.42	0.87	0.53	-0.78	-0.75	0.79	1.26	0.30	-0.79
		0.49	-0.38	1.15	0.83	-0.96	-1.57	1.21	1.36	-1.14	
42	GUINEA	0.73	-1.19	1.26	1.13	0.24	1.23	-0.42	0.60	-0.72	-0.22
		0.67	-0.24	-0.68	-0.67	-0.43	0.21	-1.38	0.37	-0.46	
43	HAITI	0.67	-1.20	1.69	0.82	-1.02	-0.57	0.30	-0.19	0.25	-0.32
		-0.86	-0.89	-1.56	0.93	-0.44	-1.54	-0.36	0.46	0.29	

Table 37 continued

44	HONDURAS	-0.45	-0.25	1.05	1.48	0.40	-0.97	0.49	-0.71	-1.13	-1.17
		0.22	-0.99	-0.01	0.36	-0.18	-1.03	1.21	0.81	0.03	
45	HUNGARY	-1.70	-1.79	-1.53	-0.12	-0.66	-0.31	0.77	-0.20	-1.12	-0.66
		0.26	-0.93	-1.26	0.63	0.48	-0.09	-0.45	0.19	-0.30	
46	ICELAND	-1.59	1.03	0.00	1.36	1.42	0.02	-1.11	-1.02	-0.77	-1.41
		-0.21	-2.47	-0.83	-1.23	-0.84	0.27	-0.59	0.98	-0.41	
47	INDIA	1.08	1.13	0.76	-2.87	-1.24	1.02	-0.28	-0.08	-1.32	-1.22
		-0.27	1.55	0.10	-0.95	-1.42	0.21	-1.81	-0.38	-1.35	
48	INDONESIA	0.84	-0.52	0.76	-2.02	-0.66	1.13	0.92	0.78	-0.39	-1.24
		0.97	-0.48	-1.92	-0.95	-1.17	1.39	1.30	0.05	-0.46	
49	IRAN	0.92	-0.43	0.04	-2.04	1.30	-0.99	-0.25	-0.12	0.59	1.10
		-0.66	0.18	-0.23	-0.83	-0.75	0.65	1.55	1.84	-0.49	
50	IRAQ	0.53	-0.31	-0.21	-0.46	0.61	-0.30	1.10	0.14	0.37	1.33
		-0.76	1.07	1.64	-0.92	1.02	-0.19	0.72	-0.41	-2.51	
51	IRELAND	-1.17	0.88	-0.51	0.52	0.01	0.16	-0.43	-1.38	0.85	-0.23
		-1.13	0.70	0.41	0.52	-0.47	0.27	-1.16	0.29	0.36	
52	ISRAEL	-1.52	0.83	0.72	0.81	-0.28	0.30	2.09	0.41	-1.07	1.04
		1.14	0.66	-0.90	-2.45	-1.74	0.90	-0.54	-0.17	0.53	
53	ITALY	-0.54	1.11	-0.41	-1.99	-1.06	0.11	1.22	-0.84	0.41	0.72
		-0.56	1.42	-0.97	1.82	0.47	0.36	0.28	0.84	1.01	
54	IVORY COAST	0.89	-0.60	0.67	0.53	0.37	1.73	0.38	0.55	-0.50	-0.21
		0.14	0.14	-2.04	0.11	0.05	0.94	0.54	1.55	0.41	
55	JAMAICA	-0.14	1.14	0.83	1.16	-1.35	0.22	0.62	0.43	-0.90	-1.42
		-0.59	-0.10	-1.14	-0.14	-0.15	-1.32	-1.63	-0.10	0.20	
56	JAPAN	-0.59	1.05	0.37	-1.79	-1.04	0.46	0.59	0.07	1.08	-0.64
		-0.45	0.33	-0.51	-1.38	1.71	-0.46	1.00	1.44	1.92	
57	JORDAN	0.63	-0.55	-0.22	0.80	0.19	-1.37	1.65	-1.23	-1.33	2.02
		-1.89	-0.05	-0.50	-0.12	0.79	0.83	0.12	-1.28	-1.03	
58	NORTH KOREA	0.05	-1.60	-1.88	0.03	-0.91	-0.10	-0.07	-0.56	0.46	1.47
		-0.18	-1.00	0.79	-0.38	1.41	-1.21	-0.73	0.77	0.87	
59	SOUTH KOREA	0.23	-0.68	0.68	-0.08	-1.88	-0.01	0.53	-0.46	1.45	2.62
		-0.17	-1.02	2.07	-0.31	-0.41	-0.14	0.22	0.44	1.19	
60	LAOS	1.37	-0.36	0.24	0.19	0.16	-1.11	-0.46	1.22	0.17	2.33
		-0.63	-0.38	-0.49	-0.19	-0.56	0.0	1.26	0.87	0.41	
61	LEBANON	-0.65	0.36	-0.55	0.53	-1.17	-0.73	1.55	0.23	1.08	-0.68
		-0.86	1.30	0.70	-0.90	-1.32	1.10	1.95	-1.74	-1.87	
62	LIBERIA	0.51	-1.01	1.61	0.86	0.47	-1.21	-1.88	1.54	0.14	-0.17
		-0.02	1.41	-0.89	0.30	-1.10	0.93	-0.58	1.58	-0.26	
63	LIBYA	0.99	0.81	0.03	0.13	1.65	-0.86	-0.74	-1.26	-1.63	0.74
		-0.60	-0.43	-0.96	0.72	0.42	0.88	-1.14	-1.20	-1.02	
64	LUXEMBOURG	-1.01	1.13	0.06	0.57	-1.05	-0.49	-0.77	-1.10	-0.78	0.24
		-0.91	0.52	-1.05	0.84	-0.18	0.37	-0.49	1.06	-0.12	
65	MALAGASY REP	1.27	1.26	-0.22	0.45	1.09	-0.55	-0.76	-0.28	-0.48	-0.24
		-1.37	-1.40	-1.34	0.74	0.43	-0.02	-1.41	-1.77	0.40	
66	MALI	1.18	-0.10	-0.88	0.34	0.57	1.60	-0.26	0.57	-0.22	-0.40
		-0.86	0.27	1.01	-0.51	0.70	-1.91	0.45	-0.99	-0.66	
67	MAURITANIA	0.97	0.14	0.06	0.90	1.22	1.29	-0.13	-0.10	-0.96	-0.23
		0.40	-0.10	0.07	-1.01	-0.25	-1.59	0.07	-1.46	0.32	
68	MEXICO	0.03	0.01	1.49	-1.24	0.65	1.36	-0.05	-0.55	1.06	1.13
		-0.12	0.60	1.39	2.11	-0.43	-1.67	-1.85	-0.29	1.51	
69	MONGOLIA	-0.68	-1.88	-0.91	1.14	1.64	-0.15	-0.25	-0.74	1.28	0.80
		0.18	-0.62	0.65	-1.19	0.06	-0.52	-1.12	-0.48	0.71	
70	MOROCCO	1.00	0.52	0.18	-0.59	0.17	1.04	0.10	-1.08	-0.81	-0.00
		-0.04	-0.76	0.80	-0.19	0.86	-0.28	-1.56	0.97	-1.88	
71	NEPAL	1.30	-0.68	-1.43	-0.82	-0.70	-1.27	-0.15	0.92	1.42	0.02
		0.63	1.71	0.36	-0.99	-0.67	-1.30	1.59	1.24	1.28	
72	NETHERLANDS	-0.95	1.11	0.05	-0.76	-1.50	-0.10	-1.28	0.69	0.13	0.96
		-0.59	0.70	-0.89	0.62	0.62	0.04	0.36	0.72	-0.61	
73	NEW ZEALAND	-2.03	0.52	1.08	0.96	1.74	0.63	-0.38	0.54	0.15	-0.16
		0.52	-0.10	-0.08	-0.88	0.25	1.28	0.27	0.39	1.52	
74	NICARAGUA	-0.00	-0.63	1.88	1.39	0.31	-1.17	-1.51	-0.61	-0.71	0.51
		0.24	-0.13	-1.65	0.88	-0.16	-1.28	1.28	0.22	-0.95	
75	NIGER	1.23	-0.13	-0.62	0.57	0.82	1.27	-0.61	0.67	-0.02	-0.60
		-1.43	-0.06	1.15	-0.68	1.33	-1.42	0.63	-1.33	0.30	
76	NIGERIA	1.63	0.80	0.54	-1.06	0.02	-1.02	0.73	0.73	-0.14	-1.20
		-0.53	0.65	0.08	0.52	0.13	0.90	-2.28	-0.37	-0.80	
77	NORWAY	-1.24	1.13	-0.12	-0.23	0.22	0.00	-1.37	-0.57	0.00	0.42
		-0.23	0.50	-0.30	-1.09	0.31	-0.18	0.28	1.90	0.28	
78	PAKISTAN	0.75	-0.99	1.71	-1.21	-0.35	0.16	0.69	0.55	-1.58	-0.32
		0.53	-1.51	1.57	-0.02	1.26	1.30	-0.98	0.99	0.77	
79	PANAMA	-0.67	0.14	1.13	1.02	-0.08	-0.34	0.33	-0.65	1.37	-1.60
		0.42	0.69	0.45	0.16	-0.41	-0.56	1.99	-1.90	0.18	
80	PARAGUAY	-0.55	-1.33	1.02	1.36	0.89	-0.94	-0.36	-0.84	0.38	0.52
		0.37	-1.56	1.04	1.04	-0.68	0.25	-0.23	1.33	1.00	
81	PERU	-0.50	-0.11	1.16	-0.50	0.96	-0.77	0.77	1.47	-0.74	-0.25
		-0.00	-0.67	0.18	0.73	-0.55	-1.26	1.01	0.92	0.38	
82	PHILIPPINES	0.23	0.91	1.43	-1.24	-1.39	1.40	-1.18	-0.72	1.76	-0.74
		-0.43	-0.46	1.16	-0.45	-1.54	0.23	-0.82	-2.85	0.19	
83	POLAND	-1.24	-1.83	-0.40	-0.64	-0.72	-0.03	0.72	-1.09	-0.09	-0.24
		0.10	0.50	-1.48	0.50	0.61	0.13	-0.52	-0.77	0.04	
84	PORTUGAL	-0.83	-1.53	0.14	0.02	-0.69	-0.81	-1.49	-1.12	0.63	0.57
		-0.89	1.02	2.06	1.63	-1.03	0.64	-0.99	0.32	0.97	
85	ROMANIA	-1.00	-1.88	-1.13	-0.67	-0.52	-0.09	-0.29	-0.47	1.39	-0.12
		0.22	-0.25	-0.51	-0.16	0.46	0.22	-0.50	0.09	-0.98	
86	RWANDA	0.62	0.04	-1.05	1.31	-1.13	0.12	0.38	-0.33	0.33	-0.55
		-0.91	-0.47	-0.92	0.33	-0.03	0.97	0.56	-1.16	0.97	

Table 37 continued

87	SAUDI ARABIA	0.71	-1.02	1.12	-0.31	1.92	-1.03	-0.73	-1.18	-1.11	-0.49
		-0.15	2.71	0.31	-1.57	-0.46	-0.17	-0.74	-1.24	1.83	
88	SENEGAL	0.93	-0.42	0.02	0.45	0.34	1.34	-0.18	-0.49	-0.27	-1.01
		-0.12	0.04	-0.03	-0.21	-0.07	0.73	-0.53	0.24	-0.28	
89	SIERRA LEONE	1.52	1.14	-0.16	0.66	-0.15	-0.12	0.50	-0.04	0.31	-0.34
		-0.19	0.08	0.66	-0.42	-0.01	-0.31	-1.22	0.07	-1.09	
90	SOMALIA	1.12	0.50	-0.57	0.71	1.47	-1.13	0.17	-1.25	-0.60	-0.14
		-1.41	-0.35	-0.87	-1.03	-0.76	1.33	-0.97	-0.34	0.24	
91	SOUTH AFRICA	-1.45	-0.50	1.10	0.03	1.64	-0.81	1.06	1.65	-0.39	-2.21
		-0.91	-0.65	1.70	-1.98	-1.24	-0.45	-1.31	-1.61	0.65	
92	SPAIN	-1.09	-1.76	0.53	-0.92	-0.35	0.03	1.11	-0.30	1.30	0.15
		-0.74	2.00	2.28	0.92	-0.76	0.71	-1.27	0.26	1.25	
93	SUDAN	0.94	-0.56	-0.84	-0.83	1.06	-0.48	0.31	-0.05	-0.18	-1.68
		0.28	-0.17	1.45	-1.07	1.30	-1.14	0.59	-0.73	-0.52	
94	SWEDEN	-1.37	1.08	-0.19	-0.44	0.29	0.14	-0.63	-0.24	0.01	0.28
		-0.12	0.66	-0.33	-0.68	0.92	0.20	0.01	1.78	-0.70	
95	SWITZERLAND	-1.25	1.13	-0.44	-0.07	-0.96	-0.96	-1.54	1.95	0.26	-0.65
		-1.20	-0.66	1.49	0.17	1.26	0.41	-0.91	0.43	-1.02	
96	SYRIA	0.03	-0.16	-0.50	0.57	0.56	-0.58	1.09	-0.81	-1.14	0.89
		-1.26	-0.89	1.55	-1.38	1.23	-0.21	1.32	-0.06	-2.13	
97	THAILAND	0.90	-0.71	0.83	-1.25	-0.42	-0.88	-0.44	-0.69	1.99	0.95
		0.78	-1.30	-0.32	-0.87	-0.42	0.71	1.77	1.07	1.75	
98	TOGO	0.95	-0.26	-0.53	1.08	-0.14	0.36	0.02	0.06	0.06	-0.84
		-0.99	-0.18	0.09	0.05	0.45	0.44	0.04	-0.31	-0.72	
99	TRINIDAD	-0.22	1.31	1.40	1.70	-1.02	0.09	1.28	0.68	-0.29	-0.06
		-0.33	0.40	-1.26	-0.83	0.00	-1.38	-1.59	-0.88	0.22	
100	TUNISIA	0.35	-0.24	-0.09	0.44	0.18	1.74	0.85	-2.26	-0.62	-0.56
		-0.13	0.15	0.18	-0.30	0.63	0.43	-1.50	1.66	-1.00	
101	TURKEY	0.52	0.83	0.89	-1.63	0.23	0.26	0.54	-1.72	1.94	1.15
		0.20	1.13	-0.19	-1.50	-1.00	-0.40	-0.50	1.24	-1.16	
102	UGANDA	1.37	0.97	-0.10	-0.00	-0.42	-0.90	0.27	0.83	0.35	-0.83
		-0.33	0.45	0.72	1.07	1.74	0.98	-1.20	-1.38	-0.32	
103	U.S.S.R.	-1.17	-1.63	0.08	1.15	1.81	-0.09	0.09	1.58	1.13	1.42
		0.64	0.78	-0.92	-0.36	0.53	-0.46	-1.16	-1.78	-1.25	
104	U.A.R.	-0.26	-1.23	0.36	-0.33	-0.61	1.18	-0.51	-1.24	-2.19	0.04
		0.51	0.18	1.97	-0.03	1.81	0.32	0.86	-0.11	-1.22	
105	UNITED KINGDOM	-1.36	0.94	0.48	-0.96	-0.81	0.01	-0.52	0.65	-0.76	0.57
		-0.69	0.29	-0.39	0.12	1.45	-0.14	0.07	-0.17	-0.31	
106	UNITED STATES	-1.43	1.25	-2.49	-1.91	1.13	0.49	-0.25	0.60	-0.73	2.06
		-0.59	-0.43	0.13	0.53	-1.71	-3.12	0.08	-0.29	0.07	
107	UPPER VOLTA	1.13	-0.23	-0.87	0.70	-0.09	1.41	-0.38	0.37	0.31	-0.61
		-0.94	-0.27	0.61	-0.38	-0.31	0.17	0.77	0.67	0.18	
108	URUGUAY	-1.10	0.94	0.02	0.43	1.56	0.20	-0.51	-0.44	0.30	-0.34
		-0.11	0.19	-0.06	0.53	-0.74	0.49	1.17	0.66	0.55	
109	VENEZUELA	-0.34	0.64	1.90	-0.00	1.95	0.25	1.76	-0.50	1.29	0.99
		2.22	0.33	-1.26	1.15	-0.15	-0.71	-0.57	0.49	-0.74	
110	NORTH VIETNAM	0.28	-1.55	-1.61	0.05	-1.25	-0.10	0.12	0.09	0.10	1.09
		0.39	-0.36	0.65	0.01	1.61	-0.47	-0.41	0.46	1.45	
111	SOUTH VIETNAM	0.44	-1.09	0.43	0.38	-1.19	0.01	0.51	0.03	1.10	1.81
		0.43	-0.87	1.13	-0.11	-0.23	0.84	1.01	-0.26	1.78	
112	YEMEN	0.87	-0.12	-1.28	0.38	1.02	-1.48	-0.45	-1.27	-0.96	-0.25
		-0.86	-0.01	0.65	-1.47	-0.91	-0.55	0.05	-1.17	0.62	
113	YUGOSLAVIA	-0.42	-1.59	0.54	-1.09	-0.34	-0.51	-1.26	2.04	0.70	1.00
		0.32	0.12	-1.35	0.47	-0.13	0.65	-0.89	-0.16	-1.86	
114	BOTSWANA	0.98	0.93	-1.60	0.30	1.25	-1.40	-0.77	0.18	-0.23	-0.67
		1.86	-0.68	-0.31	1.01	0.23	1.02	-0.68	0.17	0.28	
115	GAMBIA	1.09	1.25	-1.67	0.61	-0.55	-1.27	-1.24	-0.68	-0.14	-0.51
		1.57	-0.61	-0.19	-0.25	-0.79	0.80	0.01	-0.94	-0.18	
116	KUWAIT	-0.89	0.26	0.15	1.41	1.08	0.38	1.22	-0.49	-1.35	0.38
		2.31	0.38	2.10	-0.96	-0.15	1.09	-1.41	-0.61	-0.25	
117	LESOTHO	1.15	1.28	-1.46	1.05	-0.37	-1.33	-0.72	-0.20	0.10	0.32
		1.79	-0.67	0.20	1.43	0.80	0.53	-0.43	-0.54	-0.27	
118	MALAWI	0.76	0.68	-1.05	0.71	-0.42	0.52	0.01	0.67	-0.42	-1.04
		2.33	0.84	0.04	-0.21	-0.92	-0.63	0.27	-0.07	-0.59	
119	MALTA	-0.88	0.93	-0.71	1.29	-1.73	-0.21	-0.11	-0.25	-0.41	0.14
		1.61	0.18	0.58	0.86	-1.46	1.69	-0.18	-0.03	-0.92	
120	TANZANIA	0.99	0.81	-0.84	-0.35	0.81	0.76	0.56	-0.52	-0.45	-1.53
		1.52	0.95	0.38	-0.05	1.35	0.51	0.85	0.39	0.80	
121	ZAMBIA	0.49	1.07	-1.27	-0.01	0.85	-1.00	-0.37	0.78	-0.16	0.08
		2.92	-0.22	0.36	1.00	-0.63	0.97	-0.21	-0.37	-0.27	
122	KENYA	1.43	0.99	-0.58	-0.53	-0.55	0.03	0.90	-0.20	0.08	-0.16
		2.33	1.57	0.62	1.42	1.29	1.06	-0.67	0.09	-0.24	
123	W. SAMOA	0.56	0.27	-0.48	1.56	-1.29	1.14	-0.53	0.43	0.63	0.03
		-0.10	-0.76	0.27	0.53	-0.25	-0.05	0.12	-0.21	0.39	
124	SINGAPORE	-0.65	1.15	-0.55	1.22	-1.53	0.56	-0.05	1.15	-0.82	0.21
		2.12	0.05	0.06	-1.15	1.23	-1.35	1.08	-0.68	0.43	
125	GUYANA	0.38	0.81	0.53	1.35	1.00	-0.02	1.90	0.24	-0.21	-0.01
		-0.01	-0.78	-1.10	-0.85	-0.10	-1.25	-0.82	-0.38	0.47	
126	MALAYSIA	0.43	1.25	0.50	0.18	-0.65	-0.81	0.28	1.97	0.99	1.67
		0.52	-0.73	0.89	-0.74	0.25	-0.31	-0.64	-0.91	0.68	
127	MALDIVE ISLANDS	0.97	0.63	-1.00	1.12	-0.24	0.36	-0.70	-1.11	-0.34	0.57
		-0.96	-0.69	0.23	-0.26	-0.19	0.58	-1.24	-1.31	0.69	
128	BARBADOS	-0.27	0.96	0.18	1.70	-1.75	-0.71	-0.02	-0.03	-0.55	-1.15
		0.35	-0.36	-1.39	0.60	-0.19	-1.12	-1.97	-0.48	-0.63	
129	TAIWAN	-0.05	-0.54	2.39	-0.05	-1.78	0.96	-1.10	0.40	-1.48	2.43
		2.59	1.14	0.51	-0.87	-1.15	0.50	1.11	-0.78	1.04	

TABLE 38
Factor Scores, F = ZA, Loadings in the Range of ± .49 Set to Zero,
Regression Estimates for Missing Data

1	AFGHANISTAN	17.80	-12.78	0.10	-2.82	0.57	-2.28	-1.13	-0.05	-2.82	-0.27
		-0.50	0.81	0.60	-4.02	-1.33	-1.16	-1.69	0.72	0.99	
2	ALBANIA	3.10	-16.31	-5.31	3.92	-1.30	-1.43	-0.84	0.13	2.68	3.30
		1.08	0.81	-2.33	-1.19	1.87	0.48	-1.49	-1.00	-0.84	
3	ALGERIA	5.85	-10.22	-2.22	-1.41	4.10	0.16	0.71	-0.05	-2.83	0.85
		1.08	-0.23	-0.43	-0.31	-1.33	0.21	-1.42	-0.31	-0.64	
4	ARGENTINA	-17.50	4.83	2.55	-6.09	5.02	-0.51	1.34	-1.52	0.60	0.59
		-0.50	-1.21	1.00	3.81	1.61	0.29	2.42	0.39	-1.36	
5	AUSTRALIA	-25.60	14.24	3.44	-4.61	6.16	0.68	-1.13	0.13	0.68	0.29
		-0.50	0.81	0.35	0.16	2.43	1.34	-2.13	-0.69	0.99	
6	AUSTRIA	-22.51	13.69	-0.90	-2.71	-3.27	0.68	0.27	-1.52	-0.82	-2.07
		-0.50	0.34	-2.04	1.70	1.34	0.34	-1.49	-0.25	-0.34	
7	BELGIUM	-23.72	14.98	2.31	-3.37	-5.52	0.68	0.82	-0.05	0.19	1.83
		-0.50	0.81	-0.34	1.96	-1.33	0.07	-2.13	-0.85	0.99	
8	BOLIVIA	2.93	7.16	2.75	2.28	4.45	1.84	1.00	1.59	1.94	-1.69
		0.64	-0.76	-0.14	3.64	-1.33	-1.89	2.16	0.47	0.99	
9	BRAZIL	-7.24	4.41	4.23	-6.23	2.77	-1.25	1.08	-1.52	0.37	-0.03
		-0.50	-1.12	1.85	3.40	-0.19	-1.97	2.42	0.19	0.99	
10	BULGARIA	-12.22	-16.23	-4.00	-2.87	-1.62	-1.43	-0.56	-0.79	2.42	3.11
		1.08	0.81	-2.54	-0.90	-1.33	0.56	-1.49	-1.00	0.99	
11	BURMA	12.96	-7.90	-2.41	-2.93	0.56	0.16	-1.13	-0.05	-0.68	0.72
		1.08	-1.00	0.36	-3.73	1.83	0.71	1.76	0.57	0.99	
12	BURUNDI	13.26	-1.56	0.07	5.48	-2.94	1.84	-0.13	-0.70	0.47	-2.43
		-0.50	0.06	-0.34	0.93	-1.33	1.33	-0.78	-0.12	0.0	
13	CAMBODIA	19.07	-11.75	-3.11	1.71	-2.15	0.84	-0.21	-0.79	-0.51	2.45
		0.64	0.81	0.28	-2.54	1.78	0.95	2.19	-0.63	0.99	
14	CAMEROUN	18.31	2.10	-1.09	2.11	2.14	1.84	-0.13	1.59	-1.27	-1.07
		-0.50	-0.30	-0.55	-0.26	-1.33	1.02	0.99	-0.47	0.0	
15	CANADA	-24.51	12.68	5.54	-6.48	4.32	0.68	-1.13	1.59	0.17	1.78
		-0.50	0.81	0.46	1.69	2.52	-0.40	-2.13	-0.07	-0.84	
16	CENTRAL AFRICAN REP	17.75	-8.74	-1.62	4.03	3.52	1.84	-0.32	0.72	-0.27	-0.72
		-0.50	-0.44	1.44	0.15	-1.33	1.09	0.99	-0.69	0.13	
17	CEYLON	4.86	5.83	-1.46	-0.76	-3.88	-1.72	0.59	1.59	1.17	-3.38
		-0.50	0.81	1.01	-2.66	2.47	1.20	2.42	0.63	0.99	
18	CHAD	20.33	-3.06	-3.16	3.09	3.86	1.84	-0.47	1.59	-0.86	-2.94
		-0.50	-0.44	-0.37	-1.38	-1.33	-1.07	0.99	-0.39	0.28	
19	CHILE	-13.90	9.87	4.07	-1.32	0.06	-1.25	1.13	-1.52	-1.09	1.02
		-0.50	-0.83	1.17	3.99	0.79	-1.27	2.16	0.47	0.99	
20	CHINA, PR	2.36	-17.28	-5.31	-7.87	-0.02	-1.43	-0.84	0.31	-1.61	0.42
		1.08	-1.18	-0.91	-3.37	-1.33	-0.07	-1.49	-0.78	0.99	
21	COLOMBIA	-9.42	5.03	4.57	-2.31	1.32	-0.06	0.85	-1.52	0.35	-2.61
		-0.50	-0.92	0.87	4.40	-1.33	-2.07	1.90	0.47	0.99	
22	CONGO (BRA)	13.69	-5.87	-4.12	3.21	2.84	1.84	-0.32	1.59	-0.67	-1.44
		-0.50	-0.16	-0.37	0.69	-1.33	1.11	0.99	-0.12	0.13	
23	CONGO (LEO)	16.34	-3.51	1.02	-0.70	3.80	1.33	0.52	1.59	-0.10	0.08
		-0.50	-0.49	0.08	0.03	-1.33	1.15	2.42	-0.07	-0.56	
24	COSTA RICA	-8.40	10.48	3.33	3.80	-0.88	0.68	-1.13	-1.52	0.43	-3.82
		-0.50	-0.76	0.23	4.20	-1.33	-0.23	1.70	0.52	-1.02	
25	CUBA	-12.18	-13.40	-3.06	3.53	-1.33	-1.01	1.17	-1.52	0.26	-0.16
		1.08	0.81	-1.11	3.57	2.69	-2.20	-1.42	-0.91	-1.02	
26	CYPRUS	-4.29	8.18	-1.89	3.69	-1.20	0.69	0.27	1.59	-0.01	2.59
		-0.50	0.28	-0.55	-1.04	-1.33	0.87	-0.72	0.98	-0.34	
27	CZECHOSLOVAKIA	-17.12	-14.55	-2.50	-3.40	-2.91	-1.43	0.98	1.59	0.98	2.82
		1.08	0.31	-2.50	0.78	1.73	0.32	-1.49	-1.00	-1.19	
28	DAHOMEY	19.58	-8.28	-4.12	4.30	-0.21	1.84	-0.47	1.59	-0.44	-1.80
		-0.50	-0.13	0.91	-0.50	-1.33	0.77	1.32	0.93	0.28	
29	DENMARK	-24.54	15.27	1.63	-1.50	-2.11	0.68	-1.13	-1.52	0.82	1.44
		-0.50	0.81	0.34	0.62	1.20	0.11	-2.13	0.84	-1.19	
30	DOMINICAN REPUBLIC	-5.80	10.37	4.17	2.72	-3.13	0.68	0.71	-1.52	1.19	1.38
		-0.50	0.03	-0.64	3.88	-1.33	-2.15	0.49	-0.73	-0.64	
31	ECUADOR	-4.48	3.59	3.61	1.87	-0.22	-1.82	0.91	1.59	1.94	-0.53
		-0.50	-0.92	1.12	3.37	-1.33	-2.00	2.91	0.52	-1.02	
32	EL SALVADOR	-1.36	-7.71	2.61	3.75	-3.94	-2.28	-1.13	-1.52	1.17	-2.71
		-0.50	0.81	1.55	4.11	-1.33	-2.10	2.91	-0.38	0.99	
33	ETHIOPIA	17.39	-12.15	1.06	-1.76	1.06	-2.28	-1.13	1.59	1.20	-1.79
		-0.50	0.81	0.60	-1.33	0.88	-0.66	1.38	-0.18	-0.84	
34	FINLAND	-20.90	15.27	-0.40	-2.45	0.75	0.68	-0.21	-0.79	2.44	1.07
		-0.50	0.81	-1.53	-1.72	1.61	0.16	-2.13	0.43	0.99	
35	FRANCE	-22.54	7.47	2.53	-6.69	-0.95	1.84	-0.13	-1.52	-1.57	3.48
		-0.50	0.81	-1.87	2.06	2.69	-0.02	1.11	-0.90	0.99	
36	GABON	10.75	-7.31	-0.48	3.91	3.15	1.84	0.03	1.59	-0.73	-0.29
		-0.50	-0.44	-1.72	0.94	-1.33	1.09	1.32	-0.12	-0.16	
37	EAST RMANY	-18.95	-17.28	-1.73	-4.57	-3.81	-2.28	-0.73	0.13	-1.02	0.87
		1.08	-0.02	-2.42	0.17	2.43	0.29	-1.49	-0.52	0.51	
38	WEST GERMANY	-24.63	13.95	2.97	-6.36	-3.89	0.68	-0.73	0.13	-0.92	1.54
		-0.50	0.40	-1.69	0.94	2.88	0.21	-1.49	0.19	0.51	
39	GHANA	12.33	-12.51	0.01	-1.43	-0.78	1.10	0.67	1.59	-1.69	-3.06
		1.08	-0.04	0.69	0.65	1.32	0.66	1.11	-0.04	-0.60	
40	GREECE	-12.25	11.36	1.40	-2.80	-0.75	0.68	-1.13	-1.52	1.51	3.10
		-0.50	0.81	-1.23	-0.60	-1.33	0.60	2.65	-0.46	-1.02	
41	GUATEMALA	0.63	-0.39	3.58	2.53	-1.86	-1.82	1.26	1.59	0.57	-1.89
		-0.50	-1.15	1.46	3.46	-1.33	-2.15	2.91	-0.07	-1.02	
42	GUINEA	17.76	-12.51	-0.34	3.98	1.22	1.10	-0.13	1.59	-0.89	-0.56
		1.08	-0.21	-1.51	-1.35	-1.33	0.45	-1.49	-0.04	0.0	
43	HAITI	13.10	-11.34	2.88	4.40	-2.38	-1.90	1.26	0.13	1.94	-0.89
		-0.50	-1.06	-1.50	1.99	-1.33	-2.18	0.49	-0.72	0.99	

Table 38 continued

44	HONDURAS	1.24	0.54	3.87	4.27	1.04	-1.82	0.59	-1.52	-1.65	-2.70
		-0.50	-1.00	1.18	3.52	-1.33	-2.13	2.91	0.19	-1.02	
45	HUNGARY	-17.04	-15.04	-3.72	-3.93	-2.53	-1.43	1.10	0.13	-1.95	1.27
		1.08	-1.08	-2.32	1.67	1.68	0.40	-1.84	-0.94	-1.32	
46	ICELAND	-20.45	14.60	0.86	3.38	4.39	0.68	-0.32	-1.52	-0.47	-2.94
		-0.50	-0.72	-0.51	-0.22	-1.33	-0.10	-2.13	0.84	0.13	
47	INDIA	6.76	9.84	3.06	-7.61	-1.44	1.84	1.32	-0.05	-3.13	-2.50
		-0.50	-1.18	1.36	-3.56	-1.33	0.74	-1.49	0.0	-1.02	
48	INDONESIA	13.50	-7.01	2.18	-6.32	-0.62	1.84	1.15	-0.05	0.51	-0.54
		0.64	-1.00	-2.04	-3.20	-1.33	1.07	1.90	-0.34	-0.84	
49	IRAN	11.41	-9.93	2.01	-4.34	3.71	-1.51	-0.84	-0.05	-0.90	2.75
		-0.50	-0.92	0.04	-3.37	0.22	0.95	1.76	0.87	-0.84	
50	IRAQ	8.72	-6.69	-0.78	-2.22	2.23	-0.30	1.23	-0.05	-0.36	2.96
		-0.50	0.81	1.38	-1.90	1.87	1.05	2.05	-0.77	-1.28	
51	IRELAND	-19.90	14.09	-0.23	1.07	-0.39	0.68	-1.13	-1.52	1.94	-0.84
		-0.50	0.81	-0.15	3.07	-0.22	-0.09	-2.13	-0.18	0.99	
52	ISRAEL	-16.11	12.78	1.82	0.34	-2.18	0.68	0.82	-0.05	-1.03	3.63
		0.64	0.81	-1.89	-2.38	-1.33	0.97	-1.49	0.69	0.99	
53	ITALY	-21.49	12.98	2.84	-6.34	-3.92	0.68	1.20	-1.52	0.39	1.77
		-0.50	0.81	-2.20	3.14	2.57	0.35	-0.51	0.55	0.99	
54	IVORY COAST	17.06	-7.61	-0.52	1.84	1.66	1.84	0.03	1.59	-1.24	-0.38
		0.64	-0.24	-2.02	-0.70	-1.33	0.60	1.55	0.91	-0.16	
55	JAMAICA	-3.43	13.03	2.60	4.42	-3.59	0.68	0.03	0.13	-0.89	-3.46
		-0.50	0.18	-0.91	0.22	-1.33	-2.19	-1.42	0.19	-0.16	
56	JAPAN	-16.51	12.25	4.49	-6.43	3.03	0.68	0.59	0.13	0.49	-1.71
		-0.50	0.81	-0.29	-3.77	3.07	0.00	0.12	0.76	0.99	
57	JORDAN	10.62	-12.80	-1.94	3.02	0.05	-1.82	1.20	-1.52	-1.63	3.78
		-0.50	-0.92	-0.54	-2.56	1.01	1.02	2.05	-0.67	-1.19	
58	NORTH KOREA	6.01	-17.14	-5.31	-1.50	-2.08	-1.43	-0.73	-0.62	-0.42	3.57
		1.08	0.0	-1.14	-3.09	2.74	0.06	-1.49	-0.52	0.51	
59	SOUTH KOREA	2.97	-7.72	3.92	-1.48	-4.99	0.06	-0.73	-0.62	0.47	3.71
		0.38	0.23	1.89	-2.52	-1.33	0.11	2.05	-0.31	0.51	
60	LAOS	20.27	-10.54	3.34	2.92	1.36	-1.43	-0.32	1.59	0.47	3.32
		-0.50	-0.19	0.28	-3.00	-1.33	0.87	2.91	0.69	0.13	
61	LEBANON	-4.59	3.89	-0.44	3.15	-4.32	-0.21	0.59	0.13	1.94	-0.91
		-0.50	0.81	-0.10	0.31	-1.33	0.97	2.91	-0.25	-1.02	
62	LIBERIA	11.42	-8.80	2.21	4.59	0.80	-1.82	-1.13	1.59	1.94	-0.69
		-0.50	0.81	-1.15	-0.37	-1.33	0.60	-0.52	0.73	0.99	
63	LIBYA	10.09	3.68	-0.11	2.02	4.90	-1.25	0.41	-1.52	-1.80	0.80
		-0.50	-0.60	0.39	-0.71	0.15	0.54	-0.09	-0.28	-0.45	
64	LUXEMBOURG	-21.28	15.27	2.22	2.73	-4.24	0.68	-0.63	-1.52	-0.70	1.45
		-0.50	0.38	-1.41	2.28	0.60	0.21	-2.13	0.59	0.40	
65	MALAGASY REP	14.18	8.77	-0.83	2.21	3.68	0.68	-0.63	0.13	-1.03	-1.99
		-0.50	-0.58	-0.28	0.43	-1.33	0.13	-0.85	-0.62	0.40	
66	MALI	20.55	-4.92	-4.04	2.24	3.35	1.84	0.41	1.59	-0.95	-1.61
		-0.50	-0.36	1.76	-0.83	0.08	-1.47	1.32	-0.42	-0.45	
67	MAURITANIA	18.82	-1.55	-1.43	4.07	4.22	1.10	0.03	-0.05	-1.35	-1.99
		0.64	-0.54	0.79	-1.33	-1.33	-1.63	1.32	-0.49	-0.16	
68	MEXICO	-8.81	1.72	5.31	-3.52	1.11	1.10	0.17	-1.52	0.86	0.59
		-0.50	0.81	1.77	3.86	-1.33	-2.24	-1.49	0.39	0.99	
69	MONGOLIA	2.65	-17.14	-2.54	3.55	5.27	-1.43	-1.13	-1.52	1.08	3.16
		1.08	0.81	-0.64	-2.84	0.55	0.03	-1.49	-0.93	0.99	
70	MOROCCO	8.83	3.25	-1.10	-2.11	1.49	1.84	0.27	-0.05	-2.16	0.20
		0.64	-0.57	1.50	-0.86	1.85	0.14	-0.97	0.43	-0.34	
71	NEPAL	19.67	-11.28	-3.34	1.76	-1.27	-2.28	0.35	1.59	1.94	-0.68
		0.38	0.81	0.55	-3.90	-1.33	-0.45	2.05	0.39	0.99	
72	NETHERLANDS	-22.91	14.98	2.04	-3.27	-5.42	0.68	-1.13	0.13	0.74	2.63
		-0.50	0.81	-1.56	0.60	2.18	0.11	-2.13	0.60	-0.74	
73	NEW ZEALAND	-24.51	14.24	2.23	-0.70	3.55	0.68	-1.13	0.13	0.94	0.57
		-0.50	0.81	-0.05	0.64	0.12	1.23	-2.13	0.19	0.99	
74	NICARAGUA	0.03	-5.20	3.24	3.99	0.70	-2.28	-0.56	-1.52	-0.18	-0.63
		-0.50	0.81	-2.00	3.59	-1.33	-2.13	2.19	0.67	-1.02	
75	NIGER	21.85	-4.71	-2.82	3.64	3.81	1.84	0.47	1.59	-1.03	-2.43
		-0.50	-0.44	1.80	-1.94	1.74	-1.21	1.32	-0.55	0.28	
76	NIGERIA	15.85	3.69	0.51	-1.53	1.08	-0.98	0.74	1.59	-0.82	-3.38
		-0.50	-0.26	0.97	-0.36	0.13	0.74	-1.42	0.95	-0.68	
77	NORWAY	-23.45	15.27	1.44	-1.87	0.05	0.68	-1.13	-1.52	0.94	2.16
		-0.50	0.81	-0.50	-1.45	1.25	0.00	-2.13	0.87	0.99	
78	PAKISTAN	11.63	-10.91	2.72	-5.27	-1.32	0.16	1.30	-0.05	-2.71	-0.32
		0.38	-0.92	2.49	-1.75	2.31	0.84	-0.25	0.65	0.99	
79	PANAMA	-7.26	5.78	3.97	4.43	-0.51	-1.25	0.17	-0.79	2.57	-3.86
		-0.50	0.81	0.98	3.56	-1.33	-2.07	2.91	-0.83	0.99	
80	PARAGUAY	-2.40	-11.52	1.45	3.54	1.59	-2.28	-0.84	-1.52	1.94	1.50
		-0.50	-1.12	1.57	3.54	-1.33	-1.15	-0.85	0.80	0.99	
81	PERU	-5.60	0.70	4.40	-2.56	3.01	-0.61	0.59	1.59	-0.66	-1.28
		-0.50	-0.76	0.78	3.65	-1.33	-1.99	2.42	-0.02	0.99	
82	PHILIPPINES	-3.28	11.39	4.44	-2.02	-3.28	1.10	-1.13	-0.05	2.62	-2.45
		-0.50	-0.83	1.25	0.28	-1.33	0.60	-1.49	-0.57	0.99	
83	POLAND	-14.55	-15.09	-1.15	-6.09	-2.62	-1.82	0.87	-1.52	-0.59	2.14
		1.08	0.81	-2.25	1.89	1.07	0.29	-1.49	-1.00	-0.84	
84	PORTUGAL	-11.69	-12.05	1.22	-0.01	-3.21	-2.28	-1.13	-1.52	1.49	1.74
		-0.50	0.81	2.03	2.73	-1.33	0.03	-2.13	-0.86	0.99	
85	ROMANIA	-10.24	-16.31	-3.43	-5.46	-1.55	-1.43	-1.13	-0.79	2.60	2.26
		1.08	-0.83	-2.08	-0.44	1.94	0.45	-1.49	-1.00	-0.84	
86	RWANDA	12.45	-2.04	-4.61	5.49	-3.38	0.16	-0.32	-0.70	0.47	-2.14
		-0.50	0.09	-1.94	0.28	-1.33	1.29	0.49	-0.12	0.13	

Table 38 continued

87	SAUDI ARABIA	15.39	−10.61	1.40	−0.46	4.73	−2.28	−1.13	−1.52	−0.09	0.47
		−0.50	0.81	2.03	−2.25	−1.33	−0.32	−1.69	−0.81	0.99	
88	SENEGAL	15.68	−6.44	−3.14	2.63	0.61	1.84	0.03	−0.05	−0.95	−2.76
		0.64	−0.17	0.25	−0.83	−0.16	0.34	−0.85	0.90	−0.16	
89	SIERRA LEONE	20.24	7.95	−1.65	4.53	0.50	−0.06	0.48	1.59	−0.70	−1.61
		−0.50	−0.32	1.86	−0.88	−0.28	0.48	−0.39	1.01	−0.51	
90	SOMALIA	18.19	1.32	−2.52	3.60	4.26	−1.25	−0.47	−1.52	−1.27	−0.49
		−0.50	−0.62	−0.37	−3.05	−1.33	1.29	−0.39	0.52	0.28	
91	SOUTH AFRICA	−10.66	−0.61	2.25	−2.02	4.50	−1.43	0.03	1.59	0.47	−3.32
		−0.50	−0.70	1.78	0.21	−1.33	−0.44	−2.13	−0.88	−0.16	
92	SPAIN	−14.00	−12.35	2.52	−4.28	−1.94	−1.82	0.95	−0.79	2.36	1.83
		0.38	0.81	2.17	3.06	−0.18	−0.03	−2.13	−0.59	0.99	
93	SUDAN	17.77	−11.21	−2.01	−1.99	4.68	0.16	0.27	1.59	−1.94	−2.88
		0.64	−0.64	2.03	−1.37	1.80	−0.82	1.76	−0.65	−0.34	
94	SWEDEN	−25.21	15.27	0.96	−4.20	0.32	0.68	−0.21	−1.52	0.66	2.39
		−0.50	0.81	−0.19	−0.85	1.99	0.16	−2.13	0.76	−1.32	
95	SWITZERLAND	−23.60	15.55	1.30	−1.44	−3.20	−0.51	−1.13	1.59	0.80	−1.64
		−0.50	0.81	2.16	0.89	2.04	0.25	−2.13	0.19	−0.84	
96	SYRIA	5.90	−3.17	−2.14	−0.41	2.19	−0.27	1.05	−0.79	−2.77	2.47
		−0.50	−1.08	1.87	−1.31	1.97	0.92	2.91	−0.60	−1.19	
97	THAILAND	8.45	−12.55	1.83	−2.32	−0.37	−2.28	0.89	−0.79	2.57	1.16
		−0.50	−0.76	0.55	−3.72	−1.33	0.89	2.65	0.80	0.99	
98	TOGO	18.64	−5.50	−3.44	4.77	0.26	1.33	−0.13	1.59	−0.21	−1.61
		−0.50	−0.28	0.51	0.13	−0.30	0.74	0.49	−0.41	0.0	
99	TRINIDAD	−4.06	13.03	2.73	4.74	−3.48	0.68	0.48	0.86	0.24	−1.45
		−0.50	0.30	−1.21	−0.15	−1.33	−2.02	−1.42	0.19	−0.51	
100	TUNISIA	5.33	−2.44	−2.52	0.39	0.54	1.84	0.41	−1.52	−2.09	−1.04
		0.64	−0.40	−0.12	−1.06	0.85	0.43	−1.49	0.99	−0.45	
101	TURKEY	−2.40	6.34	3.22	−5.40	0.60	0.68	1.05	−0.79	2.46	2.23
		−0.50	0.81	−0.35	−3.26	−1.33	0.66	−0.85	0.62	−1.19	
102	UGANDA	13.65	7.57	−0.80	2.57	−0.52	−0.98	0.52	1.59	0.01	−2.43
		−0.50	−0.07	1.38	0.70	3.00	1.23	−1.42	0.19	−0.56	
103	U.S.S.R.	−16.40	−15.59	−2.51	−5.23	4.59	−1.43	0.95	0.70	2.64	3.52
		1.08	0.81	−2.43	−1.53	3.00	−0.40	−2.13	−1.00	−1.40	
104	U.A.R.	−0.70	−10.88	0.18	−5.54	−2.48	0.84	0.35	−1.52	−3.47	0.90
		1.08	−1.04	2.44	−1.13	2.38	0.82	1.13	−0.75	−1.19	
105	UNITED KINGDOM	−25.74	13.28	3.11	−5.90	−2.96	−0.06	0.76	0.13	−0.94	2.84
		−0.50	0.32	0.36	0.85	2.88	−0.05	−2.13	−0.44	−0.70	
106	UNITED STATES	−26.39	11.84	−3.63	−7.67	2.41	1.10	0.79	0.13	−1.32	3.20
		−0.50	0.18	1.01	1.19	−1.33	−2.24	−2.13	−0.18	−0.72	
107	UPPER VOLTA	21.19	−5.55	−4.61	4.40	0.66	1.84	−0.47	1.59	−0.51	−2.43
		−0.50	−0.15	0.85	−1.67	−1.33	0.60	1.32	0.95	0.28	
108	URUGUAY	−15.83	14.66	1.17	0.56	3.03	1.33	−1.13	−1.52	1.10	−1.71
		−0.50	0.81	0.21	3.48	−1.33	0.54	1.13	0.80	0.99	
109	VENEZUELA	−12.71	6.53	4.97	−1.35	4.10	1.84	1.02	−1.52	2.50	−0.34
		0.64	0.81	1.75	3.74	−1.33	−2.07	0.81	1.03	−1.36	
110	NORTH VIETNAM	9.44	−17.14	−5.31	−1.59	−3.49	−1.43	−0.73	0.10	−0.42	2.97
		1.08	0.21	−1.14	−2.29	2.74	0.79	−1.49	−0.52	0.51	
111	SOUTH VIETNAM	11.27	−12.61	1.43	−0.34	−2.98	−1.72	−0.73	0.10	0.47	3.05
		1.08	0.13	0.70	−1.35	−1.33	0.92	1.70	−0.36	0.51	
112	YEMEN	17.83	−4.80	−4.61	3.81	2.39	−0.29	−1.13	−1.52	−0.20	0.05
		0.32	−0.87	2.03	−2.39	−1.33	−0.11	0.49	−0.80	0.99	
113	YUGOSLAVIA	−6.36	−15.33	0.56	−5.41	−1.14	−1.82	−1.13	1.59	2.38	3.53
		1.08	0.81	−2.20	−0.25	0.80	0.40	−1.49	−0.23	−1.28	
114	BOTSWANA	11.28	4.17	−4.61	4.59	4.07	−0.61	−0.32	0.86	−0.00	−2.94
		0.64	−0.54	−0.37	−0.57	0.46	1.36	−0.51	0.19	0.13	
115	GAMBIA	12.99	7.56	−4.17	5.93	−1.53	−0.61	−0.63	−0.79	−0.74	−2.94
		0.64	−0.06	−0.37	−3.21	−1.33	0.40	0.55	0.19	0.40	
116	KUWAIT	−9.16	3.83	0.22	2.52	2.12	1.10	0.03	−0.79	−1.20	1.50
		0.64	−0.68	2.03	−0.24	0.25	1.13	−0.97	0.19	−0.16	
117	LESOTHO	11.84	7.89	−4.61	5.79	−0.84	−0.61	−0.47	0.86	−0.50	−1.25
		0.64	−0.11	−0.37	1.14	0.50	1.42	0.08	0.19	0.28	
118	MALAWI	12.65	4.45	−3.38	4.95	−0.59	1.10	0.27	1.59	−0.78	−2.94
		0.64	−0.09	−0.37	0.08	−1.33	−0.47	0.55	0.19	−0.34	
119	MALTA	−13.71	12.34	−1.07	5.39	−5.88	1.33	0.03	−0.79	0.47	1.41
		1.08	0.36	0.87	2.36	−1.33	0.51	−0.03	0.19	−0.16	
120	TANZANIA	12.29	6.67	−1.79	0.46	3.35	1.84	0.03	0.86	−1.20	−3.10
		0.64	−0.51	1.30	−0.30	1.63	1.29	1.70	0.19	−0.16	
121	ZAMBIA	3.55	6.69	−3.40	2.36	3.14	−0.61	0.27	1.59	0.47	−1.10
		0.64	−0.36	−0.37	0.98	−1.33	1.39	−0.51	0.19	−0.34	
122	KENYA	11.79	5.34	−0.97	0.46	−1.20	1.84	0.67	0.86	−0.95	−1.20
		0.64	0.15	1.52	−0.06	1.18	1.26	0.08	0.19	−0.60	
123	W. SAMOA	10.74	3.47	−2.63	5.40	−2.95	0.69	−0.63	0.86	0.47	−1.32
		−0.50	0.07	−0.37	1.20	−1.33	0.84	0.49	0.19	0.40	
124	SINGAPORE	−9.79	11.83	−0.24	3.15	−5.53	1.33	−0.32	1.59	−1.80	0.73
		1.08	0.26	−1.12	−1.82	1.75	−0.69	−0.51	0.19	0.13	
125	GUYANA	8.70	7.51	0.74	3.85	4.02	1.33	0.03	0.13	0.24	−0.56
		0.64	−0.66	−0.17	0.70	−1.33	−1.97	0.34	0.19	−0.16	
126	MALAYSIA	1.41	8.38	0.56	0.45	−1.60	−0.77	−0.13	1.59	0.93	1.52
		−0.50	0.03	0.83	−3.30	0.18	−0.47	−0.51	1.04	0.0	
127	MALDIVE ISLANDS	12.95	1.20	−4.61	5.49	−1.41	1.33	−0.47	−1.52	−1.10	−0.85
		−0.50	0.11	−0.37	−2.94	−1.33	1.31	−1.42	0.19	0.28	
128	BARBADOS	−7.93	11.11	−0.60	5.99	−5.24	−0.77	0.41	0.13	0.47	−2.94
		0.64	0.24	−0.78	2.00	−1.33	−2.03	−2.13	0.19	−0.45	
129	TAIWAN	−1.69	−3.87	3.47	−1.14	−5.50	0.53	0.17	−0.57	−1.64	4.00
		0.64	0.81	1.16	−3.03	−1.33	0.51	0.62	−0.31	0.99	

TABLE 39
Factor Scores, F = ZA, Loadings in the Range of ± .49 Set to Zero, Rest Set to ± 1.0, Regression Estimates for Missing Data

1	AFGHANISTAN	21.75	-16.31	-0.26	-4.43	0.81	-2.97	-1.43	0.11	-4.13	-0.36
		-0.76	1.18	0.74	-6.65	-1.65	-1.33	-3.11	1.18	1.24	
2	ALBANIA	5.06	-21.06	-7.84	5.89	-1.54	-2.08	-1.06	-0.01	3.82	4.22
		1.66	1.18	-3.13	-2.05	2.32	0.59	-2.23	-1.66	-1.05	
3	ALGERIA	7.43	-12.24	-3.05	-1.62	5.37	-0.19	0.90	0.11	-4.11	1.06
		1.66	-0.33	-0.67	-0.62	-1.65	0.08	-2.16	-0.51	-0.80	
4	ARGENTINA	-23.12	5.94	3.98	-8.85	6.86	-0.07	1.70	-2.08	0.95	0.78
		-0.76	-1.75	1.43	6.35	2.01	0.23	3.93	0.65	-1.70	
5	AUSTRALIA	-34.21	18.80	5.28	-6.80	8.37	1.16	-1.43	-0.01	1.06	0.39
		-0.76	1.18	0.62	0.23	3.01	2.19	-3.47	-1.14	1.24	
6	AUSTRIA	-29.87	17.87	-1.13	-4.10	-4.57	1.16	0.34	-2.08	-1.18	-2.59
		-0.76	0.49	-2.74	2.75	1.67	0.32	-2.23	-0.42	-0.42	
7	BELGIUM	-31.77	19.79	3.89	-4.91	-7.49	1.16	1.04	0.11	0.38	2.35
		-0.76	1.18	-0.35	3.08	-1.65	-0.18	-3.47	-1.40	1.24	
8	BOLIVIA	3.81	8.88	3.83	3.15	6.06	2.36	1.27	2.19	2.80	-2.23
		0.98	-1.11	-0.26	6.00	-1.65	-2.69	3.43	0.78	1.24	
9	BRAZIL	-10.35	5.06	6.29	-9.07	3.68	-1.44	1.37	-2.08	0.63	-0.03
		-0.76	-1.63	2.50	5.64	-0.22	-2.83	3.93	0.31	1.24	
10	BULGARIA	-15.11	-20.91	-5.94	-4.33	-1.85	-2.08	-0.71	-0.99	3.47	3.97
		1.66	1.18	-3.41	-1.20	-1.65	0.74	-2.23	-1.66	1.24	
11	BURMA	17.37	-9.85	-3.46	-4.02	1.29	-0.19	-1.43	0.11	-0.62	0.99
		1.66	-1.45	0.47	-6.30	2.26	1.00	3.01	0.94	1.24	
12	BURUNDI	17.17	-1.75	-0.28	8.03	-4.03	2.36	-0.16	-1.06	0.61	-3.16
		-0.76	0.08	-0.43	1.40	-1.65	2.16	-1.50	-0.20	0.0	
13	CAMBODIA	25.21	-14.79	-4.49	2.18	-2.93	0.51	-0.27	-0.99	-0.38	3.14
		0.98	1.18	0.32	-4.29	2.20	1.45	3.22	-1.05	1.24	
14	CAMEROUN	24.40	2.48	-1.67	2.61	2.73	2.36	-0.16	2.19	-1.90	-1.40
		-0.76	-0.44	-0.70	-0.60	-1.65	1.58	1.31	-0.78	0.0	
15	CANADA	-32.80	16.62	8.05	-9.49	5.55	1.16	-1.43	2.19	0.35	2.30
		-0.76	1.18	0.75	2.83	3.13	-1.05	-3.47	-0.12	-1.05	
16	CENTRAL AFRICAN REP	23.85	-10.69	-2.52	5.75	4.50	2.36	-0.40	1.09	-0.48	-0.97
		-0.76	-0.64	1.91	0.25	-1.65	1.72	1.31	-1.14	0.16	
17	CEYLON	7.63	6.65	-2.06	-1.19	-4.88	-1.92	0.75	2.19	1.73	-4.30
		-0.76	1.18	1.43	-4.69	3.06	1.92	3.93	1.05	1.24	
18	CHAD	27.16	-3.93	-4.68	4.18	5.04	2.36	-0.59	2.19	-1.35	-3.80
		-0.76	-0.64	-0.50	-2.31	-1.65	-1.17	1.31	-0.65	0.35	
19	CHILE	-18.65	13.00	5.95	-1.42	-0.17	-1.44	1.43	-2.08	-1.95	1.30
		-0.76	-1.20	1.56	6.70	0.99	-1.83	3.43	0.78	1.24	
20	CHINA, PR	1.68	-22.92	-7.84	-11.48	0.89	-2.08	-1.06	0.44	-1.89	0.59
		1.66	-1.71	-1.32	-5.60	-1.65	-0.44	-2.23	-1.29	1.24	
21	COLOMBIA	-13.16	6.17	6.60	-2.82	1.87	-0.22	1.08	-2.08	0.60	-3.28
		-0.76	-1.34	1.14	7.27	-1.65	-3.02	2.97	0.78	1.24	
22	CONGO (BRA)	18.78	-7.43	-5.80	4.18	3.56	2.36	-0.40	2.19	-1.02	-1.89
		-0.76	-0.52	-0.50	1.02	-1.65	1.76	1.31	-0.20	-0.16	
23	CONGO (LEO)	21.46	-4.73	1.18	-0.61	5.08	1.82	0.66	2.19	-0.23	0.05
		-0.76	-0.71	0.08	-0.13	-1.65	1.82	3.93	-0.35	-0.69	
24	COSTA RICA	-10.54	13.39	4.69	5.31	-1.45	1.16	-1.43	-2.08	0.71	-4.89
		-0.76	-1.11	0.31	7.00	-1.65	-0.73	2.61	0.86	-1.28	
25	CUBA	-16.28	-17.15	-4.04	-5.44	-2.05	-1.45	1.48	-2.08	0.48	-0.23
		1.66	1.18	-1.52	6.21	3.33	-3.28	-2.16	-1.51	-1.28	
26	CYPRUS	-3.96	9.52	-2.89	4.97	-1.88	0.78	0.34	2.19	-0.15	3.33
		-0.76	0.41	-0.77	-1.53	-1.65	1.31	-1.42	1.61	-0.42	
27	CZECHOSLOVAKIA	-21.89	-18.58	-3.74	-4.83	-3.68	-2.08	1.24	2.19	1.47	3.60
		1.66	1.18	-3.36	1.20	2.15	0.28	-2.23	-1.66	-1.48	
28	DAHOMEY	26.26	-10.20	-5.78	6.07	-0.47	2.36	-0.59	2.19	-0.74	-2.35
		-0.76	-0.19	1.21	-0.85	-1.65	1.13	1.93	1.53	0.35	
29	DENMARK	-32.94	20.19	2.87	-2.58	-2.85	1.16	-1.43	-2.08	1.26	1.83
		-0.76	1.18	0.61	1.58	1.49	-0.11	-3.47	1.39	-1.48	
30	DOMINICAN REPUBLIC	-8.34	12.97	5.93	4.16	-4.36	1.16	0.90	-2.08	1.61	1.76
		-0.76	0.04	-0.90	6.45	-1.65	-3.20	0.79	-1.21	-0.80	
31	ECUADOR	-6.07	3.87	5.17	2.81	-0.36	-2.49	1.16	2.19	2.80	-0.67
		-0.76	-1.34	1.44	5.52	-1.65	-2.90	4.54	0.86	-1.28	
32	EL SALVADOR	-1.56	-9.48	3.75	5.47	-5.42	-2.97	-1.43	-2.08	1.73	-3.41
		-0.76	1.18	2.06	6.76	-1.65	-3.09	4.54	-0.62	1.24	
33	ETHIOPIA	21.46	-15.52	1.17	-2.60	1.49	-2.97	-1.43	2.19	1.77	-2.22
		-0.76	1.18	0.74	-2.88	1.09	-0.40	2.34	-0.30	-1.05	
34	FINLAND	-27.67	20.19	-0.41	-4.34	1.03	1.16	-0.27	-0.99	3.50	1.34
		-0.76	1.18	-2.08	-2.42	2.00	-0.00	-3.47	0.71	1.24	
35	FRANCE	-30.12	9.94	4.19	-9.70	-1.18	2.36	1.72	-2.08	-1.83	4.46
		-0.76	1.18	-2.51	3.43	3.33	-0.34	1.48	-1.48	1.24	
36	GABON	15.07	-9.06	-1.00	5.41	3.91	2.36	0.04	2.19	-1.11	-0.42
		-0.76	-0.64	-2.28	1.20	-1.65	1.72	1.93	-0.20	-0.20	
37	EAST GERMANY	-24.73	-22.92	-2.38	-6.89	-4.91	-2.97	-0.93	-0.01	-1.45	1.13
		1.66	-0.03	-3.25	0.34	3.01	0.23	-2.23	-0.86	0.64	
38	WEST GERMANY	-32.70	18.39	4.78	-9.23	-5.04	1.16	-0.93	-0.01	-1.32	1.96
		-0.76	0.57	-2.26	1.49	3.57	0.08	-2.23	0.31	0.64	
39	GHANA	16.24	-15.84	-0.07	-2.30	-1.09	0.98	0.85	2.19	-2.47	-3.87
		1.66	-0.05	0.91	1.05	1.64	0.91	1.48	-0.07	-0.75	
40	GREECE	-15.54	14.84	2.35	-4.19	-1.21	1.16	-1.43	-2.08	2.21	3.97
		-0.76	1.18	-1.65	-0.67	-1.65	0.80	4.03	-0.75	-1.28	
41	GUATEMALA	0.21	-0.74	5.12	3.84	-2.47	-2.49	1.60	2.19	0.91	-2.39
		-0.76	-1.67	1.91	5.78	-1.65	-3.17	4.54	-0.12	-1.28	
42	GUINEA	23.79	-15.84	-0.80	5.79	1.45	0.98	-0.16	2.19	-1.41	-0.76
		1.66	-0.30	-2.00	-2.26	-1.65	0.53	-2.23	-0.07	0.0	
43	HAITI	16.57	-14.43	3.86	6.57	-3.01	-2.26	1.60	-0.01	2.80	-1.06
		-0.76	-1.53	-2.03	3.28	-1.65	-3.22	0.79	-1.18	1.24	

Table 39 continued

44	HONDURAS	1.83	0.00	5.47	6.29	1.19	-2.49	0.75	-2.08	-1.96	-3.41
		-0.76	-1.45	1.56	5.75	-1.65	-3.13	4.54	0.31	-1.28	
45	HUNGARY	-22.52	-18.97	-5.51	-6.02	-3.20	-2.08	1.40	-0.01	-2.34	1.64
		1.66	-1.57	-3.12	2.83	2.08	0.44	-2.90	-1.56	-1.65	
46	ICELAND	-25.72	19.14	1.33	4.30	5.64	1.16	-0.40	-2.08	-0.69	-3.80
		-0.76	-1.04	-0.59	0.07	-1.65	-0.50	-3.47	1.39	0.16	
47	INDIA	8.10	12.11	4.26	-11.07	-1.06	2.36	1.67	0.11	-4.58	-3.13
		-0.76	-1.71	1.88	-6.11	-1.65	1.06	-2.23	0.0	-1.28	
48	INDONESIA	17.39	-8.42	2.86	-9.25	-0.07	2.36	1.45	0.11	0.82	-0.56
		0.98	-1.45	-2.74	-5.47	-1.65	1.68	2.97	-0.56	-1.05	
49	IRAN	14.47	-12.59	3.25	-6.29	5.37	-1.91	-1.06	0.11	1.36	3.52
		-0.76	-1.34	-0.01	-5.49	0.29	1.45	3.01	1.44	-1.05	
50	IRAQ	12.27	-9.33	-1.23	-3.31	3.33	-0.31	1.56	0.11	-0.94	3.77
		-0.76	1.18	1.77	-3.28	2.32	1.65	3.49	-1.26	-1.59	
51	IRELAND	-26.27	18.40	-0.19	1.58	-0.97	1.16	-1.43	-2.08	2.80	-1.09
		-0.76	1.18	-0.13	5.16	-0.28	-0.47	-3.47	-0.30	1.24	
52	ISRAEL	-19.23	16.51	2.40	0.43	-2.89	1.16	1.04	0.11	-1.06	4.65
		0.98	1.18	-2.54	-4.08	-1.65	1.50	-2.23	1.14	1.24	
53	ITALY	-29.07	16.86	4.58	-9.19	-5.16	1.16	1.52	-2.08	0.65	2.27
		-0.76	1.18	-2.96	5.17	3.18	0.36	-0.49	0.91	1.24	
54	IVORY COAST	22.69	-9.28	-0.77	2.40	2.05	2.36	0.04	2.19	-1.87	-0.54
		0.98	-0.36	-2.67	-1.24	-1.65	0.80	2.33	1.51	-0.20	
55	JAMAICA	-4.14	16.60	3.45	6.58	-4.69	1.16	0.04	-0.01	-1.28	-4.49
		-0.76	0.26	-1.22	0.19	-1.65	-3.25	-2.16	0.31	-0.20	
56	JAPAN	-22.03	15.47	6.73	-9.42	-3.61	1.16	0.75	-0.01	0.79	-2.15
		-0.76	1.18	-0.33	-6.30	3.81	-0.30	-0.25	1.25	1.24	
57	JORDAN	15.49	-16.74	-2.93	4.27	-0.19	-2.49	1.52	-2.08	-2.51	4.83
		-0.76	-1.34	-0.83	-4.39	1.25	1.58	3.49	-1.10	-1.48	
58	NORTH KOREA	7.86	-22.65	-7.84	-2.12	-2.77	-2.00	-0.93	-0.95	-0.62	4.57
		1.66	0.0	-1.55	-5.10	3.39	-0.20	-2.23	-0.86	0.64	
59	SOUTH KOREA	4.03	-9.94	5.67	-1.20	-6.50	-0.29	-0.93	-0.95	0.61	4.75
		0.57	0.33	2.55	-4.19	-1.65	-0.11	3.49	-0.51	0.64	
60	LAOS	26.95	-13.83	-0.10	4.15	1.54	-2.08	-0.40	2.19	0.61	4.26
		-0.76	-0.27	0.32	-5.06	-1.65	1.31	4.54	1.14	0.16	
61	LEBANON	-4.90	4.66	-0.80	4.80	-5.72	-0.14	0.75	-0.01	2.80	-0.93
		-0.76	1.18	-0.20	0.42	-1.65	1.50	4.54	-0.42	-1.28	
62	LIBERIA	14.54	-10.81	2.64	6.71	0.78	-2.49	-1.43	2.19	2.80	-0.94
		-0.76	1.18	-1.53	-0.44	-1.65	0.80	-1.50	1.21	1.24	
63	LIBYA	14.30	4.48	-0.25	2.36	6.45	-1.44	0.52	-2.08	-2.70	1.07
		-0.76	-0.88	0.46	-1.21	0.20	0.70	-0.24	-0.46	-0.56	
64	LUXEMBOURG	-28.28	20.19	3.77	3.92	-5.88	1.16	-0.79	-2.08	-1.02	1.87
		-0.76	0.55	-1.94	3.67	0.75	0.08	-3.47	0.97	0.50	
65	MALAGASY REP	18.78	10.85	-1.35	2.81	4.82	1.16	-0.79	-0.01	-1.56	-2.59
		-0.76	-0.85	-0.35	0.63	-1.65	0.73	-1.16	-1.02	0.50	
66	MALI	27.29	-6.06	-5.82	2.90	4.36	2.36	0.52	2.19	-1.49	-2.12
		-0.76	-0.52	2.34	-1.32	0.08	-1.91	1.93	-0.70	-0.56	
67	MAURITANIA	25.62	-2.18	-2.41	5.76	5.46	0.98	0.04	0.11	-2.08	-2.59
		0.98	-0.78	1.05	-2.03	-1.65	-2.22	1.93	-0.81	-0.20	
68	MEXICO	-13.09	2.22	7.70	-4.60	1.27	0.98	0.22	-2.08	1.31	0.70
		-0.76	1.18	2.39	6.45	-1.65	-3.34	-2.23	0.65	1.24	
69	MONGOLIA	5.16	-22.65	-4.10	5.04	6.93	-2.08	-1.43	-2.08	1.61	4.08
		1.66	1.18	-0.91	-4.55	0.69	-0.24	-2.23	-1.53	1.24	
70	MOROCCO	11.28	4.19	-1.66	-3.20	2.17	2.36	0.34	0.11	-3.18	0.32
		0.98	-0.82	2.04	-1.58	2.29	-0.04	-1.32	0.71	-0.42	
71	NEPAL	24.91	-14.13	-4.83	3.08	-1.69	-2.97	0.44	2.19	2.80	-0.91
		0.57	1.18	0.67	-6.43	-1.65	-0.38	3.49	0.65	1.24	
72	NETHERLANDS	-31.02	19.79	3.53	-4.76	-7.32	1.16	-1.43	-0.01	1.15	3.37
		-0.76	1.18	-2.06	0.82	2.71	-0.11	-3.47	1.00	-0.93	
73	NEW ZEALAND	-32.31	18.80	3.64	-1.13	4.54	1.16	-1.43	-0.01	1.42	0.72
		-0.76	1.18	0.05	1.05	0.17	1.97	-3.47	0.31	1.24	
74	NICARAGUA	-0.06	-6.01	4.61	5.85	0.75	-2.97	-0.71	-2.08	0.10	-0.74
		-0.76	1.18	-2.67	5.98	-1.65	-3.13	3.22	1.10	-1.28	
75	NIGER	29.09	-5.82	-4.24	5.17	4.95	2.36	-0.59	2.19	-1.61	-3.16
		-0.76	-0.64	2.39	-3.09	2.15	-1.42	1.93	-0.91	0.35	
76	NIGERIA	19.83	4.17	0.65	-1.73	1.61	-0.03	0.94	2.19	-1.30	-4.38
		-0.76	-0.38	1.29	-0.59	0.14	1.06	-2.16	1.57	-0.84	
77	NORWAY	-31.50	20.19	2.56	-3.23	0.13	1.16	-1.43	-2.08	1.42	2.77
		-0.76	1.18	-0.60	-2.01	1.55	-0.30	-3.47	1.44	1.24	
78	PAKISTAN	14.67	-14.45	4.04	-7.60	-1.07	-0.19	1.64	0.11	-4.02	-0.34
		0.57	-1.34	3.35	-2.86	2.86	1.26	-0.31	1.08	1.24	
79	PANAMA	-8.79	6.52	5.60	6.55	-1.11	-1.44	0.22	-0.99	3.67	-4.94
		-0.76	1.18	1.37	6.04	-1.65	-3.02	4.54	-1.37	1.24	
80	PARAGUAY	-3.29	-14.96	2.00	5.13	2.32	-2.97	-1.06	-2.08	2.80	1.93
		-0.76	-1.63	2.08	5.75	-1.65	-1.59	-1.16	1.32	1.24	
81	PERU	-7.56	0.77	6.42	-3.87	4.23	-0.78	0.75	2.19	-1.35	-1.54
		-0.76	-1.11	1.01	5.99	-1.65	-2.88	3.93	-0.03	1.24	
82	PHILIPPINES	-4.66	14.62	6.52	-2.13	-4.22	0.98	-1.43	0.11	3.74	-3.07
		-0.76	-1.20	1.75	0.32	-1.65	0.96	-2.23	-0.94	1.24	
83	POLAND	-19.11	-19.35	-1.80	-8.98	-3.44	-2.49	1.10	-2.08	-0.48	2.74
		1.66	1.18	-3.02	3.00	1.32	0.23	-2.23	-1.66	-1.05	
84	PORTUGAL	-16.67	-14.43	2.13	0.66	-4.67	-2.97	-1.43	-2.08	2.18	2.24
		-0.76	1.18	2.71	4.53	-1.65	-0.24	-3.47	-1.43	1.24	
85	ROMANIA	-13.27	-21.06	-5.04	-0.63	-2.00	-2.08	-1.43	-0.99	3.72	2.90
		1.66	-1.20	-2.77	-0.64	2.41	0.53	-2.23	-1.66	-1.05	
86	RWANDA	16.39	-3.20	-6.56	8.06	-4.57	0.21	-0.40	-1.06	0.61	-2.79
		-0.76	0.14	-2.58	0.24	-1.65	2.08	0.79	-0.20	0.16	

Table 39 continued

87	SAUDI ARABIA	19.30	-13.40	1.90	-0.27	6.30	-2.97	-1.43	-2.08	0.21	0.62
		-0.76	1.18	2.69	-3.53	-1.65	-0.08	-3.11	-1.35	1.24	
88	SENEGAL	20.29	-8.26	-4.52	3.89	0.64	2.36	0.04	0.11	-1.49	-3.48
		0.98	-0.25	0.35	-1.12	-0.21	0.32	-1.16	1.48	-0.20	
89	SIERRA LEONE	26.60	9.19	-2.50	6.69	0.52	-0.22	0.61	2.19	-1.13	-2.12
		-0.76	-0.46	2.47	-1.44	-0.36	0.59	-0.35	1.67	-0.64	
90	SOMALIA	24.44	0.85	-3.82	5.16	5.55	-1.44	-0.59	-2.08	-1.97	-0.70
		-0.76	-0.90	-0.50	-4.98	-1.65	2.08	-0.03	0.86	0.35	
91	SOUTH AFRICA	-15.17	-1.27	3.18	-2.59	6.27	-2.08	0.04	2.19	0.61	-4.20
		-0.76	-1.01	2.35	0.38	-1.65	-0.00	-3.47	-1.45	-0.20	
92	SPAIN	-19.72	-14.98	3.50	-5.95	-2.74	-2.49	1.20	-0.99	3.39	2.31
		0.57	1.18	2.89	4.93	-0.24	-0.36	-3.47	-0.97	1.24	
93	SUDAN	23.12	-14.77	-2.93	-3.15	6.27	-0.19	0.34	2.19	-2.88	-3.63
		0.98	-0.93	2.69	-2.28	2.23	-0.71	3.01	-1.08	-0.42	
94	SWEDEN	-34.00	20.19	1.57	-6.35	0.48	1.16	-0.27	-2.08	1.04	3.07
		-0.76	1.18	-0.15	-0.97	2.47	-0.00	-3.47	1.25	-1.65	
95	SWITZERLAND	-31.59	20.48	2.05	-2.43	-4.27	-0.07	-1.43	2.19	1.23	-2.00
		-0.76	1.18	2.92	1.48	2.53	0.16	-3.47	0.31	-1.05	
96	SYRIA	8.26	-4.43	-3.09	-1.11	3.13	-0.26	1.33	-0.99	-4.05	3.17
		-0.76	-1.57	2.50	-2.20	2.44	1.39	4.54	-1.00	1.48	
97	THAILAND	10.55	-16.12	2.90	-2.83	-0.03	-2.97	1.13	-0.99	3.67	1.46
		-0.76	-1.11	0.83	-6.25	-1.65	1.35	4.03	1.32	1.24	
98	TOGO	24.84	-7.25	-4.98	6.72	0.12	1.82	-0.16	2.19	-0.40	-2.12
		-0.76	-0.41	0.62	0.07	-0.39	1.06	0.79	-0.67	0.0	
99	TRINIDAD	-4.31	16.00	4.10	7.02	-4.51	1.16	0.61	1.08	0.21	-1.93
		-0.76	0.44	-1.67	-0.43	-1.65	-2.93	-2.16	0.31	-0.64	
100	TUNISIA	7.15	-2.39	-3.51	0.18	0.78	2.36	0.52	-2.08	-3.11	-1.33
		0.98	-0.57	-0.23	-1.88	1.06	0.49	-2.23	1.64	-0.56	
101	TURKEY	-4.02	7.43	4.80	-7.73	0.76	1.16	0.91	-0.99	3.52	2.82
		-0.76	1.18	-0.42	-5.32	-1.65	-0.74	-1.16	1.02	-1.48	
102	UGANDA	17.41	8.86	-1.25	4.15	-0.74	-0.93	0.66	2.19	-0.06	-3.16
		-0.76	-0.11	1.83	1.09	3.72	1.97	-2.16	0.31	-0.69	
103	U.S.S.R.	-21.83	-19.93	-3.67	-6.71	6.55	-2.08	1.20	1.06	3.77	4.51
		1.66	1.18	-3.25	-2.59	3.72	-1.05	-3.47	-1.66	-1.74	
104	U.A.R.	-0.68	-13.08	0.13	-8.20	-3.22	0.51	0.44	-2.08	-5.04	1.21
		1.66	-1.50	3.28	-1.69	2.95	1.21	1.26	-1.24	-1.48	
105	UNITED KINGDOM	-34.80	17.42	4.98	-8.48	-3.76	-0.22	0.97	-0.01	-1.34	3.65
		-0.76	0.46	0.62	1.30	3.57	-0.42	-3.47	-0.73	-0.87	
106	UNITED STATES	-35.40	15.09	-4.79	-11.05	3.21	0.98	1.00	-0.01	-1.86	4.11
		-0.76	0.26	-1.38	2.05	-1.65	-3.34	-3.47	-0.30	-0.90	
107	UPPER VOLTA	28.16	-6.92	-6.56	6.34	0.78	2.36	-0.59	2.19	-0.85	-3.16
		-0.76	-0.22	1.13	-2.61	-1.65	0.80	1.93	1.57	0.35	
108	URUGUAY	-20.69	18.91	1.80	0.53	3.96	1.82	-1.43	-2.08	1.64	-2.15
		-0.76	1.18	0.28	5.76	-1.65	0.70	1.26	1.32	1.24	
109	VENEZUELA	-17.08	8.46	7.24	-1.47	5.65	2.36	1.29	-2.08	3.58	-0.40
		0.98	1.18	-2.37	6.18	-1.65	-3.02	1.96	1.70	-1.70	
110	NORTH VIETNAM	12.39	-22.65	-7.84	-2.27	-4.54	-2.08	-0.93	0.13	-0.62	3.80
		1.66	0.30	-1.55	-3.94	3.39	1.17	-2.23	-0.86	0.64	
111	SOUTH VIETNAM	14.71	-16.44	2.03	0.12	-3.89	-1.92	-0.93	0.13	0.61	3.91
		1.66	0.19	0.97	-2.34	-1.65	1.39	2.61	-0.59	0.64	
112	YEMEN	23.89	-6.89	-6.56	5.59	3.16	-0.28	-1.43	-2.08	0.07	0.08
		0.50	-1.26	2.69	-3.80	-1.65	0.31	0.79	-1.32	1.24	
113	YUGOSLAVIA	-7.92	-19.83	0.79	-7.89	-1.23	-2.49	-1.43	2.19	3.41	4.53
		1.66	1.18	-2.97	-0.58	1.00	0.44	-2.23	-0.38	-1.59	
114	BOTSWANA	14.75	5.37	-6.56	6.57	5.22	-0.78	-0.40	1.08	-0.05	-3.80
		0.98	-0.78	-0.50	-0.94	0.57	2.21	-0.49	0.31	0.16	
115	GAMBIA	17.81	9.32	-5.94	8.68	-2.35	-0.78	-0.79	-0.99	-1.19	-3.80
		0.98	-0.08	-0.50	-5.23	-1.65	0.44	0.80	0.31	0.50	
116	KUWAIT	-10.21	4.97	0.17	3.56	2.58	0.98	0.04	-0.99	-1.86	1.91
		0.98	-0.98	2.69	-0.18	0.29	1.78	-1.32	0.31	-0.20	
117	LESOTHO	16.09	9.95	-6.56	8.50	-1.40	-0.78	-0.59	1.08	-0.78	-1.65
		0.98	-0.16	-0.50	1.75	0.63	2.32	0.64	0.31	0.35	
118	MALAWI	16.88	5.52	-4.80	7.20	-0.95	0.98	0.34	2.19	-1.24	-3.80
		0.98	-0.14	-0.50	0.15	-1.65	-0.05	0.80	0.31	-0.42	
119	MALTA	-17.17	16.03	-1.91	7.80	-7.96	1.82	0.04	-0.99	0.61	1.80
		1.66	0.52	1.24	3.78	-1.65	0.64	-0.37	0.31	-0.20	
120	TANZANIA	15.49	8.60	-2.61	0.63	4.44	2.36	0.04	1.08	-1.81	-4.03
		0.98	-0.74	1.72	-0.59	2.03	2.08	2.61	0.31	-0.20	
121	ZAMBIA	3.84	8.70	-4.68	3.10	4.04	-0.78	0.34	2.19	0.61	-1.46
		0.98	-0.52	-0.50	1.56	-1.65	2.27	-0.49	0.31	-0.42	
122	KENYA	15.80	6.80	-1.47	0.60	-1.66	2.36	0.85	1.08	-1.43	-1.61
		0.98	0.22	2.02	-0.12	1.46	2.03	0.64	0.31	-0.75	
123	W. SAMOA	14.53	4.48	-4.09	7.98	-4.14	0.78	-0.79	1.08	0.61	-1.75
		-0.76	0.11	-0.50	1.96	-1.65	1.26	0.79	0.31	0.50	
124	SINGAPORE	-11.74	15.66	-0.55	4.16	-7.33	1.82	-0.40	2.19	-2.60	0.91
		1.66	0.38	-1.48	-2.67	2.16	-0.47	0.49	0.31	0.16	
125	GUYANA	12.09	10.08	0.68	5.35	5.21	1.82	0.04	-0.01	0.21	-0.76
		0.98	-0.96	-0.28	1.09	-1.65	-2.83	1.14	0.31	-0.20	
126	MALAYSIA	2.19	10.02	0.82	1.24	-1.91	-0.34	-0.16	2.19	1.25	1.93
		-0.76	0.04	1.14	-5.75	0.20	-0.43	-0.49	1.72	0.0	
127	MALDIVE ISLANDS	17.22	1.26	-6.56	8.11	-2.50	1.82	-0.59	-2.08	-1.72	-1.13
		-0.76	0.16	-0.50	-4.74	-1.65	2.13	-2.16	0.31	0.35	
128	BARBADOS	-9.39	14.41	-1.20	8.78	-6.96	-0.34	0.52	-0.01	0.61	-3.80
		0.98	0.36	-1.02	3.43	-1.65	-2.96	-3.47	0.31	-0.56	
129	TAIWAN	-2.44	-3.95	5.12	-0.94	-7.30	-0.07	0.22	-0.66	-2.49	5.13
		0.98	1.18	1.64	-5.09	-1.65	0.64	0.35	-0.51	1.24	

TABLE 40
Correlations of Table 37 Factor Scores with Estimate Deck

	1	2	3	4	5	6	7	8	9	10	11	12	13	14	15	16	17	18	19
VAR 1	-0.47	-0.29	-0.02	-0.29	-0.09	-0.13	0.07	-0.09	0.15	0.76	0.02	0.29	0.03	-0.11	0.00	0.00	0.04	0.15	0.02
VAR 2	-0.19	0.05	-0.01	-0.11	-0.90	-0.01	-0.03	0.09	0.06	0.14	0.06	0.09	-0.05	0.02	0.12	0.02	0.07	0.07	0.09
VAR 3	0.91	-0.13	-0.07	0.21	0.05	0.10	-0.08	0.06	-0.12	-0.22	-0.07	-0.10	0.05	-0.16	0.02	-0.04	0.04	-0.10	0.06
VAR 4	-0.52	-0.04	0.13	-0.78	-0.08	0.07	0.12	0.09	0.13	0.17	-0.00	0.15	0.06	0.05	0.13	-0.05	-0.00	0.15	0.10
VAR 5	0.84	-0.24	0.08	-0.01	-0.04	0.02	-0.06	-0.04	-0.11	-0.05	0.05	-0.08	0.18	-0.18	-0.07	-0.18	0.07	-0.07	0.09
VAR 6	0.70	-0.43	0.01	0.07	0.07	0.07	-0.08	-0.09	0.10	-0.04	-0.12	-0.15	0.04	-0.17	-0.16	-0.07	0.06	0.05	-0.09
VAR 7	0.38	-0.04	0.32	0.30	0.02	0.13	0.07	0.02	0.05	0.13	0.23	-0.24	0.28	-0.11	0.05	-0.26	0.14	-0.32	-0.01
VAR 8	0.56	-0.09	-0.08	-0.03	0.21	0.17	0.06	-0.12	-0.36	-0.08	-0.09	0.16	0.07	-0.47	-0.01	0.13	-0.14	-0.19	-0.31
VAR 9	-0.39	0.08	0.26	0.04	-0.12	-0.04	0.20	0.03	0.12	-0.06	-0.02	-0.01	-0.06	0.77	-0.06	-0.02	0.06	-0.09	-0.03
VAR 10	-0.62	0.16	0.22	0.07	-0.14	-0.05	0.05	-0.05	0.14	-0.14	-0.06	0.06	-0.12	0.60	-0.05	-0.08	0.07	0.11	-0.08
VAR 11	0.71	-0.03	0.07	0.03	-0.15	0.03	0.05	-0.00	-0.05	-0.07	-0.12	-0.11	0.17	-0.49	-0.02	0.03	0.04	0.04	-0.06
VAR 12	0.12	0.03	0.22	0.13	0.19	0.05	0.19	-0.29	0.08	-0.14	0.02	0.06	0.41	0.46	-0.17	-0.27	0.21	-0.03	-0.11
VAR 13	-0.31	0.30	-0.12	0.02	-0.62	0.08	-0.06	-0.09	0.01	0.18	-0.15	0.44	0.08	0.19	0.10	0.04	0.02	0.06	0.09
VAR 14	-0.10	-0.26	-0.33	0.07	-0.08	-0.07	0.17	0.02	0.03	0.09	-0.01	0.08	-0.68	0.03	-0.10	-0.11	-0.07	0.02	-0.12
VAR 15	-0.13	0.31	-0.14	0.13	0.68	0.03	0.06	0.05	-0.08	-0.03	-0.12	0.03	-0.02	-0.35	0.08	0.05	-0.05	0.11	-0.08
VAR 16	-0.27	0.00	-0.03	-0.15	-0.87	-0.08	0.06	0.00	0.09	0.08	0.01	0.14	-0.04	0.00	0.05	-0.02	-0.03	0.05	0.03
VAR 17	0.18	-0.12	0.08	-0.59	0.65	0.10	0.06	0.08	-0.04	-0.00	-0.01	-0.06	0.10	-0.03	0.03	-0.09	0.01	0.03	0.10
VAR 18	-0.91	0.18	0.06	-0.07	-0.01	-0.04	0.03	-0.04	0.03	0.17	0.05	0.15	-0.13	0.18	0.00	0.00	-0.07	0.16	-0.03
VAR 19	-0.20	-0.26	-0.05	-0.24	-0.06	-0.06	-0.02	0.01	-0.10	0.75	0.00	-0.07	-0.04	-0.12	0.05	0.09	0.12	0.02	-0.09
VAR 20	-0.42	-0.01	-0.10	-0.03	-0.17	0.12	-0.10	0.09	0.13	0.13	0.09	-0.02	-0.75	0.06	-0.01	0.17	-0.04	0.06	-0.02
VAR 21	-0.34	-0.24	-0.10	-0.16	-0.08	-0.05	0.06	-0.06	0.14	0.83	0.03	0.27	-0.03	-0.10	0.01	0.02	-0.05	0.11	0.04
VAR 22	0.16	-0.36	-0.43	0.01	-0.09	0.05	-0.06	0.05	-0.26	-0.18	0.09	0.13	-0.04	-0.33	0.28	0.06	-0.24	-0.06	-0.34
VAR 23	0.64	-0.36	-0.19	0.21	0.09	0.12	0.03	0.11	-0.10	-0.12	-0.09	-0.11	-0.09	-0.03	0.08	0.12	-0.09	-0.32	-0.16
VAR 24	0.20	0.47	0.20	0.08	0.02	0.02	-0.01	0.06	0.02	-0.04	0.08	-0.11	-0.04	-0.04	-0.09	0.14	0.01	0.54	-0.01
VAR 25	-0.80	0.14	0.06	-0.29	-0.04	0.08	-0.04	0.03	0.23	0.26	0.07	0.02	0.04	0.05	-0.00	0.04	0.12	0.02	0.04
VAR 26	-0.85	0.23	0.02	-0.10	-0.18	-0.03	0.06	0.01	0.15	0.22	0.08	0.05	-0.17	0.09	0.05	0.03	-0.03	0.03	0.04
VAR 27	-0.86	0.21	-0.02	-0.13	-0.17	0.00	-0.03	0.01	0.23	0.14	0.00	-0.01	-0.19	0.16	0.08	0.04	-0.02	0.10	0.11
VAR 28	0.23	-0.08	0.15	0.34	-0.00	-0.09	0.01	-0.03	0.46	0.11	0.01	0.06	-0.09	-0.20	-0.11	0.07	-0.10	-0.26	-0.06
VAR 29	0.04	-0.00	0.12	-0.24	-0.05	0.05	0.69	0.01	-0.16	0.05	0.08	-0.03	-0.06	0.20	0.04	-0.09	0.06	-0.01	-0.18
VAR 30	0.25	-0.12	0.03	-0.08	0.15	-0.04	0.24	-0.06	-0.26	-0.21	-0.01	-0.64	0.13	0.05	-0.12	-0.02	0.09	0.02	-0.08
VAR 31	0.17	0.16	0.01	-0.07	-0.03	0.22	0.04	-0.11	0.05	-0.52	0.11	-0.19	-0.02	0.06	-0.11	0.14	0.11	-0.37	-0.18
VAR 32	0.24	-0.15	-0.05	0.04	0.02	0.08	0.06	-0.25	-0.71	0.01	0.02	-0.05	-0.05	-0.31	0.05	-0.01	-0.04	-0.16	0.01
VAR 33	-0.09	-0.05	-0.08	-0.01	0.04	0.01	0.12	0.05	-0.00	0.07	0.04	-0.03	-0.07	0.05	0.17	0.03	0.04	-0.03	-0.73
VAR 34	0.24	-0.04	0.13	-0.03	-0.02	0.06	0.12	-0.04	-0.52	0.12	-0.05	-0.08	0.27	-0.13	0.12	0.02	-0.03	-0.04	-0.19
VAR 35	-0.15	0.13	0.79	0.03	0.02	-0.01	0.05	0.12	0.17	-0.04	-0.05	0.01	-0.04	0.17	-0.16	-0.16	0.02	0.08	0.15
VAR 36	-0.09	0.14	0.84	-0.07	0.03	-0.06	0.15	0.04	0.02	0.08	0.05	-0.02	0.06	0.14	-0.19	-0.12	0.14	0.07	0.04
VAR 37	-0.24	0.09	0.35	-0.46	0.06	0.14	0.22	-0.08	-0.04	0.03	0.21	0.39	-0.11	0.10	-0.00	-0.07	0.03	-0.11	-0.27
VAR 38	0.38	0.04	-0.48	0.31	-0.20	0.12	0.09	0.08	-0.05	-0.10	0.04	-0.23	0.04	-0.11	0.15	0.16	-0.32	-0.10	0.21
VAR 39	0.22	-0.40	-0.57	0.12	-0.05	-0.00	0.03	0.12	-0.11	-0.17	0.03	0.02	-0.09	-0.29	0.15	0.12	-0.34	-0.23	-0.19
VAR 40	-0.45	0.24	0.62	-0.43	-0.04	-0.01	0.15	0.02	0.05	0.16	-0.00	0.19	0.09	0.13	-0.01	0.07	0.05	0.16	0.04
VAR 41	-0.49	0.25	0.60	-0.41	-0.04	-0.03	0.09	0.08	0.11	0.02	-0.01	0.21	-0.05	0.21	0.07	0.00	-0.00	0.15	0.10
VAR 42	-0.45	-0.04	-0.18	-0.60	-0.09	0.09	-0.03	-0.00	-0.12	-0.01	-0.04	0.03	-0.15	-0.04	0.33	0.04	0.09	0.29	-0.05
VAR 43	-0.34	-0.02	-0.30	-0.59	-0.12	0.03	0.02	-0.07	-0.15	0.02	-0.07	0.04	-0.11	-0.10	0.32	0.04	0.05	0.29	-0.15
VAR 44	-0.28	-0.05	-0.16	-0.40	-0.03	-0.04	-0.02	0.04	-0.05	0.09	0.02	0.09	0.02	0.04	0.79	0.08	0.09	0.10	-0.07
VAR 45	-0.20	-0.05	-0.22	-0.28	-0.15	0.04	0.02	0.07	-0.04	0.06	0.02	0.16	0.12	-0.00	0.81	0.01	-0.04	0.01	-0.10
VAR 46	0.15	0.18	-0.06	0.87	0.08	0.03	-0.11	-0.06	-0.15	-0.19	-0.09	-0.09	-0.11	-0.06	-0.13	0.02	0.03	-0.04	-0.15
VAR 47	0.72	-0.15	-0.12	0.25	-0.01	0.05	-0.13	0.19	-0.10	-0.23	0.13	-0.12	0.02	-0.14	-0.10	0.14	-0.03	-0.09	0.11
VAR 48	-0.77	0.38	0.09	-0.11	-0.06	-0.09	0.07	0.05	-0.05	0.08	0.01	0.12	0.03	0.17	0.07	0.02	-0.06	-0.02	0.11
VAR 49	0.87	-0.25	-0.08	0.15	0.04	0.02	-0.07	-0.03	-0.05	-0.12	-0.02	-0.26	0.06	-0.09	-0.07	0.07	0.05	-0.06	0.09
VAR 50	0.56	0.01	-0.16	0.73	0.06	-0.06	-0.09	-0.07	-0.06	-0.20	-0.01	-0.14	-0.08	-0.01	-0.09	0.02	0.05	-0.12	-0.07
VAR 51	0.77	-0.31	-0.00	0.33	0.02	0.01	-0.03	-0.11	-0.00	-0.13	-0.14	-0.18	0.10	-0.06	-0.02	-0.03	0.11	-0.05	-0.02
VAR 52	0.13	-0.79	-0.11	-0.11	0.02	-0.20	0.05	-0.11	-0.03	0.16	0.27	0.08	0.05	-0.05	0.08	0.08	-0.00	0.09	0.03
VAR 53	0.89	-0.13	-0.05	0.20	0.15	-0.01	-0.04	0.07	-0.12	-0.16	-0.05	-0.04	0.06	-0.03	-0.06	0.05	-0.06	-0.18	0.02
VAR 54	0.18	0.04	-0.15	0.17	-0.10	-0.01	0.03	0.71	0.09	-0.15	-0.08	-0.12	-0.09	0.03	0.17	-0.08	-0.04	-0.03	0.03
VAR 55	0.05	0.06	0.27	0.15	0.12	-0.10	0.12	0.15	-0.06	-0.18	0.05	-0.02	0.02	0.11	-0.10	-0.76	0.07	-0.15	0.04
VAR 56	0.41	-0.10	0.02	-0.13	0.02	0.15	0.00	0.63	-0.03	-0.14	0.09	0.04	0.14	-0.27	-0.17	0.09	0.03	0.00	-0.22
VAR 57	0.57	0.09	-0.14	0.39	-0.05	0.30	0.19	0.05	-0.31	-0.18	0.11	-0.21	0.03	-0.14	0.04	0.30	-0.20	-0.35	-0.10
VAR 58	0.87	-0.12	-0.08	0.25	0.13	0.07	0.02	0.03	-0.13	-0.10	0.05	-0.13	0.11	-0.13	0.09	0.08	0.07	-0.21	-0.00
VAR 59	0.70	0.07	-0.02	0.39	0.03	0.04	-0.08	0.20	-0.17	-0.21	0.11	-0.03	-0.10	-0.08	-0.04	0.10	0.04	-0.24	0.20
VAR 60	0.02	0.46	0.30	0.02	0.12	-0.10	-0.09	0.00	0.08	-0.04	-0.63	0.03	0.07	0.09	-0.11	-0.01	0.17	0.04	0.05
VAR 61	0.07	-0.93	-0.13	-0.10	-0.06	-0.05	-0.13	0.07	0.08	0.12	0.00	0.04	0.02	-0.10	0.05	-0.03	0.04	0.04	0.07
VAR 62	0.44	-0.14	0.08	0.04	0.04	-0.10	0.40	0.02	0.11	0.00	0.05	-0.22	0.08	0.07	0.06	-0.08	0.56	-0.05	-0.04
VAR 63	0.29	-0.84	-0.07	-0.08	0.06	-0.06	-0.10	0.09	-0.01	0.06	0.02	-0.01	0.08	-0.15	-0.00	-0.06	0.22	-0.01	0.12
VAR 64	0.29	-0.85	-0.16	0.04	-0.00	0.12	-0.11	0.03	-0.01	0.06	0.04	0.06	0.05	-0.06	0.02	0.04	-0.02	-0.02	0.10
VAR 65	0.29	-0.83	-0.31	0.01	0.03	0.01	-0.15	0.02	0.03	0.02	0.05	0.00	0.06	-0.09	0.08	-0.01	-0.08	-0.03	0.05
VAR 66	0.31	-0.38	0.11	0.02	0.09	-0.19	0.28	0.36	0.04	0.11	-0.24	-0.02	0.23	0.01	-0.19	-0.04	0.20	-0.11	0.04
VAR 67	-0.31	-0.03	-0.04	0.47	-0.24	0.17	0.04	-0.36	0.13	-0.06	0.15	0.07	-0.01	-0.02	-0.04	-0.13	0.03	0.26	0.20
VAR 68	0.66	-0.50	-0.21	0.26	-0.01	-0.08	-0.10	0.08	-0.01	-0.07	-0.04	-0.11	-0.01	-0.07	0.05	0.03	-0.11	-0.17	-0.00
VAR 69	-0.26	0.71	0.15	0.08	0.03	0.12	-0.18	-0.07	-0.07	-0.09	-0.16	0.03	-0.21	0.05	0.04	-0.12	-0.02	0.05	0.14
VAR 70	-0.66	0.33	0.00	0.01	-0.02	-0.06	-0.11	-0.15	-0.02	0.07	0.44	0.16	-0.09	0.33	0.03	0.00	0.00	0.17	-0.04
VAR 71	-0.47	0.49	0.02	0.08	-0.11	0.08	-0.34	-0.10	-0.08	0.07	0.29	0.18	-0.17	0.26	0.03	0.08	-0.15	0.05	-0.09
VAR 72	0.14	-0.68	-0.11	-0.09	0.00	0.38	-0.19	-0.07	0.09	0.12	-0.13	-0.04	0.06	-0.05	0.04	-0.09	-0.03	-0.16	0.15
VAR 73	-0.19	-0.27	0.06	-0.03	-0.09	-0.82	-0.09	0.05	0.25	0.08	-0.10	0.01	0.01	0.06	0.04	-0.10	0.00	0.08	0.13
VAR 74	0.45	-0.77	-0.05	0.07	0.09	0.08	0.00	-0.02	0.07	-0.02	-0.17	-0.09	0.05	-0.04	0.09	0.01	0.02	-0.08	0.16
VAR 75	0.44	0.11	0.06	0.15	0.13	-0.20	0.35	-0.02	0.13	0.01	0.04	-0.10	0.09	0.11	-0.04	-0.05	0.59	-0.01	-0.09
VAR 76	-0.26	0.11	-0.19	-0.11	-0.02	-0.02	-0.08	-0.04	0.00	0.02	0.07	0.01	-0.08	-0.02	0.05	0.04	-0.81	-0.04	-0.01
VAR 77	0.00	0.70	0.01	0.10	0.02	0.50	0.11	-0.04	-0.04	-0.16	-0.19	-0.04	-0.17	0.02	0.05	0.24	-0.02	0.05	-0.09
VAR 78	-0.46	0.28	0.11	0.11	-0.05	-0.42	0.01	-0.01	0.45	0.09	-0.11	-0.13	-0.01	0.16	-0.14	-0.07	-0.02	0.02	0.06
VAR 79	-0.08	-0.34	0.08	0.29	-0.20	0.20	0.14	-0.19	0.00	-0.08	-0.21	0.24	-0.04	-0.34	-0.17	-0.19	0.25	0.22	0.06
VAR 80	0.34	-0.85	-0.05	-0.02	0.07	0.02	0.02	0.03	-0.00	0.07	-0.06	0.02	0.00	-0.10	0.06	0.01	0.09	0.04	0.04
VAR 81	0.32	-0.85	-0.06	0.01	-0.00	-0.10	0.01	-0.04	0.05	0.10	-0.01	0.01	-0.01	-0.05	0.10	-0.07	0.06	-0.00	0.04
VAR 82	-0.14	0.43	0.23	-0.32	0.11	-0.30	-0.17	0.06	-0.02	0.21	-0.17	0.01	-0.13	0.22	0.04	-0.16	-0.16	0.02	-0.04
VAR 83	-0.29	0.81	0.05	0.05	-0.07	-0.11	0.02	0.02	0.15	0.01	-0.04	-0.04	-0.11	0.15	-0.05	0.01	-0.06	-0.04	-0.08
VAR 84	0.81	-0.26	-0.15	0.06	0.17	-0.05	0.01	0.01	-0.00	-0.14	-0.01	-0.00	-0.02	-0.07	0.01	0.06	0.15	-0.12	0.02
VAR 85	-0.05	0.65	-0.16	0.14	-0.06	0.40	-0.09	0.05	-0.08	-0.10	-0.13	0.06	-0.19	0.07	0.03	0.10	-0.25	-0.05	0.06
VAR 86	-0.25	0.79	-0.04	0.05	-0.15	0.08	-0.11	0.01	-0.09	0.03	0.16	0.05	-0.05	0.07	0.06	0.01	-0.15	-0.10	0.12
VAR 87	0.30	0.62	0.37	0.13	0.10	0.10	0.01	-0.07	-0.19	-0.11	-0.12	0.07	0.19	0.01	-0.15	0.11	0.21	0.08	0.05
VAR 88	0.47	0.05	-0.24	0.09	0.08	-0.05	-0.01	0.11	-0.13	-0.09	0.06	0.02	0.09	-0.32	0.09	0.53	0.04	-0.23	0.05
VAR 89	0.22	0.17	0.42	0.42	0.18	0.03	0.11	0.13	-0.02	-0.32	0.20	-0.24	0.03	0.42	-0.05	-0.07	0.14	-0.09	0.17
VAR 90	-0.06	0.14	0.35	0.37	0.21	0.12	0.15	0.06	-0.05	-0.37	-0.02	-0.12	-0.04	0.58	-0.06	-0.12	-0.02	0.09	-0.08
VAR 91	-0.38	-0.27	0.08	-0.76	-0.09	-0.09	0.08	-0.02	0.13	0.45	0.04	0.11	0.05	-0.01	0.05	-0.06	0.08	0.17	0.11

TABLE 41
Correlations of Table 38 Factor Scores with Estimate Deck

	1	2	3	4	5	6	7	8	9	10	11	12	13	14	15	16	17	18	19
VAR 1	-0.46	-0.18	0.11	-0.49	-0.22	-0.28	-0.01	-0.34	0.21	0.95	0.19	0.39	-0.25	-0.13	0.30	0.12	-0.23	-0.20	-0.02
VAR 2	-0.27	0.09	0.09	-0.17	-0.92	-0.07	0.04	-0.06	0.06	0.20	0.12	0.26	-0.19	-0.09	0.20	0.07	-0.13	0.02	0.05
VAR 3	0.94	-0.36	-0.39	0.49	0.24	0.13	-0.08	0.41	-0.27	-0.43	0.02	-0.34	0.28	-0.40	-0.29	0.11	0.39	0.08	0.08
VAR 4	-0.63	0.11	0.41	-0.93	-0.17	-0.06	0.19	-0.18	0.10	0.38	0.01	0.21	-0.08	0.08	0.42	-0.04	-0.28	-0.12	-0.07
VAR 5	0.83	-0.45	-0.20	0.25	0.12	-0.03	0.04	0.22	-0.24	-0.26	0.10	-0.35	0.38	-0.41	-0.28	0.00	0.41	0.06	0.13
VAR 6	0.77	-0.57	-0.29	0.25	0.24	-0.04	-0.04	0.22	-0.09	-0.13	0.08	-0.30	0.14	-0.37	-0.31	0.11	0.38	-0.04	0.05
VAR 7	0.41	-0.14	0.08	0.32	0.09	0.04	0.08	0.10	-0.08	-0.14	0.01	-0.20	0.32	-0.06	-0.25	-0.16	0.41	0.02	-0.04
VAR 8	0.59	-0.27	-0.34	0.11	0.29	0.21	0.05	0.16	-0.41	-0.07	0.13	-0.20	0.19	-0.54	-0.05	0.30	0.12	-0.04	-0.17
VAR 9	-0.43	0.23	0.41	-0.05	-0.20	-0.06	0.25	-0.14	0.21	-0.08	-0.17	0.02	-0.08	0.88	-0.05	-0.32	0.06	-0.14	-0.06
VAR 10	-0.66	0.38	0.46	-0.10	-0.26	-0.06	0.09	-0.27	0.32	-0.04	-0.24	0.15	-0.17	0.85	0.01	-0.39	-0.10	-0.06	-0.08
VAR 11	0.70	-0.24	-0.17	0.16	0.05	0.09	0.05	0.30	-0.23	-0.16	-0.08	-0.30	0.34	-0.64	-0.16	0.22	0.33	0.12	-0.01
VAR 12	0.11	0.04	0.24	0.18	0.20	0.03	0.19	-0.17	0.04	-0.28	-0.21	-0.25	0.48	0.59	-0.26	-0.40	0.41	0.01	-0.06
VAR 13	-0.40	0.35	0.10	-0.05	-0.72	0.10	-0.08	-0.13	0.02	0.19	-0.10	0.52	-0.12	0.13	0.17	0.03	-0.25	0.03	0.05
VAR 14	-0.04	-0.24	-0.32	0.03	-0.11	-0.09	0.05	0.01	0.08	0.22	0.31	0.14	-0.88	-0.03	0.00	-0.02	-0.12	-0.14	-0.16
VAR 15	-0.07	0.31	-0.07	0.05	0.68	0.24	-0.11	0.08	-0.07	-0.01	-0.22	0.04	0.06	-0.13	0.06	0.06	-0.12	0.17	-0.12
VAR 16	-0.34	0.08	0.11	-0.23	-0.91	-0.14	0.09	-0.10	0.09	0.22	0.13	0.29	-0.20	-0.02	0.19	0.04	-0.21	-0.07	0.01
VAR 17	0.14	-0.18	0.11	-0.50	0.71	0.04	0.15	0.09	-0.13	0.00	-0.01	-0.27	0.22	-0.11	0.10	-0.06	0.10	-0.07	-0.01
VAR 18	-0.92	0.41	0.37	-0.35	-0.21	-0.05	-0.02	-0.41	0.24	0.36	-0.01	0.40	-0.32	0.45	0.26	-0.19	-0.42	-0.02	-0.06
VAR 19	-0.21	-0.24	-0.01	-0.39	-0.10	-0.14	-0.00	-0.18	-0.03	0.87	0.23	0.08	-0.21	-0.23	0.27	0.22	-0.06	-0.24	-0.11
VAR 20	-0.43	0.14	-0.05	-0.16	-0.24	0.08	-0.09	-0.09	0.23	0.27	0.17	0.36	-0.90	0.03	0.14	0.13	-0.38	-0.04	-0.08
VAR 21	-0.32	-0.17	-0.04	-0.32	-0.14	-0.17	-0.04	-0.24	0.15	0.94	0.20	0.36	-0.30	-0.18	0.24	0.18	-0.27	-0.17	-0.01
VAR 22	0.24	-0.40	-0.59	-0.05	-0.02	-0.06	0.00	0.15	-0.27	0.10	0.51	0.07	-0.14	-0.44	0.35	0.31	-0.25	-0.25	-0.24
VAR 23	0.73	-0.54	-0.52	0.33	0.21	0.02	0.00	0.33	-0.21	-0.17	0.21	-0.26	-0.02	-0.31	-0.11	0.21	0.24	-0.20	-0.11
VAR 24	0.03	0.41	0.21	0.16	0.03	0.19	-0.02	0.07	-0.01	-0.22	-0.27	-0.04	0.10	-0.03	-0.14	0.02	0.07	0.99	0.05
VAR 25	-0.83	0.31	0.36	-0.52	-0.21	-0.04	0.01	-0.34	0.30	0.40	-0.04	0.31	-0.21	0.26	0.26	-0.05	-0.27	-0.09	-0.01
VAR 26	-0.89	0.42	0.30	-0.38	-0.35	-0.02	0.02	-0.34	0.31	0.40	0.01	0.40	-0.33	0.30	0.27	-0.08	-0.40	-0.10	-0.03
VAR 27	-0.90	0.42	0.30	-0.37	-0.34	-0.03	-0.08	-0.32	0.35	0.31	-0.03	0.37	-0.38	0.38	0.26	-0.11	-0.41	-0.04	0.03
VAR 28	0.35	-0.17	-0.19	0.47	0.06	-0.07	-0.13	0.15	0.50	-0.02	0.04	0.08	-0.02	-0.16	-0.36	0.06	0.07	-0.01	0.04
VAR 29	-0.02	0.03	0.17	-0.20	-0.02	0.07	0.99	0.03	-0.19	-0.02	0.02	-0.25	0.03	0.19	0.06	-0.18	0.24	-0.02	-0.27
VAR 30	0.40	-0.15	-0.08	0.11	0.32	0.05	0.25	0.10	-0.37	-0.30	0.02	-1.00	0.29	-0.00	-0.18	-0.03	0.34	0.04	-0.11
VAR 31	0.32	0.17	-0.18	0.34	0.11	0.40	0.09	0.20	-0.24	-0.67	-0.01	-0.34	0.08	0.05	-0.17	0.08	0.07	0.08	-0.15
VAR 32	0.47	-0.18	-0.38	0.20	0.11	0.23	0.10	0.05	-0.81	-0.09	0.22	-0.33	0.11	-0.35	-0.05	0.18	0.04	-0.05	-0.06
VAR 33	-0.04	-0.00	-0.07	-0.08	0.06	0.06	0.27	0.12	-0.11	0.05	0.08	-0.11	-0.13	0.06	0.17	0.02	0.03	-0.05	-1.00
VAR 34	0.49	-0.05	-0.30	0.28	0.16	0.34	0.18	0.24	-0.73	-0.20	0.05	-0.37	0.26	-0.28	-0.06	0.18	0.09	0.08	-0.12
VAR 35	-0.19	0.25	0.83	-0.01	-0.01	-0.05	0.08	-0.06	0.27	-0.13	-0.36	0.06	0.16	0.37	-0.27	-0.40	0.12	0.16	0.13
VAR 36	-0.16	0.22	0.88	-0.11	-0.00	-0.08	0.20	-0.15	0.15	-0.04	-0.32	-0.09	0.28	0.32	-0.26	-0.39	0.29	0.19	0.02
VAR 37	-0.35	0.16	0.41	-0.49	-0.00	0.04	0.49	-0.18	0.06	0.16	-0.04	0.09	-0.03	0.20	0.18	-0.16	-0.08	-0.10	-0.29
VAR 38	0.42	-0.10	-0.56	0.38	-0.08	0.19	-0.18	0.33	-0.29	-0.20	0.30	-0.12	-0.07	-0.34	-0.01	0.30	-0.12	0.04	0.10
VAR 39	0.36	-0.48	-0.77	0.13	0.04	-0.09	-0.08	0.28	-0.19	0.02	0.52	-0.02	-0.24	-0.43	0.18	0.35	-0.22	-0.32	-0.17
VAR 40	-0.59	0.39	0.86	-0.55	-0.15	-0.01	0.19	-0.25	0.14	0.21	-0.28	0.17	0.13	0.29	0.14	-0.16	-0.07	0.13	-0.03
VAR 41	-0.63	0.42	0.84	-0.53	-0.14	-0.03	0.17	-0.23	0.22	0.11	-0.32	0.24	0.02	0.37	0.19	-0.24	-0.17	0.12	0.01
VAR 42	-0.50	0.07	0.12	-0.77	-0.16	0.05	0.07	-0.14	-0.15	0.31	0.08	0.15	-0.24	-0.03	0.52	0.12	-0.25	-0.10	-0.13
VAR 43	-0.40	0.03	-0.04	-0.71	-0.12	0.03	0.12	-0.12	-0.18	0.33	0.13	0.07	-0.21	-0.13	0.55	0.20	-0.25	-0.09	-0.16
VAR 44	-0.34	0.00	-0.05	-0.51	-0.06	-0.10	0.05	-0.07	-0.11	0.31	0.18	0.16	-0.11	-0.09	0.96	0.21	-0.20	-0.13	-0.17
VAR 45	-0.26	-0.02	-0.16	-0.41	-0.16	-0.02	0.07	0.03	-0.15	0.25	0.23	0.19	-0.05	-0.13	0.96	0.20	-0.26	-0.13	-0.17
VAR 46	0.28	0.16	-0.26	0.88	0.09	0.18	-0.23	0.09	-0.02	-0.33	-0.14	-0.04	-0.02	0.11	-0.37	-0.05	0.14	0.16	0.06
VAR 47	0.78	-0.31	-0.40	0.51	0.16	0.10	-0.12	0.44	-0.17	-0.37	0.14	-0.26	0.19	-0.32	-0.32	0.21	0.25	0.07	0.13
VAR 48	-0.84	0.55	0.38	-0.33	-0.26	-0.00	0.04	-0.27	0.13	0.21	-0.19	0.32	-0.12	0.40	0.26	-0.13	-0.39	0.06	0.06
VAR 49	0.91	-0.47	-0.38	0.42	0.23	-0.00	-0.04	0.30	-0.22	-0.32	0.10	-0.44	0.26	-0.38	-0.32	0.19	0.43	0.04	0.15
VAR 50	0.66	-0.14	-0.42	0.89	0.16	0.05	-0.19	0.22	-0.06	-0.39	-0.00	-0.20	0.08	-0.07	-0.41	0.03	0.33	0.10	0.07
VAR 51	0.86	-0.49	-0.32	0.52	0.17	-0.07	-0.04	0.20	-0.17	-0.33	0.01	-0.36	0.25	-0.24	-0.32	0.05	0.48	-0.02	0.04
VAR 52	0.29	-0.83	-0.30	-0.15	0.03	-0.42	0.03	-0.05	-0.02	0.31	0.55	-0.06	-0.09	-0.27	0.09	0.20	0.08	-0.33	-0.03
VAR 53	0.91	-0.37	-0.37	0.48	0.31	0.07	-0.03	0.39	-0.23	-0.39	0.01	-0.35	0.28	-0.30	-0.34	0.14	0.35	0.07	0.04
VAR 54	0.21	0.01	-0.19	0.21	0.02	0.07	-0.03	0.86	0.01	-0.26	-0.01	-0.02	-0.09	-0.05	0.05	0.02	0.02	0.03	-0.12
VAR 55	0.09	0.04	0.24	0.19	0.15	-0.14	0.15	0.06	0.05	-0.33	-0.14	-0.14	0.19	0.32	-0.30	-0.88	0.24	-0.03	0.05
VAR 56	0.45	-0.20	-0.17	0.06	0.17	0.13	0.10	0.79	-0.10	-0.18	0.11	-0.16	0.18	-0.33	-0.09	0.18	0.17	0.10	-0.07
VAR 57	0.63	-0.08	-0.42	0.49	0.09	0.39	0.07	0.34	-0.42	-0.31	0.24	-0.34	0.13	-0.29	-0.16	0.33	0.16	0.05	-0.05
VAR 58	0.91	-0.36	-0.41	0.49	0.27	0.11	0.02	0.35	-0.30	-0.32	0.08	-0.38	0.27	-0.40	-0.19	0.21	0.43	0.08	0.02
VAR 59	0.73	-0.13	-0.31	0.60	0.15	0.14	-0.12	0.42	-0.17	-0.43	-0.01	-0.22	0.16	-0.25	-0.35	0.12	0.29	0.14	0.15
VAR 60	-0.09	0.47	0.42	0.09	0.11	0.06	-0.02	-0.06	0.12	-0.22	-1.00	0.02	0.27	0.21	-0.21	-0.12	0.15	0.27	0.08
VAR 61	0.29	-0.92	-0.32	-0.13	-0.04	-0.41	-0.13	0.10	0.03	0.30	0.44	0.01	-0.13	-0.26	0.08	0.07	0.04	-0.45	0.04
VAR 62	0.44	-0.27	0.04	0.13	0.12	-0.05	0.25	0.11	-0.03	-0.14	0.02	-0.38	0.20	0.01	-0.13	-0.11	0.88	-0.00	-0.04
VAR 63	0.51	-0.90	-0.35	0.01	0.14	-0.40	-0.06	0.15	-0.08	0.14	0.37	-0.22	0.04	-0.30	-0.04	0.03	0.27	-0.33	0.04
VAR 64	0.47	-0.89	-0.45	0.05	0.05	-0.25	-0.10	0.14	-0.08	0.17	0.44	-0.05	-0.10	-0.28	-0.04	0.17	0.07	-0.37	0.04
VAR 65	0.47	-0.88	-0.57	0.04	0.10	-0.31	-0.15	0.15	-0.07	0.14	0.48	-0.07	-0.12	-0.32	0.05	0.18	0.01	-0.38	0.02
VAR 66	0.42	-0.43	-0.01	0.13	0.15	-0.25	0.08	0.27	0.09	0.03	-0.11	-0.23	0.24	-0.04	-0.22	-0.07	0.44	-0.14	0.07
VAR 67	-0.22	0.10	-0.00	0.28	-0.31	0.03	-0.18	-0.30	0.14	-0.02	0.09	0.21	-0.22	0.18	-0.15	-0.13	-0.12	0.07	0.12
VAR 68	0.78	-0.67	-0.55	0.40	0.14	-0.22	-0.09	0.32	-0.15	-0.15	0.23	-0.23	0.04	-0.37	-0.16	0.16	0.20	-0.17	0.04
VAR 69	-0.38	0.78	0.32	0.06	-0.02	0.33	-0.13	-0.14	0.02	-0.22	-0.47	0.20	-0.05	0.22	-0.01	-0.16	-0.22	0.28	0.12
VAR 70	-0.74	0.50	0.25	-0.15	-0.19	0.03	-0.02	-0.36	0.15	0.15	0.06	0.28	-0.22	0.46	0.18	-0.16	-0.31	0.10	-0.03
VAR 71	-0.57	0.61	0.22	-0.02	-0.23	0.23	-0.16	-0.20	0.09	0.03	-0.07	0.33	-0.21	0.31	0.10	-0.03	-0.43	0.14	-0.04
VAR 72	0.28	-0.72	-0.33	-0.12	0.04	-0.07	-0.16	0.06	-0.05	0.19	0.31	-0.04	-0.12	-0.23	0.02	0.08	-0.07	-0.39	0.01
VAR 73	-0.14	-0.23	0.11	-0.07	-0.12	-0.94	-0.08	-0.14	0.36	0.15	-0.09	0.11	-0.03	0.07	0.07	-0.17	-0.02	-0.08	0.04
VAR 74	0.61	-0.85	-0.38	0.13	0.18	-0.25	-0.07	0.17	-0.05	-0.01	0.22	-0.20	0.00	-0.25	-0.09	0.09	0.25	-0.26	0.08
VAR 75	0.39	-0.01	0.08	0.28	0.18	-0.03	0.28	0.09	0.04	-0.21	-0.14	-0.34	0.26	0.10	-0.23	-0.12	0.86	0.18	-0.11
VAR 76	-0.36	0.18	-0.14	-0.20	-0.09	-0.01	-0.13	-0.08	0.05	0.18	0.23	0.22	-0.29	-0.05	0.25	0.09	-0.90	-0.01	-0.05
VAR 77	-0.14	0.69	0.12	0.11	0.05	0.81	0.02	0.04	-0.11	-0.26	-0.33	0.06	-0.05	0.10	-0.02	0.12	-0.08	0.33	-0.06
VAR 78	-0.55	0.41	0.37	-0.02	-0.17	-0.27	-0.08	-0.24	0.48	0.05	-0.25	0.20	-0.10	0.39	-0.06	-0.26	-0.14	0.12	0.12
VAR 79	0.08	-0.27	-0.02	0.16	-0.18	-0.07	-0.04	-0.09	0.08	0.06	0.01	0.05	-0.08	-0.05	-0.16	-0.14	0.16	-0.10	0.12
VAR 80	0.53	-0.90	-0.36	0.03	0.13	-0.27	-0.01	0.14	-0.08	0.15	0.40	-0.16	-0.02	-0.29	-0.01	0.07	0.21	-0.30	0.02
VAR 81	0.47	-0.92	-0.34	-0.01	0.05	-0.45	-0.00	0.05	0.01	0.23	0.45	-0.14	-0.03	-0.25	0.01	0.02	0.22	-0.37	-0.01
VAR 82	-0.34	0.39	0.39	-0.29	0.06	-0.09	0.05	-0.13	0.11	0.07	-0.36	0.13	0.03	0.14	0.13	-0.18	-0.14	0.19	-0.00
VAR 83	-0.42	0.83	0.33	0.04	-0.10	0.24	0.00	-0.06	0.19	-0.13	-0.41	0.15	-0.01	0.24	-0.02	-0.09	-0.11	0.33	0.01
VAR 84	0.84	-0.49	-0.38	0.32	0.33	-0.06	0.01	0.32	-0.17	-0.28	0.09	-0.32	0.19	-0.35	-0.23	0.16	0.46	0.00	-0.02
VAR 85	-0.16	0.66	-0.04	0.19	-0.04	0.57	-0.13	0.12	-0.10	-0.23	-0.24	0.16	-0.12	0.06	-0.00	0.11	-0.31	0.31	0.04
VAR 86	-0.39	0.81	0.15	0.06	-0.18	0.38	-0.10	-0.04	-0.04	-0.13	-0.25	0.23	-0.02	0.11	0.08	0.02	-0.31	0.23	0.12
VAR 87	0.17	0.54	0.39	0.28	0.12	0.39	0.09	0.04	-0.16	-0.38	-0.58	-0.13	0.49	0.09	-0.33	-0.04	0.32	0.47	0.13
VAR 88	0.50	-0.11	-0.41	0.21	0.15	0.13	-0.14	0.32	-0.20	-0.11	0.03	-0.13	0.14	-0.49	-0.01	0.72	0.12	0.00	0.03
VAR 89	0.24	0.16	0.26	0.47	0.19	0.08	0.11	0.14	0.05	-0.57	-0.18	-0.30	0.25	0.50	-0.37	-0.35	0.38	0.24	0.16
VAR 90	-0.04	0.22	0.28	0.34	0.17	0.13	0.17	0.05	0.06	-0.49	-0.24	-0.20	0.11	0.77	-0.27	-0.46	0.16	0.19	-0.03
VAR 91	-0.45	-0.18	0.29	-0.88	-0.15	-0.28	0.16	-0.27	0.10	0.63	0.16	0.17	-0.11	-0.09	0.36	0.02	-0.15	-0.22	-0.01

TABLE 42
Correlations of Table 39 Factor Scores with Estimate Deck

	1	2	3	4	5	6	7	8	9	10	11	12	13	14	15	16	17	18	19
VAR 1	-0.46	-0.18	0.12	-0.48	-0.21	-0.30	-0.01	-0.34	0.23	0.95	0.19	0.39	-0.25	-0.12	0.30	0.06	-0.23	-0.20	-0.02
VAR 2	-0.27	0.09	0.10	-0.16	-0.90	-0.06	0.04	-0.06	0.07	0.21	0.12	0.26	-0.19	-0.10	0.20	0.04	-0.13	0.02	0.05
VAR 3	0.93	-0.36	-0.41	0.49	0.23	0.10	-0.08	0.42	-0.29	-0.43	0.02	-0.34	0.27	-0.40	-0.29	0.19	0.40	0.08	0.08
VAR 4	-0.65	0.11	0.43	-0.92	-0.14	-0.06	0.19	-0.18	0.12	0.39	0.01	0.21	-0.08	0.08	0.42	-0.10	-0.28	-0.12	-0.07
VAR 5	0.82	-0.46	-0.22	0.26	0.13	-0.09	0.04	0.23	-0.25	-0.26	0.10	-0.35	0.37	-0.41	-0.29	0.06	0.41	0.06	0.13
VAR 6	0.77	-0.58	-0.32	0.26	0.25	-0.09	-0.04	0.24	-0.11	-0.13	0.08	-0.30	0.13	-0.38	-0.31	0.14	0.38	-0.04	0.05
VAR 7	0.41	-0.14	0.06	0.33	0.08	0.01	0.08	0.11	-0.09	-0.14	0.01	-0.20	0.32	-0.07	-0.25	-0.11	0.42	0.02	-0.04
VAR 8	0.59	-0.27	-0.35	0.10	0.30	0.17	0.05	0.19	-0.43	-0.07	0.13	-0.20	0.19	-0.54	-0.05	0.34	0.12	-0.04	-0.17
VAR 9	-0.43	0.23	0.41	-0.04	-0.22	-0.03	0.25	-0.15	0.22	-0.08	-0.17	0.02	-0.08	0.87	-0.05	-0.35	0.06	-0.14	-0.06
VAR 10	-0.66	0.39	0.47	-0.10	-0.27	-0.01	0.09	-0.29	0.34	-0.04	-0.24	0.15	-0.16	0.86	0.01	-0.43	-0.11	-0.06	-0.08
VAR 11	0.70	-0.24	-0.19	0.17	0.07	0.07	0.05	0.31	-0.25	-0.16	-0.08	-0.30	0.33	-0.65	-0.16	0.27	0.34	0.12	-0.01
VAR 12	0.10	0.04	0.23	0.19	0.18	0.02	0.19	-0.17	0.03	-0.28	-0.21	-0.25	0.47	0.59	-0.26	-0.38	0.41	0.01	-0.06
VAR 13	-0.40	0.36	0.12	-0.05	-0.75	0.13	-0.08	-0.14	0.03	0.19	-0.10	0.52	-0.11	0.13	0.17	-0.01	-0.26	0.03	0.05
VAR 14	-0.03	-0.24	-0.33	0.02	-0.11	-0.10	0.05	0.00	0.09	0.22	0.31	0.14	-0.89	-0.04	0.00	-0.04	-0.11	-0.14	-0.16
VAR 15	-0.07	0.31	-0.06	0.03	0.67	0.28	-0.11	0.07	-0.08	-0.01	-0.22	0.04	0.06	-0.12	0.06	0.08	-0.12	0.17	-0.12
VAR 16	-0.34	0.08	0.12	-0.22	-0.89	-0.12	0.09	-0.11	0.10	0.22	0.13	0.29	-0.19	-0.02	0.19	0.00	-0.21	-0.07	0.01
VAR 17	0.13	-0.19	0.11	-0.50	0.73	0.02	0.15	0.10	-0.12	0.01	-0.01	-0.27	0.22	-0.11	0.10	-0.04	0.10	-0.07	-0.01
VAR 18	-0.92	0.42	0.39	-0.35	-0.22	-0.02	-0.02	-0.42	0.26	0.37	-0.01	0.40	-0.31	0.46	0.26	-0.26	-0.43	-0.02	-0.06
VAR 19	-0.20	-0.23	-0.00	-0.39	-0.08	-0.16	-0.00	-0.17	-0.02	0.87	0.23	0.08	-0.21	-0.24	0.27	0.18	-0.06	-0.24	-0.11
VAR 20	-0.43	0.14	-0.04	-0.17	-0.24	0.11	-0.09	-0.10	0.24	0.26	0.17	0.36	-0.89	0.03	0.14	0.08	-0.38	-0.04	-0.08
VAR 21	-0.32	-0.17	-0.03	-0.32	-0.14	-0.19	-0.04	-0.24	0.16	0.94	0.20	0.36	-0.29	-0.18	0.24	0.14	-0.27	-0.17	-0.01
VAR 22	0.25	-0.40	-0.60	-0.06	-0.01	-0.09	0.00	0.15	-0.26	0.10	0.51	0.07	-0.14	-0.43	0.35	0.31	-0.24	-0.25	-0.24
VAR 23	0.73	-0.54	-0.54	0.33	0.21	-0.02	0.00	0.34	-0.23	-0.18	0.21	-0.26	-0.03	-0.32	-0.11	0.26	0.24	-0.20	-0.11
VAR 24	0.03	0.41	0.22	0.16	0.03	0.23	-0.02	0.08	-0.02	-0.22	-0.27	-0.04	0.10	-0.03	-0.14	0.02	0.07	0.99	0.05
VAR 25	-0.82	0.32	0.38	-0.51	-0.20	-0.02	0.01	-0.34	0.32	0.41	-0.04	0.31	-0.20	0.27	0.26	-0.11	-0.27	-0.09	-0.01
VAR 26	-0.88	0.43	0.32	-0.38	-0.34	0.02	0.02	-0.35	0.33	0.40	0.01	0.40	-0.32	0.31	0.27	-0.15	-0.40	-0.10	-0.03
VAR 27	-0.89	0.43	0.32	-0.37	-0.34	0.01	-0.08	-0.34	0.37	0.31	-0.03	0.37	-0.37	0.38	0.26	-0.18	-0.41	-0.04	0.03
VAR 28	0.36	-0.18	-0.21	0.49	0.05	-0.07	-0.13	0.16	0.47	-0.03	0.04	0.08	-0.02	-0.16	-0.36	0.09	0.08	-0.01	0.04
VAR 29	-0.03	0.03	0.17	-0.20	-0.00	0.06	0.99	0.04	-0.20	-0.02	0.02	-0.25	0.02	0.18	0.06	-0.18	0.25	-0.02	-0.27
VAR 30	0.39	-0.16	-0.10	0.10	0.33	0.03	0.25	0.11	-0.38	-0.30	0.02	-1.00	0.28	-0.01	-0.18	-0.00	0.36	0.04	-0.11
VAR 31	0.32	0.17	-0.19	0.34	0.11	0.40	0.09	0.20	-0.26	-0.67	-0.01	-0.34	0.08	0.04	-0.17	0.11	0.08	0.08	-0.15
VAR 32	0.47	-0.18	-0.39	0.20	0.11	0.18	0.10	0.06	-0.79	-0.10	0.22	-0.33	0.11	-0.36	-0.05	0.21	0.03	-0.05	-0.06
VAR 33	-0.04	-0.00	-0.07	-0.09	0.06	0.06	0.27	0.11	-0.11	0.05	0.08	-0.11	-0.13	0.06	0.17	0.01	0.04	-0.05	-1.00
VAR 34	0.50	-0.05	-0.31	0.27	0.15	0.31	0.18	0.25	-0.77	-0.21	0.05	-0.37	0.25	-0.29	-0.06	0.22	0.09	0.08	-0.12
VAR 35	-0.20	0.25	0.81	-0.00	-0.01	-0.02	0.08	-0.06	0.27	-0.13	-0.36	0.06	0.16	0.37	-0.27	-0.41	0.11	0.16	0.13
VAR 36	-0.16	0.22	0.86	-0.09	0.00	-0.06	0.20	-0.14	0.15	-0.04	-0.32	-0.09	0.28	0.32	-0.26	-0.39	0.28	0.19	0.02
VAR 37	-0.35	0.16	0.41	-0.48	0.01	0.03	0.49	-0.17	0.07	0.17	-0.04	0.09	-0.03	0.20	0.18	-0.19	-0.09	-0.10	-0.29
VAR 38	0.42	-0.10	-0.57	0.37	-0.09	0.18	-0.18	0.32	-0.30	-0.21	0.30	-0.12	-0.07	-0.35	-0.01	0.32	-0.10	0.04	0.10
VAR 39	0.37	-0.49	-0.79	0.12	0.04	-0.12	-0.08	0.28	-0.19	0.02	0.52	-0.02	-0.25	-0.44	0.18	0.37	-0.21	-0.32	-0.17
VAR 40	-0.60	0.40	0.87	-0.54	-0.13	0.02	0.19	-0.25	0.15	0.22	-0.28	0.17	0.14	0.29	0.14	-0.20	-0.08	0.13	-0.03
VAR 41	-0.64	0.42	0.85	-0.52	-0.13	0.01	0.17	-0.23	0.23	0.11	-0.32	0.24	0.03	0.37	0.19	-0.29	-0.17	0.12	0.01
VAR 42	-0.51	0.07	0.14	-0.80	-0.14	0.06	0.07	-0.14	-0.13	0.31	0.08	0.15	-0.23	-0.02	0.52	0.07	-0.26	-0.10	-0.13
VAR 43	-0.41	0.03	-0.02	-0.74	-0.10	0.03	0.12	-0.12	-0.16	0.33	0.13	0.07	-0.21	-0.12	0.55	0.15	-0.25	-0.09	-0.16
VAR 44	-0.34	0.01	-0.03	-0.53	-0.05	-0.09	0.05	-0.08	-0.10	0.31	0.18	0.16	-0.11	-0.09	0.96	0.18	-0.19	-0.13	-0.17
VAR 45	-0.26	-0.02	-0.14	-0.42	-0.15	-0.02	0.07	0.03	-0.14	0.25	0.23	0.19	-0.05	-0.12	0.96	0.16	-0.25	-0.13	-0.17
VAR 46	0.30	0.16	-0.27	0.86	0.05	0.20	-0.23	0.08	-0.03	-0.33	-0.14	-0.04	-0.02	0.12	-0.37	-0.01	0.13	0.16	0.06
VAR 47	0.78	-0.32	-0.42	0.51	0.16	0.05	-0.12	0.44	-0.18	-0.38	0.14	-0.26	0.19	-0.33	-0.32	0.27	0.24	0.07	0.13
VAR 48	-0.84	0.55	0.41	-0.34	-0.27	0.05	0.04	-0.29	0.14	0.21	-0.19	0.32	-0.11	0.40	0.26	-0.17	-0.40	0.06	0.06
VAR 49	0.91	-0.48	-0.40	0.42	0.23	-0.05	-0.04	0.31	-0.23	-0.32	0.10	-0.44	0.25	-0.39	-0.32	0.24	0.44	0.04	0.15
VAR 50	0.68	-0.15	-0.44	0.88	0.13	0.05	-0.19	0.22	-0.07	-0.39	-0.00	-0.20	0.07	-0.07	-0.41	0.08	0.33	0.10	0.07
VAR 51	0.86	-0.49	-0.35	0.52	0.17	-0.11	-0.04	0.21	-0.18	-0.33	0.01	-0.36	0.24	-0.25	-0.32	0.10	0.48	-0.02	0.04
VAR 52	0.29	-0.83	-0.31	-0.14	0.04	-0.49	0.03	-0.04	-0.01	0.31	0.55	-0.06	-0.10	-0.27	0.09	0.20	0.08	-0.33	-0.03
VAR 53	0.91	-0.37	-0.39	0.48	0.30	0.03	-0.03	0.40	-0.25	-0.39	0.01	-0.35	0.27	-0.31	-0.34	0.22	0.36	0.07	0.04
VAR 54	0.21	0.01	-0.19	0.20	0.02	0.08	-0.03	0.82	-0.01	-0.27	-0.01	-0.02	-0.09	-0.06	0.05	0.06	0.02	0.03	-0.12
VAR 55	0.08	0.04	0.23	0.20	0.15	-0.14	0.15	0.06	0.05	-0.33	-0.14	-0.14	0.18	0.32	-0.30	-0.81	0.24	-0.03	0.05
VAR 56	0.45	-0.21	-0.18	0.06	0.18	0.09	0.10	0.82	-0.12	-0.18	0.11	-0.16	0.18	-0.34	-0.10	0.21	0.17	0.10	-0.07
VAR 57	0.65	-0.09	-0.44	0.49	0.08	0.36	0.07	0.35	-0.45	-0.32	0.24	-0.34	0.12	-0.30	-0.16	0.38	0.17	0.05	-0.05
VAR 58	0.91	-0.37	-0.43	0.49	0.27	0.08	0.02	0.36	-0.32	-0.33	0.08	-0.38	0.26	-0.40	-0.19	0.29	0.44	0.08	0.02
VAR 59	0.73	-0.13	-0.33	0.61	0.13	0.13	-0.12	0.42	-0.18	-0.43	-0.01	-0.22	0.15	-0.25	-0.35	0.19	0.29	0.14	0.15
VAR 60	-0.09	0.47	0.43	0.09	0.09	0.13	-0.02	-0.06	0.11	-0.22	-1.00	0.02	0.27	0.21	-0.21	-0.10	0.13	0.27	0.08
VAR 61	0.28	-0.91	-0.33	-0.13	-0.02	-0.49	-0.13	0.11	0.04	0.31	0.44	0.01	-0.13	-0.27	0.08	0.07	0.03	-0.45	0.04
VAR 62	0.45	-0.28	0.03	0.13	0.13	-0.07	0.25	0.12	-0.04	-0.14	0.02	-0.38	0.19	-0.00	-0.13	-0.07	0.91	-0.00	-0.04
VAR 63	0.50	-0.90	-0.36	0.02	0.15	-0.47	-0.06	0.17	-0.08	0.14	0.37	-0.22	0.03	-0.30	-0.04	0.06	0.27	-0.33	0.04
VAR 64	0.46	-0.88	-0.47	0.06	0.05	-0.33	-0.10	0.15	-0.08	0.17	0.44	-0.05	-0.11	-0.28	-0.04	0.18	0.06	-0.37	0.04
VAR 65	0.47	-0.88	-0.58	0.05	0.10	-0.38	-0.15	0.16	-0.07	0.14	0.48	-0.07	-0.13	-0.32	0.05	0.19	0.01	-0.38	0.02
VAR 66	0.42	-0.43	-0.03	0.14	0.15	-0.27	0.08	0.28	0.07	0.03	-0.11	-0.23	0.23	-0.05	-0.22	-0.04	0.44	-0.14	0.07
VAR 67	-0.22	0.10	-0.00	0.27	-0.33	0.05	-0.18	-0.32	0.15	-0.02	0.09	0.21	-0.22	0.19	-0.15	-0.14	-0.13	0.07	0.12
VAR 68	0.78	-0.68	-0.57	0.41	0.13	-0.26	-0.09	0.33	-0.17	-0.16	0.23	-0.23	0.03	-0.38	-0.16	0.21	0.20	-0.17	0.04
VAR 69	-0.39	0.79	0.34	0.05	-0.04	0.39	-0.13	-0.15	0.02	-0.22	-0.47	0.20	-0.04	0.23	-0.01	-0.17	-0.22	0.28	0.12
VAR 70	-0.74	0.51	0.27	-0.16	-0.20	0.07	-0.02	-0.37	0.16	0.14	0.06	0.28	-0.21	0.48	0.18	-0.20	-0.31	0.10	-0.03
VAR 71	-0.57	0.63	0.24	-0.03	-0.25	0.27	-0.16	-0.21	0.10	0.03	-0.07	0.33	-0.20	0.32	0.10	-0.07	-0.44	0.14	-0.04
VAR 72	0.28	-0.71	-0.34	-0.12	0.04	-0.15	-0.16	0.07	-0.05	0.19	0.31	-0.04	-0.12	-0.23	0.02	0.07	-0.08	-0.39	0.01
VAR 73	-0.15	-0.24	0.10	-0.06	-0.12	-0.89	-0.08	-0.14	0.37	0.16	-0.09	0.11	-0.03	0.07	0.07	-0.17	-0.01	-0.08	0.04
VAR 74	0.61	-0.85	-0.40	0.13	0.18	-0.31	-0.07	0.17	-0.05	-0.01	0.22	-0.20	-0.01	-0.26	-0.09	0.11	0.25	-0.26	0.08
VAR 75	0.39	-0.02	0.07	0.28	0.17	-0.02	0.28	0.10	0.03	-0.21	-0.14	-0.34	0.25	0.10	-0.23	-0.07	0.88	0.18	-0.11
VAR 76	-0.37	0.18	-0.13	-0.20	-0.09	0.01	-0.13	-0.08	0.05	0.18	0.23	0.22	-0.29	-0.04	0.25	0.06	-0.86	-0.01	-0.05
VAR 77	-0.13	0.70	0.13	0.10	0.04	0.89	0.02	0.04	-0.12	-0.26	-0.33	0.06	-0.04	0.10	-0.02	0.12	-0.08	0.33	-0.06
VAR 78	-0.55	0.41	0.38	-0.01	-0.18	-0.21	-0.08	-0.26	0.48	0.05	-0.25	0.20	-0.09	0.40	-0.06	-0.28	-0.13	0.12	0.12
VAR 79	0.08	-0.26	-0.03	0.16	-0.19	-0.09	-0.04	-0.09	0.08	0.06	0.01	0.05	-0.08	-0.04	-0.16	-0.15	0.14	-0.10	0.12
VAR 80	0.53	-0.90	-0.38	0.04	0.14	-0.34	-0.01	0.15	-0.07	0.15	0.40	-0.16	-0.03	-0.29	-0.01	0.08	0.21	-0.30	0.02
VAR 81	0.46	-0.92	-0.36	-0.00	0.06	-0.51	-0.00	0.05	0.01	0.23	0.45	-0.14	-0.04	-0.25	0.01	0.03	0.22	-0.37	-0.01
VAR 82	-0.35	0.39	0.41	-0.29	0.06	-0.05	0.05	-0.13	0.11	0.08	-0.36	0.13	0.03	0.14	0.13	-0.19	-0.14	0.19	-0.00
VAR 83	-0.41	0.82	0.34	0.04	-0.11	0.32	0.00	-0.07	0.18	-0.13	-0.41	0.15	-0.00	0.24	-0.02	-0.10	-0.10	0.33	0.01
VAR 84	0.83	-0.49	-0.40	0.32	0.33	-0.10	0.01	0.33	-0.19	-0.28	0.09	-0.32	0.18	-0.35	-0.23	0.22	0.47	0.00	-0.02
VAR 85	-0.15	0.66	-0.03	0.18	-0.06	0.61	-0.13	0.11	-0.11	-0.24	-0.24	0.16	-0.11	0.06	-0.00	0.11	-0.32	0.31	0.04
VAR 86	-0.38	0.81	0.17	0.06	-0.20	0.43	-0.10	-0.05	-0.05	-0.14	-0.25	0.23	-0.00	0.11	0.08	0.01	-0.31	0.23	0.12
VAR 87	0.17	0.54	0.40	0.29	0.11	0.43	0.09	0.05	-0.18	-0.38	-0.58	-0.13	0.49	0.09	-0.33	0.00	0.31	0.47	0.13
VAR 88	0.50	-0.12	-0.41	0.21	0.15	0.12	-0.14	0.33	-0.22	-0.11	0.03	-0.13	0.13	-0.50	-0.01	0.81	0.12	0.00	0.03
VAR 89	0.23	0.15	0.25	0.48	0.17	0.10	0.11	0.13	0.04	-0.57	-0.18	-0.30	0.25	0.50	-0.37	-0.30	0.39	0.24	0.16
VAR 90	-0.04	0.22	0.28	0.33	0.15	0.16	0.17	0.04	0.05	-0.49	-0.24	-0.20	0.11	0.77	-0.27	-0.46	0.16	0.19	-0.03
VAR 91	-0.46	-0.18	0.31	-0.87	-0.12	-0.30	0.16	-0.27	0.12	0.64	0.16	0.17	-0.10	-0.09	0.36	-0.03	-0.15	-0.22	-0.01

TABLE 43
Intercorrelations of Table 37 Factor Scores

X	Y	R	X	Y	R	X	Y	R	X	Y	R
1	2	-0.00	3	4	0.04	6	7	0.06	10	11	0.02
1	3	-0.03	3	5	0.07	6	8	-0.09	10	12	0.08
1	4	0.05	3	6	0.07	6	9	-0.08	10	13	0.03
1	5	0.01	3	7	0.06	6	10	-0.08	10	14	0.03
1	6	0.01	3	8	0.08	6	11	-0.02	10	15	-0.02
1	7	-0.07	3	9	0.00	6	12	0.02	10	16	-0.04
1	8	-0.02	3	10	0.05	6	13	0.00	10	17	0.04
1	9	-0.08	3	11	0.04	6	14	-0.03	10	18	0.03
1	10	-0.09	3	12	0.06	6	15	0.00	10	19	0.06
1	11	-0.02	3	13	-0.01	6	16	0.04	11	12	0.06
1	12	-0.06	3	14	0.08	6	17	-0.01	11	13	0.02
1	13	0.02	3	15	-0.10	6	18	-0.00	11	14	0.06
1	14	-0.09	3	16	-0.02	6	19	-0.01	11	15	0.02
1	15	0.03	3	17	-0.00	7	8	0.02	11	16	0.03
1	16	0.01	3	18	0.02	7	9	-0.01	11	17	0.04
1	17	-0.00	3	19	0.05	7	10	0.00	11	18	0.02
1	18	-0.12	4	5	-0.00	7	11	0.01	11	19	-0.00
1	19	-0.02	4	6	0.02	7	12	0.03	12	13	0.01
2	3	0.02	4	7	-0.01	7	13	0.03	12	14	-0.01
2	4	0.02	4	8	-0.06	7	14	-0.00	12	15	0.03
2	5	-0.00	4	9	-0.09	7	15	-0.04	12	16	-0.02
2	6	0.02	4	10	-0.10	7	16	0.02	12	17	0.06
2	7	0.01	4	11	-0.00	7	17	0.00	12	18	0.03
2	8	-0.02	4	12	-0.12	7	18	-0.02	12	19	-0.06
2	9	-0.08	4	13	-0.02	7	19	-0.01	13	14	-0.03
2	10	-0.05	4	14	-0.02	8	9	0.06	13	15	0.06
2	11	-0.01	4	15	-0.04	8	10	-0.02	13	16	-0.02
2	12	-0.02	4	16	0.03	8	11	0.01	13	17	-0.00
2	13	-0.08	4	17	-0.04	8	12	-0.04	13	18	-0.04
2	14	0.04	4	18	-0.14	8	13	-0.03	13	19	0.04
2	15	-0.01	4	19	-0.07	8	14	-0.04	14	15	0.10
2	16	0.03	5	6	-0.00	8	15	-0.02	14	16	0.04
2	17	-0.03	5	7	0.04	8	16	-0.01	14	17	0.00
2	18	-0.04	5	8	-0.03	8	17	0.04	14	18	-0.00
2	19	-0.07	5	9	-0.09	8	18	-0.02	14	19	-0.01
			5	10	-0.02	8	19	-0.04	15	16	-0.01
			5	11	-0.00	9	10	0.13	15	17	0.07
			5	12	-0.07	9	11	-0.01	15	18	-0.02
			5	13	0.02	9	12	0.05	15	19	0.00
			5	14	-0.06	9	13	0.09	16	17	-0.01
			5	15	-0.05	9	14	0.11	16	18	0.00
			5	16	0.01	9	15	-0.06	16	19	-0.04
			5	17	-0.02	9	16	-0.00	17	18	0.14
			5	18	-0.01	9	17	0.12	17	19	0.08
			5	19	-0.02	9	18	0.04	18	19	0.02
						9	19	0.07			

TABLE 44
Intercorrelations of Table 38 Factor Scores

X	Y	R	X	Y	R	X	Y	R	X	Y	R
1	2	-0.48	3	4	-0.29	6	7	0.06	10	11	0.22
1	3	-0.43	3	5	-0.07	6	8	0.11	10	12	0.30
1	4	0.49	3	6	-0.02	6	9	-0.29	10	13	-0.27
1	5	0.27	3	7	0.17	6	10	-0.21	10	14	-0.19
1	6	0.04	3	8	-0.21	6	11	-0.06	10	15	0.28
1	7	-0.02	3	9	0.23	6	12	-0.05	10	16	0.18
1	8	0.38	3	10	0.01	6	13	-0.00	10	17	-0.20
1	9	-0.27	3	11	-0.42	6	14	-0.00	10	18	-0.21
1	10	-0.35	3	12	0.08	6	15	-0.05	10	19	-0.04
1	11	0.08	3	13	0.20	6	16	0.16	11	12	-0.02
1	12	-0.39	3	14	0.41	6	17	-0.01	11	13	-0.26
1	13	0.27	3	15	-0.10	6	18	0.19	11	14	-0.20
1	14	-0.45	3	16	-0.37	6	19	-0.05	11	15	0.20
1	15	-0.31	3	17	0.10	7	8	0.03	11	16	0.11
1	16	0.17	3	18	0.21	7	9	-0.19	11	17	-0.15
1	17	0.03	3	19	0.06	7	10	-0.02	11	18	-0.27
1	18	0.03	4	5	0.03	7	11	0.02	11	19	-0.08
1	19	0.03	4	6	0.09	7	12	-0.24	12	13	-0.28
2	3	0.40	4	7	-0.20	7	13	0.02	12	14	0.00
2	4	0.03	4	8	0.16	7	14	0.18	12	15	0.18
2	5	-0.08	4	9	0.01	7	15	0.06	12	16	0.03
2	6	0.44	4	10	-0.42	7	16	-0.17	12	17	-0.34
2	7	0.03	4	11	-0.09	7	17	0.23	12	18	-0.04
2	8	-0.10	4	12	-0.10	7	18	-0.02	12	19	0.11
2	9	0.03	4	13	0.07	7	19	-0.27	13	14	0.00
2	10	-0.21	4	14	0.05	8	9	-0.04	13	15	-0.08
2	11	-0.47	4	15	-0.47	8	10	-0.27	13	16	-0.06
2	12	0.15	4	16	-0.03	8	11	0.05	13	17	0.29
2	13	0.04	4	17	0.23	8	12	-0.10	13	18	0.09
2	14	0.29	4	18	0.15	8	13	0.04	13	19	0.13
2	15	-0.01	4	19	0.08	8	14	-0.21	14	15	-0.11
2	16	-0.08	5	6	0.10	8	15	-0.01	14	16	-0.47
2	17	-0.17	5	7	-0.01	8	16	0.11	14	17	0.05
2	18	0.40	5	8	0.10	8	17	0.10	14	18	-0.03
2	19	0.00	5	9	-0.09	8	18	0.07	14	19	-0.05
			5	10	-0.16	8	19	-0.11	15	16	0.21
			5	11	-0.10	9	10	0.12	15	17	-0.23
			5	12	-0.31	9	11	-0.12	15	18	-0.13
			5	13	0.20	9	12	0.37	15	19	-0.17
			5	14	-0.04	9	13	-0.17	16	17	-0.11
			5	15	-0.11	9	14	0.22	16	18	0.02
			5	16	-0.03	9	15	-0.13	16	19	-0.01
			5	17	0.14	9	16	-0.13	17	18	0.06
			5	18	0.03	9	17	-0.01	17	19	-0.02
			5	19	-0.05	9	18	-0.00	18	19	0.04
						9	19	0.10			

TABLE 45
Intercorrelations of Table 39 Factor Scores

X	Y	R		X	Y	R
1,	2	-0.48		7,	8	0.04
1,	3	-0.47		7,	9	-0.19
1,	4	0.51		7,	10	-0.01
1,	5	0.26		7,	11	0.02
1,	6	0.01		7,	12	-0.24
1,	7	-0.02		7,	13	0.02
1,	8	0.40		7,	14	0.17
1,	9	-0.30		7,	15	0.06
1,	10	-0.35		7,	16	-0.17
1,	11	0.09		7,	17	0.25
1,	12	-0.39		7,	18	-0.02
1,	13	0.25		7,	19	-0.27
1,	14	-0.46		8,	9	-0.07
1,	15	-0.31		8,	10	-0.27
1,	16	0.25		8,	11	0.06
1,	17	0.45		8,	12	-0.10
1,	18	0.03		8,	13	0.05
1,	19	0.03		8,	14	-0.23
2,	3	0.42		8,	15	-0.02
2,	4	0.02		8,	16	0.16
2,	5	-0.10		8,	17	0.11
2,	6	0.52		8,	18	0.07
2,	7	0.02		8,	19	-0.11
2,	8	-0.12		9,	10	0.13
2,	9	0.02		9,	11	-0.11
2,	10	-0.20		9,	12	0.38
2,	11	-0.46		9,	13	-0.18
2,	12	0.16		9,	14	0.23
2,	13	0.05		9,	15	-0.12
2,	14	0.29		9,	16	-0.16
2,	15	-0.00		9,	17	-0.02
2,	16	-0.09		9,	18	-0.01
2,	17	-0.17		9,	19	0.11
2,	18	0.40		10,	11	0.22
2,	19	0.00		10,	12	0.30
3,	4	-0.30		10,	13	-0.27
3,	5	-0.07		10,	14	-0.19
3,	6	0.01		10,	15	0.29
3,	7	0.17		10,	16	0.13
3,	8	-0.22		10,	17	-0.20
3,	9	0.23		10,	18	-0.22
3,	10	0.03		10,	19	-0.05
3,	11	-0.43		11,	12	-0.02
3,	12	0.09		11,	13	-0.27
3,	13	0.20		11,	14	-0.21
3,	14	0.42		11,	15	0.20
3,	15	-0.08		11,	16	0.10
3,	16	-0.39		11,	17	-0.12
3,	17	0.08		11,	18	-0.27
3,	18	0.21		11,	19	-0.08
3,	19	0.07		12,	13	-0.27
4,	5	-0.01		12,	14	0.00
4,	6	0.08		12,	15	0.17
4,	7	-0.19		12,	16	0.00
4,	8	0.16		12,	17	-0.35
4,	9	0.01		12,	18	-0.04
4,	10	-0.43		12,	19	0.11
4,	11	-0.09		13,	14	0.00
4,	12	-0.10		13,	15	-0.07
4,	13	0.08		13,	16	-0.02
4,	14	0.05		13,	17	0.27
4,	15	-0.49		13,	18	0.09
4,	16	0.00		13,	19	0.13
4,	17	0.23		14,	15	-0.10
4,	18	0.15		14,	16	-0.50
4,	19	0.09		14,	17	0.05
5,	6	0.08		14,	18	-0.03
5,	7	-0.00		14,	19	-0.05
5,	8	0.12		15,	16	0.17
5,	9	-0.10		15,	17	-0.22
5,	10	-0.15		15,	18	-0.13
5,	11	-0.09		15,	19	-0.17
5,	12	-0.33		16,	17	-0.07
5,	13	0.19		16,	18	0.01
5,	14	-0.06		16,	19	-0.00
5,	15	-0.09		17,	18	0.07
5,	16	0.00		17,	19	-0.03
5,	17	0.14		18,	19	0.04
5,	18	0.03				
5,	19	-0.06				
6,	7	0.06				
6,	8	0.10				
6,	9	-0.27				
6,	10	-0.23				
6,	11	-0.13				
6,	12	-0.02				
6,	13	-0.00				
6,	14	0.01				
6,	15	-0.05				
6,	16	0.16				
6,	17	-0.03				
6,	18	0.23				
6,	19	-0.05				

Appendixes

APPENDIX A
Sources of Variables

<pre>
01   MANY INHABITANTS PER PHYSICIAN (1)
02   BIG GROSS NATIONAL PRODUCT (1)
03   LARGE PERCENTAGE INCREASE IN POPULATION (1)
04   LARGE PERCENTAGE CHRISTIANS OF TOTAL POPULATION(1)
05   BIG POPULATION PER SQUARE KILOMETER (1)
06   BIG AREA (1)
07   LARGE PERCENTAGE EXPORT OF TOTAL TO US (2)
08   LARGE PERCENTAGE IMPORT OF TOTAL TO US (2)
09   LARGE NUMBER OF PERSONS IN UN MISSIONS (3)
10   RECENTLY ENTERED UN (4)
11   ALLIED WITH USSR (5)
12   LARGE DOLLAR IMPORT FROM US (6)
13   LARGE DOLLAR EXPORT TO US (6)
14   SMALL POPULATION (7)
15   SMALL PER CAPITA GROSS NATIONAL PRODUCT (7)
16   NON-WESTERN (7)
17   POLITICALLY NOT MODERN (7)
18   LARGE DISTANCE FROM US (8)
19   LARGE DISTANCE FROM USSR (8)
20   LARGE DISTANCE FROM CHINA (8)
21   MANY MEN UNDER ARMS (9)
</pre>

The following resources were used following the above code:

1. Bruce M. Russett, et al. *World Handbook of Political and Social Indicators* (New Haven: Yale University Press, 1964).
2. Dan Golenpaul, ed., *Information Please Almanac, 1968* (New York: Simon and Schuster, 1967).
3. Permanent Missions to the United Nations, United Nations Document, March, 1968.
4. Golenpaul, *op. cit.*
5. S. H. Steinberg, ed., *The Statesman's Yearbook, 1966–67* (New York: St. Martin's Press, 1966).
   States coded 1 = allied with U.S.   2 = neutral   3 = allied with U.S.S.R.
6. *Direction of Trade Annual 1962–66* (Washington: International Monetary Fund and International Bank for Reconstruction and Development, 1967).
7. Arthur S. Banks and Robert Textor, *A Cross Polity Survey* (Cambridge: M.I.T. Press, 1963).
8. Great Circle Distance on Standard Globe Used. States with common border originally coded 0.
9. *Reader's Digest Almanac and Yearbook 1968* (New York: Reader's Digest Association, 1957).

APPENDIX B
Final Scale Value of States on 21 Variables

01 ALBANIA
```
 0.61  -1.49   1.27  -0.35   0.44  -1.58  -1.62  -1.62  -1.45
 1.50   1.66  -1.62  -1.62   1.24   0.36   0.86  -0.65   0.26
-0.48  -0.13  -0.61
```

02 ARGENTINA
```
-1.49   0.79  -0.57   0.61  -1.40   1.14  -0.22   0.61   0.26
-0.61  -0.88   0.61   0.53  -1.06  -0.40   0.18  -0.65  -0.13
 1.67   1.58   1.05
```

03 AUSTRALIA
```
-0.96   1.05   0.18   0.0   -1.62   1.32   0.53   0.53  -0.13
-0.61  -0.88   1.14   1.23   0.05  -1.40  -1.22  -0.65   1.67
 0.88  -0.13  -0.26
```

04 BELGIUM
```
-1.14   0.96  -1.63   1.27   1.49  -1.49   0.22  -0.61   0.62
-0.61  -0.88   1.23   1.32   0.05  -1.40  -1.22  -0.65  -0.70
-0.09   0.70   0.22
```

05 BRAZIL
```
 0.09   1.23   1.41   0.70  -1.27   1.40   1.05   0.79   1.23
-0.61  -0.88   1.05   1.40  -1.06   0.36   0.18  -0.65  -0.97
 0.97   1.14   0.88
```

06 BULGARIA
```
-1.32  -0.09  -1.23   0.09   0.70  -0.79  -1.49  -1.49  -0.65
 1.50   1.66  -1.32  -1.32   0.05  -0.40  -0.50  -0.65   0.53
-0.88  -0.39   0.53
```

07 CAMBODIA
```
 1.67  -1.14   1.58  -1.49   0.17  -0.44  -0.88  -1.23  -0.97
 1.50   0.78  -1.40  -1.40   1.24   1.31   1.36   1.73   1.10
 0.66  -1.23  -0.88
```

08 CANADA
```
-0.88   1.40   0.18   0.31  -1.49   1.58   1.58   1.67   0.93
-0.61  -0.88   1.67   1.67  -1.06  -1.40  -1.22  -0.65  -1.54
 0.09   0.39   0.22
```

09 CEYLON
```
 0.79  -0.61   0.61  -0.53   1.32  -1.05   0.09  -1.05  -1.63
 1.50   0.78  -1.23  -0.70   0.05   1.31   0.18   1.73   1.40
 0.44  -1.05  -1.27
```

10 CHILE
```
-0.04  -0.17   0.44   1.27  -1.01   0.61   0.96   1.14  -0.13
-0.61  -0.88   0.70   0.79   0.05  -0.40   0.18  -0.65  -0.53
 1.59   1.67  -0.35
```

APPENDIX B continued

11	COLOMBIA	0.17	0.09	0.18	0.88	-0.70	0.79	1.23	1.40	0.62
		-0.61	-0.88	0.88	0.88	0.05	0.36	0.18	1.08	-1.23
		1.06	1.32	0.96						
12	COSTA RICA	0.31	-1.32	1.67	1.27	0.0	-1.14	1.40	1.23	0.26
		-0.61	-0.88	-0.44	-0.35	1.24	0.36	0.18	-0.65	-1.23
		1.41	1.40	-1.67						
13	CZECHOSLOVAKIA	-1.58	0.61	-1.45	0.44	1.14	-0.61	-1.10	-0.96	0.93
		-0.61	1.66	-1.05	-0.79		-0.86	-1.22	-0.65	0.0
		-1.37	-0.39	0.70						
14	DENMARK	-1.05	0.26	-1.45	1.27	1.05	-1.23	0.22	-0.70	0.62
		-0.61	-0.88	0.35	0.70	1.24	-0.86	-1.22	-0.65	-0.53
		-0.62	0.17	0.0						
15	ECUADOR	0.31	-0.79	1.14	0.53	-0.39	-0.17	1.32	1.05	-1.32
		-0.61	-0.88	-0.26	0.09	1.24	0.36	0.18	1.08	-1.05
		1.50	1.49	-1.05						
16	FRANCE	-0.70	1.49	-1.14	1.05	0.88	0.44	0.0	-0.26	1.41
		-0.61	-0.88	1.40	1.49	-1.06	-1.40	-1.22	-0.65	-0.88
		-0.18	0.53	1.49						
17	INDIA	0.88	1.32	0.18	-1.23	1.23	1.23	0.79	0.96	1.23
		-0.61	0.78	1.32	1.14	-1.66	1.31	0.18	-0.65	0.79
		-1.37	-1.54	1.40						
18	IRAN	0.70	-0.26	-0.57	-1.49	-0.70	0.88	-0.48	0.35	-0.48
		-0.61	-0.88	0.44	0.26	-1.06	1.31	1.36	1.08	1.10
		-1.37	-1.14	0.61						
19	IRAQ	0.96	-0.70	1.27	-1.14	-0.39	0.17	-1.10	0.09	-0.48
		-0.61	0.78	-0.70	-0.96	0.05	0.36	1.36	-0.65	1.32
		-0.88	-0.88	0.09						
20	LEBANON	-0.48	-0.96	0.61	-0.26	1.40	-1.67	-0.79	0.26	-0.13
		-0.61	0.78	-0.17	-1.14	1.24	-0.40	0.18	-0.65	1.23
		-0.35	-0.79	-1.14						
21	LIBERIA	1.49	-1.58	-0.92	-0.61	-0.88	0.88	1.14	1.49	-1.45
		-0.61	0.78	-0.96	-0.44	1.24	1.31	0.86	1.73	0.61
		0.79	0.92	-1.40						
22	LIBYA	1.05	-1.67	-0.26	-0.88	-1.62	0.96	-0.70	0.17	-0.48
		1.50	0.78	-0.53	-0.53	1.24	1.31	1.36	1.73	0.44
		0.18	0.04	-1.27						
23	MEXICO	-0.04	0.53	1.05	1.67	-0.17	1.05	1.49	1.58	-0.97
		-0.61	-0.88	1.49	1.58	-1.06	0.36	0.18	-0.65	-1.54
		1.14	1.05	-0.17						
24	MOROCCO	1.32	-0.53	0.79	-0.88	0.09	0.09	-1.27	-0.17	-0.97
		1.94	0.78	-0.35	-1.05	0.05	1.31	1.36	-0.65	-0.35
		0.35	0.79	-0.75						
25	NEPAL	1.58	-1.23	-0.26	-1.67	0.61	-0.53	-1.40	-1.14	-1.63
		1.50	0.78	-1.49	-1.49	0.05	1.31	1.72	1.73	0.88
		-0.26	-1.54	-0.44						
26	NETHERLANDS	-0.35	0.70	-0.92	0.31	1.67	-1.40	-0.61	-0.35	0.26
		-0.61	-0.88	1.58	1.05	0.05	-0.86	-1.22	-0.65	-0.53
		-0.48	0.39	0.44						
27	NEW ZEALAND	-1.40	0.0	-0.09	0.18	-1.27	-0.26	0.61	-0.09	-0.97
		-0.61	-0.88	-0.09	0.61	1.24	-1.40	-1.22	-0.65	1.49
		1.23	0.92	-0.09						
28	NORWAY	-0.79	0.17	-1.32	0.79	-1.01	0.0	0.44	-0.79	-0.97
		-0.61	-0.88	0.17	0.44	1.24	-1.40	-1.22	-0.65	-0.79
		-1.37	-0.70	-0.75						
29	PANAMA	0.53	-1.40	0.61	1.54	-0.53	-0.96	1.67	1.32	1.23
		-0.61	-0.88	0.09	-0.26	1.24	0.36	0.18	1.08	-1.23
		1.32	1.23	-1.53						
30	POLAND	-0.48	1.14	0.79	0.97	0.96	-0.09	0.35	-0.88	0.26
		-0.61	1.66	-0.61	0.0	-1.06	-0.40	-1.22	-0.65	-0.13
		-1.37	-0.53	0.79						
31	SWEDEN	-0.61	0.88	-1.63	1.54	-0.26	0.26	-0.35	-0.48	0.26
		1.05	0.78	0.96	0.96	0.05	-1.40	-1.22	-0.65	-0.26
		-0.88	0.04	-0.53						
32	THAILAND	1.14	-0.35	0.97	-1.49	0.53	0.35	-0.09	0.44	-0.97
		1.05	-0.88	0.0	-0.09	-1.06	1.31	0.86	1.08	0.97
		0.53	-1.32	0.35						
33	TUNISIA	1.23	-0.88	-0.57	-0.88	0.26	-0.70	-1.27	-1.40	-0.48
		1.94	0.78	-0.79	-1.23	1.24	0.36	1.36	-0.65	0.18
		0.0	0.61	-0.96						
34	TURKEY	0.44	0.44	0.88	-1.32	0.35	0.70	0.70	0.70	1.06
		-0.61	-0.88	0.79	0.17	-1.06	0.36	-0.50	-0.65	0.70
		-1.37	-0.96	1.23						
35	USSR	-1.67	1.58	-0.26	-0.70	-1.14	1.67	-0.96	-1.32	1.67
		-0.61	1.66	-0.88	-0.61	-1.66	-0.86	-0.50	-0.65	-1.54
		-1.37	-1.54	1.67						
36	UNITED STATES	-1.23	1.67	-0.57	-0.18	-0.09	1.49	-1.62	-1.62	1.59
		-0.61	-0.88	-1.62	-1.62	-1.66	-1.40	-1.22	-0.65	-1.54
		-1.37	-0.61	1.58						
37	YUGOSLAVIA	-0.17	0.35	-1.05	-0.09	0.79	-0.35	-0.22	0.0	-0.97
		-0.61	0.78	0.26	-0.17	-1.06	-0.40	-0.50	-0.65	0.09
		-0.70	-0.26	1.14						
38	KENYA	1.40	-1.05	0.18	-0.44	-0.70	0.53	-0.48	-0.48	0.79
		2.10	0.78	-1.14	-0.88	0.05	0.36	1.36	1.73	1.58
		0.66	0.26	-1.53						
39	TAIWAN	-0.26	-0.44	1.49	-1.05	1.58	-1.32	0.88	0.88	1.50
		-0.61	-0.88	0.53	0.35	0.05	1.31	0.86	1.73	0.35
		0.26	-1.54	1.32						

APPENDIX C

		01	02	03	04	05	06	07	08	09	10
01	ALBANIA	-1.14	-0.26	0.79	1.58	1.42	1.56	0.59	1.69	1.70	0.24
02	ARGENTINA	0.44	-0.53	-0.22	-0.53	-0.46	0.14	-0.69	-1.95	-0.73	0.24
03	AUSTRALIA	0.79	-1.49	-0.35	-0.61	-1.37	-1.06	-1.51	-0.62	-0.73	-1.42
04	BELGIUM	1.32	-0.79	0.04	0.13	-0.46	-1.06	-0.69	-0.62	-0.73	-1.42
05	BRAZIL	-0.44	0.18	-0.44	-0.79	-0.46	0.14	0.59	-0.62	-0.73	0.24
06	BULGARIA	0.61	-0.44	0.53	1.32	0.64	1.56	-0.69	1.69	1.70	0.24
07	CAMBODIA	-1.40	1.32	1.32	0.26	1.42	0.96	1.51	1.08	1.21	1.71
08	CANADA	1.58	-1.49	0.66	0.0	-1.37	-1.06	-0.69	-0.62	-0.73	-1.42
09	CEYLON	-1.05	0.39	-1.05	-1.49	0.64	-0.27	0.59	-0.62	-0.73	0.24
10	CHILE	0.17	0.04	-0.09	-0.17	-0.46	-1.06	-0.69	-0.62	-0.73	0.24
11	COLOMBIA	0.26	0.39	-1.14	-1.23	-0.46	-1.06	0.59	-0.62	-0.73	0.24
12	COSTA RICA	-0.26	-0.35	-1.58	-1.62	1.42	-1.06	-0.69	-0.62	-0.73	0.24
13	CZECHOSLOVAKIA	1.01	-1.32	0.35	1.05	-0.46	1.56	-0.69	1.69	1.70	0.24
14	DENMARK	1.40	-0.96	-0.22	0.13	-0.46	-1.06	-1.51	-0.62	-0.73	-1.42
15	ECUADOR	-0.96	0.61	-0.79	-0.92	0.64	-1.06	0.59	-0.62	-0.73	0.24
16	FRANCE	1.14	-0.96	1.14	1.49	-1.37	0.14	-0.69	-0.62	-0.73	-1.42
17	INDIA	-1.58	1.49	-0.88	-1.36	-1.37	-1.06	0.59	-0.62	-0.73	0.24
18	IRAN	-0.35	1.14	1.05	0.70	0.64	0.96	1.51	1.08	0.58	1.71
19	IRAQ	-1.23	0.88	1.40	0.88	0.64	0.55	0.59	0.72	0.88	1.37
20	LEBANON	-0.70	0.04	-0.57	-1.36	0.64	0.14	0.59	-0.62	-0.73	0.24
21	LIBERIA	-0.09	1.05	-1.40	-0.35	1.42	0.14	1.51	1.08	1.21	0.24
22	LIBYA	-0.61	0.26	0.96	-0.92	1.42	0.14	1.51	-0.62	0.58	0.24
23	MEXICO	0.0	1.40	-1.32	0.61	-0.46	-1.06	0.59	-0.62	0.58	0.24
24	MOROCCO	-0.88	0.79	0.66	-1.14	0.64	0.14	0.59	-0.62	0.58	0.24
25	NEPAL	-1.67	1.67	-1.49	-0.44	1.42	0.96	1.51	1.08	1.21	1.71
26	NETHERLANDS	1.01	-0.96	0.44	0.79	-0.46	-1.06	-0.69	-0.62	-0.73	-1.42
27	NEW ZEALAND	0.88	-1.49	-0.70	-0.26	-0.46	-1.06	-1.51	-0.62	-0.73	-1.42
28	NORWAY	1.23	-1.18	0.26	0.44	-0.46	-1.06	-1.51	-0.62	-0.73	-1.42
29	PANAMA	0.35	-0.09	-1.67	-1.62	1.42	-1.06	-0.69	-0.62	-0.73	0.24
30	POLAND	0.53	-0.61	0.17	0.53	-0.46	0.96	-0.69	1.69	1.21	0.24
31	SWEDEN	1.49	-1.67	0.88	0.35	-0.46	-1.06	-1.51	-0.62	-0.73	-1.42
32	THAILAND	-1.49	1.23	0.04	-0.09	0.64	0.96	0.59	1.08	1.21	1.71
33	TUNESIA	-0.53	0.53	-0.96	-1.05	0.64	0.96	0.59	-0.62	0.58	0.24
34	TURKEY	-0.79	0.70	-0.57	1.23	-0.46	0.14	0.59	-0.62	-0.73	0.24
35	USSR	0.70	-0.70	1.58	1.14	-1.69	0.96	-0.69	1.69	1.70	0.24
36	UNITED STATES	1.67	-1.18	1.49	0.96	-1.69	-1.06	-1.51	-0.62	-0.73	-1.42
37	YUGOSLAVIA	0.09	-0.18	1.23	1.40	-0.46	0.96	0.59	1.69	1.21	0.24
38	KENYA	-1.32	0.96	-1.23	-0.70	1.42	1.56	1.51	-0.62	-0.73	1.47
39	TAIWAN	-0.17	1.58	1.67	1.67	-1.37	1.56	-0.09	1.08	-1.85	-0.88

Code:

01. MANY RADIOS PER 1000 POPULATION.
02. LARGE NUMBER OF INHABITANTS PER HOSPITAL BED.
03. LARGE DEFENSE EXPENDITURES AS A % OF GNP.
04. LARGE % MILITARY PERSONNEL AS A % OF POP AGED 15-64.
05. VERY LOW INTERNATIONAL FINANCIAL STATUS.
06. CENSORSHIP.
07. VERY LOW NEWSPAPER CIRCULATION PER 1000 POPULATION.
08. NO EFFECTIVE CONSTITUTIONAL LIMITATIONS.
09. OPPOSITION GROUPS NOT TOLERATED.
10. TRADITIONAL BUREAUCRACY.

APPENDIX D

VARIABLE LIST FOR NATIONAL ATTRIBUTE STUDY

1. Military Personnel as a Percentage of Total Population (1)
2. Population per 1,000 Hectares of Agricultural Lands (1)
3. Inhabitants per Physician (1)
4. Gross National Product (1)
5. Inhabitants per Hospital Bed (1)
6. Percentage of Labor Force Employed in Agriculture (1)
7. Annual Percentage Rate of Increase in Population (1)
8. Moslems as a Percentage of Population (1)
9. Roman Catholics as a Percentage of Total Population (1)
10. All Christians as a Percentage of Total Population (1)
11. Russett's Asian Region (5)
12. Russett's Latin American Region (5)
13. Russett's Western European Region (5)
14. Russett's Eastern European Region (5)
15. Russett's Anglo-Saxon Region (5)
16. Population per Square Kilometer (1)
17. Area in Square Kilometers (1)
18. Radios per 1,000 Population (1)
19. Defense Expenditures as a Percentage of Gross National Product (1)
20. Votes in National Elections as a Percentage of Voting Age Population (1)
21. Military Personnel as a Percentage of Population Age 15–64 (1)
22. Alker & Russett's Cold War Factor Scores (3)
23. Alker & Russett's Self-Determination Factor Scores (3)
24. Alker & Russett's U.N. Supranationalism Factor Scores (3)
25. Students Enrolled in Higher Education per 100,000 Population (1)
26. Primary and Secondary School Pupils as a Percentage of Population Age 5–19 (1)
27. Percentage Literate of Population Age 15 and over (1)
28. Percentage Literate of Population Age 15 and over, Average Annual Increase (1)
29. Rummel's Turmoil Factor Scores (4)
30. Rummel's Revolutionary Factor Scores (4)
31. Rummel's Subversive Factor Scores (4)
32. Rummel's War Factor Scores (4)
33. Rummel's Diplomatic Factor Scores (4)
34. Rummel's Belligerent Factor Scores (4)
35. Percentage Export of Total Exports to U.S. (6)
36. Percentage Imports of Total Imports from U.S. (6)
37. Number in Permanent Missions at the United Nations (7)
38. Year Entered the United Nations (8)
39. Military Alliances (12)
40. Imports from the United States (10)
41. Exports to the United States (10)
42. Exports to the U.S.S.R. (10)
43. Imports from the U.S.S.R. (10)
44. Exports to Communist China (10)
45. Imports from Communist China (10)
46. Population (2)
47. Urbanization (2)
48. Agricultural Population (2)
49. Per Capita Gross National Product (2)
50. International Financial Status (2)
51. Economic Development Status (2)
52. Freedom of the Press (2)
53. Newspaper Circulation per 1,000 Population (2)
54. Religious Homogeneity (2)
55. Racial Homogeneity (2)
56. Linguistic Homogeneity (2)
57. Date of Independence (2)
58. Westernization (2)
59. Political Modernization (2)
60. System Style (2)
61. Constitutional Status of Present Regime (2)

Appendix D continued

62. Governmental Stability (2)
63. Representative Character of
    Current Regime (2)
64. Current Electoral System (2)
65. Freedom of Group Opposition (2)
66. Political Inculturation (2)
67. Sectionalism (2)
68. Interest Articulation by
    Associational Groups (2)
69. Interest Articulation by
    Institutional Groups (2)
70. Interest Articulation by Non-
    Associational Groups (2)
71. Interest Articulation by Anomic
    Groups (2)
72. Interest Articulation by Political
    Parties (2)
73. Interest Aggregation by Executive
    (2)
74. Interest Aggregation by
    Legislature (2)
75. Stability of Party System (2)
76. Personalissimo (2)
77. Political Leadership (2)
78. Leadership Charisma (2)
79. Vertical Power Distribution (2)
80. Horizontal Power Distribution (2)
81. Current Status of Legislature (2)
82. Character of Legislature (2)
83. Current Status of Executive (2)
84. Character of Bureaucracy (2)
85. Political Participation by
    Military (2)
86. Role of Police (2)
87. Communist Bloc (2)
88. Distance from U.S. (11)
89. Distance from U.S.S.R. (11)
90. Distance from China (11)
91. Military Strength (9)

The following are primary sources (following above code): (1) Bruce M. Russett, *et al., World Handbook of Political and Social Indicators* (New Haven: Yale University Press, 1964); (2) Arthur S. Banks and Robert Textor, *A Cross Polity Survey* (Cambridge: M.I.T. Press, 1963); (3) Hayward R. Alker, Jr. and Bruce M. Russett, *World Politics in the General Assembly* (New Haven: Yale University Press, 1965); (4) Rudolph J. Rummel, "Dimensions of Conflict Behavior Within and Between Nations," *General Systems,* Yearbook of the Society for the Advancement of General Systems Theory (Ann Arbor, 1963); (5) Bruce M. Russett, "Delineating International Regimes," Carnegie-IDRC Joint Study Group on Measurement Problems, Paper No. G60 (Indiana University, February, 1965); (6) Dan Golenpaul, ed., *Information Please Almanac 1968* (New York: Simon and Schuster, 1967); (7) *Permanent Missions to the United Nations,* United Nations Document, March 1968; (8) Golenpaul, *op. cit.*; (9) *Readers' Digest Almanac and Yearbook 1968* (New York: Readers' Digest Association, 1957); (10) *Direction of Trade Annual 1962–66* (Washington: International Monetary Fund and International Bank for Reconstruction and Development, 1967); (11) Great circle distance on a standard globe used. States with common border originally coded 0; (12) S. H. Steinberg, ed., *The Statesmen's Yearbook 1966–67* (New York: St. Martin's Press, 1966). States coded "1" allied with U.S., "2" neutral, and "3" allied with U.S.S.R.

Additional resources include: Louis Barron, ed., *Worldmark Encyclopedia of the Nations: Africa* (New York: Worldmark Press, Inc., 1963); Louis Barron, ed., *Worldmark Encyclopedia of the Nations: Africa; Asia; Australasia; Europe; the Americas* (New York: Worldmark Press, Inc., 1965); British Information Services, *Batswana* (London: Curwen Press, 1963); British Information Services, *Kenya* (London: F. Mildner & Sons, 1963); British Information Services, *Lesotho* (Norwich, England: Page Bros., Ltd., 1966); British Information Services, *Malta* (Manchester, England: H. M. Stationery Office Press, 1964); British Information Services, *Zambia* (Wilts, England: Swindon Press, Ltd., 1964); *Gallatin Business Intelligence: Africa: Asia: Oceania: the Americas* (New York: Copley International Corporation, 1967); Norton Ginsburg, *Atlas of Economic Development* (Chicago: University of Chicago Press, 1964); Colin Legum, ed., *Africa: Hand-*

book to the Continent (New York: Frederick A. Praeger, Inc., 1966); *Information Please Almanac and Yearbook 1967*, 22nd ed. (New York: Simon and Schuster, 1966); *The Europa Yearbook 1967*, Volumes I, II (London: Europa Publications, Ltd., 1967); *The Gallatin Annual of International Business* (New York: American Heritage Publishing Co., Inc., 1965); *The Middle East*, 10th ed. (London: Europa Publications, Ltd., 1963); *The U.S. Book of Facts, Statistics and Information* (New York: Washington Square Press, Inc., 1966); United Nations Department of Economic and Social Affairs, *United Nations Statistical Yearbook* (New York: U.N. Statistical Office, 1967); United Nations Office of Public Information, *Everyman's United Nations*, 7th ed., (New York: U.N. Office of Public Information, 1964); United States Department of State Publications, *Background Notes* (Washington, D.C.: U.S. Government Printing Office, 1967); *Whitaker's Almanac 1967 and 1968* (London: William Clowes & Sons, Ltd., 1967); *World Almanac and Book of Facts 1967 and 1968* (New York: Newspaper Enterprise Association, Inc., 1966, 1967); *World Economic Atlas,* 3rd ed. (London: Oxford University Press, 1965); *The Annual Register: World Events in 1966* (New York: St. Martin's Press, 1967); *The International Yearbook and Statesmen's Who's Who* (London: Burke's Peerage, Ltd., 1967).

APPENDIX E

Zero Values for Subjects

W. SAMOA	48.00	MALAYSIA	12.00
MALDIVE ISLANDS	43.00	PAKISTAN	11.00
RWANDA	35.00	CHINA. PR	10.00
YEMEN	35.00	HONDURAS	10.00
BOTSWANA	34.00	UNITED KINGDOM	10.00
LESOTHO	34.00	ICELAND	10.00
BURUNDI	33.00	IRAN	10.00
NORTH KOREA	32.00	ROMANIA	9.00
NORTH VIETNAM	32.00	TUNISIA	9.00
TOGO	30.00	LIBERIA	9.00
GAMBIA	30.00	CAMBODIA	8.00
ZAMBIA	29.00	JORDAN	8.00
SOUTH KOREA	27.00	AUSTRIA	8.00
KUWAIT	27.00	GUATEMALA	8.00
CONGO (BRA)	26.00	HAITI	8.00
CONGO (LEO)	26.00	THAILAND	8.00
GABON	26.00	U.A.R.	7.00
MALAWI	26.00	UNITED STATES	7.00
MALI	25.00	ARGENTINA	7.00
NIGER	25.00	INDONESIA	7.00
UGANDA	25.00	LEBANON	7.00
CHAD	24.00	URUGUAY	6.00
UPPER VOLTA	24.00	PARAGUAY	6.00
SOUTH VIETNAM	24.00	TAIWAN	6.00
DAHOMEY	23.00	SWITZERLAND	6.00
SOMALIA	23.00	HUNGARY	5.00
SUDAN	23.00	NICARAGUA	5.00
ALGERIA	22.00	COLOMBIA	5.00
CENTRAL AFRICAN REP	22.00	PORTUGAL	5.00
SINGAPORE	22.00	BOLIVIA	5.00
IVORY COAST	21.00	ALBANIA	5.00
CYPRUS	21.00	BULGARIA	4.00
CAMEROUN	21.00	SPAIN	4.00
LAOS	20.00	CZECHOSLOVAKIA	4.00
LIBYA	20.00	U.S.S.R.	4.00
NIGERIA	20.00	VENEZUELA	4.00
SENEGAL	20.00	COSTA RICA	4.00
BARBADOS	20.00	ISRAEL	4.00
MAURITANIA	19.00	MEXICO	3.00
MONGOLIA	19.00	GREECE	3.00
GUINEA	18.00	TURKEY	3.00
IRAQ	18.00	BRAZIL	3.00
MALAGASY REP	18.00	ECUADOR	3.00
SAUDI ARABIA	18.00	AUSTRALIA	3.00
TRINIDAD	18.00	FRANCE	3.00
MALTA	18.00	NEW ZEALAND	3.00
SIERRA LEONE	17.00	CHILE	3.00
KENYA	17.00	CEYLON	2.00
JAMAICA	16.00	IRELAND	2.00
NEPAL	16.00	JAPAN	2.00
BURMA	16.00	DENMARK	2.00
TANZANIA	15.00	PHILIPPINES	2.00
EAST GERMANY	15.00	NETHERLANDS	2.00
SOUTH AFRICA	15.00	EL SALVADOR	2.00
DOMINICAN REPUBLIC	15.00	SWEDEN	2.00
GHANA	14.00	NORWAY	2.00
SYRIA	14.00	YUGOSLAVIA	2.00
GUYANA	14.00	ITALY	1.00
LUXEMBOURG	14.00	CANADA	1.00
CANADA	13.00	PANAMA	1.00
ETHIOPIA	13.00	FINLAND	1.00
PANAMA	13.00	BELGIUM	1.00
WEST GERMANY	12.00	POLAND	1.00
MOROCCO	12.00	INDIA	1.00
PERU	12.00		
CUBA	12.00		
AFGHANISTAN	12.00		

Zero Values for Variables

VARIABLE	28	89.00	VARIABLE	18	11.00
VARIABLE	73	64.00	VARIABLE	82	11.00
VARIABLE	29	60.00	VARIABLE	54	9.00
VARIABLE	30	60.00	VARIABLE	50	7.00
VARIABLE	31	60.00	VARIABLE	43	7.00
VARIABLE	32	60.00	VARIABLE	67	7.00
VARIABLE	33	60.00	VARIABLE	5	7.00
VARIABLE	34	60.00	VARIABLE	84	7.00
VARIABLE	11	53.00	VARIABLE	85	7.00
VARIABLE	15	52.00	VARIABLE	68	6.00
VARIABLE	12	51.00	VARIABLE	55	6.00
VARIABLE	13	51.00	VARIABLE	60	5.00
VARIABLE	14	51.00	VARIABLE	51	5.00
VARIABLE	53	50.00	VARIABLE	1	5.00
VARIABLE	19	49.00	VARIABLE	45	4.00
VARIABLE	21	43.00	VARIABLE	8	4.00
VARIABLE	75	38.00	VARIABLE	42	4.00
VARIABLE	64	34.00	VARIABLE	57	4.00
VARIABLE	20	33.00	VARIABLE	44	4.00
VARIABLE	24	33.00	VARIABLE	4	3.00
VARIABLE	62	33.00	VARIABLE	79	3.00
VARIABLE	23	31.00	VARIABLE	58	3.00
VARIABLE	25	30.00	VARIABLE	3	3.00
VARIABLE	2	27.00	VARIABLE	70	2.00
VARIABLE	63	27.00	VARIABLE	48	1.00
VARIABLE	26	26.00	VARIABLE	16	1.00
VARIABLE	27	25.00	VARIABLE	10	1.00
VARIABLE	83	25.00	VARIABLE	17	1.00
VARIABLE	61	24.00	VARIABLE	40	1.00
VARIABLE	72	24.00	VARIABLE	56	1.00
VARIABLE	74	24.00	VARIABLE	41	1.00
VARIABLE	81	24.00	VARIABLE	59	0-0
VARIABLE	6	23.00	VARIABLE	37	0-0
VARIABLE	66	22.00	VARIABLE	46	0-0
VARIABLE	76	22.00	VARIABLE	7	0-0
VARIABLE	77	19.00	VARIABLE	39	0-0
VARIABLE	52	18.00	VARIABLE	35	0-0
VARIABLE	65	18.00	VARIABLE	49	0-0
VARIABLE	71	18.00	VARIABLE	36	0-0
VARIABLE	80	18.00	VARIABLE	38	0-0
VARIABLE	69	17.00	VARIABLE	87	0-0
VARIABLE	78	17.00	VARIABLE	88	0-0
VARIABLE	26	15.00	VARIABLE	89	0-0
VARIABLE	86	14.00	VARIABLE	90	0-0
VARIABLE	47	11.00	VARIABLE	91	0-0
VARIABLE	9	11.00			

APPENDIX F

Ranked Factor Scores

1	AFGHANISTAN	128	35	73	20	68	1	2	37	11	57	48	117	51	8	52	7	41	105	44
2	ALBANIA	57	9	37	118	44	43	30	66	124	112	104	52	30	38	104	38	21	10	12
3	ALGERIA	71	29	40	31	116	96	108	43	36	43	114	119	90	63	3	96	24	117	38
4	ARGENTINA	16	60	82	25	128	44	124	41	109	32	99	5	102	121	101	129	124	54	28
5	AUSTRALIA	5	96	118	49	129	77	10	93	60	65	105	25	66	46	127	120	62	9	123
6	AUSTRIA	28	116	18	36	26	49	38	12	55	18	32	62	32	101	90	65	46	68	49
7	BELGIUM	9	103	61	15	4	58	52	110	18	86	11	110	60	102	1	94	61	43	55
8	BOLIVIA	73	78	94	71	113	82	102	97	81	38	97	31	26	100	18	16	86	129	41
9	BRAZIL	61	76	96	7	114	17	110	54	68	80	85	14	109	125	80	26	85	37	78
10	BULGARIA	18	5	12	40	58	52	75	61	103	70	103	91	12	36	7	62	36	34	42
11	BURMA	114	65	38	10	62	32	18	55	48	19	118	3	31	15	78	116	110	15	63
12	BURUNDI	93	58	79	123	33	112	54	39	104	82	42	32	54	96	77	104	52	47	104
13	CAMBODIA	126	39	49	68	51	99	8	26	25	120	100	100	22	60	125	39	126	27	109
14	CAMEROUN	119	81	36	57	77	104	34	100	90	75	27	45	34	111	43	118	73	57	35
15	CANADA	11	90	117	13	123	103	4	123	62	104	64	47	103	94	107	53	58	52	56
016	CENTRAL AFRICAN REP	86	25	47	100	96	127	61	86	84	68	7	80	96	82	37	111	90	64	90
17	CEYLON	90	91	58	59	6	5	96	105	100	7	35	50	52	11	129	88	108	75	114
18	CHAD	104	57	15	70	90	119	41	118	54	64	1	59	55	69	29	30	98	56	65
19	CHILE	36	77	119	75	91	46	121	46	26	109	60	78	110	123	98	49	91	81	99
20	CHINA, PR	63	13	2	4	50	54	74	49	34	34	54	1	43	6	10	5	26	102	128
21	COLOMBIA	55	72	116	19	98	85	76	36	79	14	66	41	44	127	14	29	78	12	107
22	CONGO (BRA)	77	42	19	93	92	116	63	114	45	50	3	57	48	103	44	122	109	86	70
23	CONGO (LEO)	113	67	84	41	106	56	125	107	102	66	36	106	29	85	41	126	88	39	92
24	COSTA RICA	43	82	104	111	53	106	14	3	114	3	86	11	53	107	67	54	114	46	15
25	CUBA	32	19	11	33	55	97	100	13	76	6	119	112	71	120	122	10	94	32	19
26	CYPRUS	41	83	29	104	40	53	113	99	92	52	12	85	87	10	5	101	49	107	14
27	CZECHOSLOVAKIA	4	7	30	37	34	48	84	124	91	69	111	116	23	84	60	66	59	25	5
28	DAHOMEY	97	43	28	95	63	123	47	111	96	84	8	48	85	90	40	92	95	108	50
29	DENMARK	8	106	67	69	39	91	13	22	98	47	82	105	99	28	84	71	89	126	11
030	DOMINICAN REPUBLIC	69	86	126	101	25	94	57	4	107	116	40	67	21	114	66	4	42	7	45
31	ECUADOR	52	71	109	77	67	22	81	125	66	85	46	34	92	97	33	6	115	59	17
32	EL SALVADOR	46	23	108	119	21	13	27	51	51	13	73	75	108	106	69	24	123	67	108
33	ETHIOPIA	122	26	88	23	76	3	3	113	44	11	38	127	36	52	103	22	93	94	34
34	FINLAND	10	108	33	52	78	90	85	50	99	31	51	103	62	2	74	48	65	121	57
35	FRANCE	25	75	89	14	72	108	122	16	6	119	30	128	2	77	113	47	120	31	121
36	GABON	70	30	90	112	95	121	69	120	46	76	20	69	24	93	58	127	99	72	81
037	EAST GERMANY	2	1	31	72	22	30	123	90	97	16	55	21	28	68	115	102	28	95	120
038	WEST GERMANY	27	119	53	27	12	69	80	91	63	102	14	10	47	112	123	91	29	60	103
39	GHANA	92	18	99	50	52	111	64	92	43	4	112	109	68	65	117	70	79	106	54
40	GREECE	38	99	32	44	57	80	58	6	106	111	44	113	46	34	25	68	129	104	39
41	GUATEMALA	53	40	111	91	24	29	104	112	74	20	102	39	113	95	22	11	113	118	13
42	GUINEA	88	17	105	106	70	120	33	101	38	61	108	54	64	61	42	72	18	98	36
43	HAITI	89	21	123	102	15	37	88	62	39	33	22	18	6	109	45	15	39	96	67
44	HONDURAS	50	47	114	124	74	18	83	33	9	10	81	23	50	71	64	23	112	93	47
45	HUNGARY	3	4	6	55	36	51	115	65	10	17	78	16	10	89	88	50	56	63	33
46	ICELAND	7	113	71	127	121	67	20	24	21	36	58	2	18	17	36	52	30	119	30
47	INDIA	118	126	100	1	9	110	50	59	8	9	52	6	77	26	9	69	4	38	8
48	INDONESIA	102	38	107	5	31	113	103	87	33	8	110	36	1	13	15	123	102	49	32
49	IRAN	111	50	78	3	118	21	60	53	110	113	26	74	38	24	30	105	111	124	43
50	IRAQ	87	61	55	43	93	45	120	68	80	121	19	107	125	19	102	42	80	24	1
51	IRELAND	20	100	43	87	61	79	40	7	94	40	13	84	89	88	38	67	12	84	71
52	ISRAEL	13	98	103	105	48	88	129	73	14	118	115	83	27	1	8	117	40	36	76
53	ITALY	40	112	56	6	16	75	117	27	88	96	31	121	14	126	95	78	72	97	111
54	IVORY COAST	99	32	92	82	79	129	72	95	52	55	77	90	3	87	76	109	92	125	98
55	JAMAICA	58	120	102	115	13	81	106	81	24	21	37	65	15	59	75	12	5	65	95
56	JAPAN	42	111	93	9	17	93	90	64	122	27	39	66	35	5	126	40	107	114	127
57	JORDAN	91	45	52	96	81	4	127	10	3	127	2	53	40	55	97	107	66	11	10
58	NORTH KOREA	60	14	3	53	32	87	71	21	120	117	101	12	119	49	116	25	38	100	101
059	SOUTH KOREA	74	46	85	47	1	65	109	38	126	128	61	8	128	41	31	57	71	89	106
60	LAOS	124	54	62	60	69	23	43	108	87	126	45	40	33	67	26	59	122	85	72
61	LEBANON	39	73	41	86	10	33	119	71	108	15	24	114	107	21	11	119	127	4	4
62	LIBERIA	78	24	121	103	80	16	1	122	35	46	69	115	46	75	17	110	45	120	29
63	LIBYA	117	88	63	54	119	26	17	17	16	100	41	89	16	110	87	115	20	17	25
64	LUXEMBOURG	24	121	68	92	19	39	44	14	23	77	16	92	17	99	54	73	31	116	40
65	MALAGASY REP	125	129	51	79	107	40	39	60	49	94	6	27	7	119	93	80	11	6	83
66	MALI	105	53	17	76	89	128	37	102	58	62	29	94	104	42	94	2	96	26	64
67	MAURITANIA	106	66	57	94	101	122	42	85	30	91	89	77	73	33	48	8	75	16	100
68	MEXICO	65	64	124	24	102	117	68	29	117	115	70	86	118	129	59	3	3	40	122
69	MONGOLIA	47	6	24	99	112	74	21	8	123	74	88	35	116	23	79	36	14	53	73
70	MOROCCO	115	74	77	35	83	109	62	28	56	73	71	56	88	54	106	41	6	113	6
71	NEPAL	120	36	8	30	38	2	45	96	121	60	91	122	81	32	28	32	121	111	115
72	NETHERLANDS	29	115	76	34	5	57	15	88	65	107	34	101	25	78	99	60	84	91	37
73	NEW ZEALAND	1	79	115	110	124	100	48	78	59	45	94	49	61	20	85	121	82	69	124
74	NICARAGUA	62	31	125	120	84	12	5	42	12	92	74	43	4	104	61	14	117	74	20
75	NIGER	112	55	26	88	97	125	36	116	95	83	5	73	111	43	110	9	103	21	94
76	NIGERIA	129	92	75	11	73	8	99	106	86	22	50	120	65	108	71	99	2	66	52
77	NORWAY	19	125	66	51	75	63	9	32	47	87	56	81	37	12	86	44	74	127	66
78	PAKISTAN	101	28	127	21	42	66	105	77	7	37	80	7	123	58	118	124	13	101	82
79	PANAMA	37	68	112	107	66	50	79	34	119	1	98	87	83	66	39	34	128	2	51
80	PARAGUAY	44	16	106	116	108	24	56	18	64	90	92	4	112	115	35	98	60	122	110
81	PERU	48	52	120	45	110	36	98	119	20	49	75	44	72	92	49	20	101	110	61
82	PHILIPPINES	72	107	122	12	7	118	11	35	128	23	53	37	115	37	4	74	33	1	58
83	POLAND	21	3	46	32	28	63	97	9	42	26	79	88	5	80	91	55	37	20	69
84	PORTUGAL	34	12	72	65	27	20	7	15	93	89	23	108	127	122	19	89	15	82	102
85	ROMANIA	31	2	21	29	43	71	51	30	125	35	95	58	69	48	92	56	43	77	21
86	RWANDA	75	63	22	126	41	73	91	48	75	72	28	17	39	83	68	113	83	28	117
87	SAUDI ARABIA	95	27	97	38	122	6	46	11	4	24	63	129	67	14	32	82	53	8	129
88	SENEGAL	103	44	60	80	87	115	67	67	40	29	68	71	58	56	70	93	51	103	62
89	SIERRA LEONE	127	114	45	81	60	47	94	75	105	71	72	95	86	62	62	43	22	78	27
90	SOMALIA	109	69	23	78	117	25	77	23	72	56	9	79	59	44	27	125	48	44	80
91	SOUTH AFRICA	14	37	110	64	126	31	114	126	22	2	25	64	122	3	20	28	9	13	112
92	SPAIN	26	8	83	28	47	60	107	56	113	67	17	126	129	105	23	95	10	88	116
93	SUDAN	110	41	27	26	111	27	87	76	83	5	76	102	114	27	114	21	81	29	60
94	SWEDEN	12	117	64	46	85	72	35	45	57	79	62	97	41	25	105	64	63	123	23
95	SWITZERLAND	17	118	48	61	14	19	12	128	82	25	4	29	121	70	109	84	19	61	18
96	SYRIA	66	59	44	89	94	34	112	25	13	105	10	22	120	7	111	46	119	42	2
97	THAILAND	108	34	98	16	46	9	28	31	129	101	109	13	57	22	55	86	125	109	125
98	TOGO	94	51	35	108	65	78	55	82	61	39	21	55	70	76	82	90	77	55	31
99	TRINIDAD	54	127	113	125	18	64	116	98	37	44	59	93	19	29	73	13	8	18	91
100	TUNISIA	81	49	59	74	82	126	95	1	70	28	65	99	79	51	96	79	7	128	24
101	TURKEY	85	95	101	8	86	83	93	2	127	114	83	111	63	4	21	37	50	115	16
102	UGANDA	123	105	54	62	59	10	86	109	78	53	57	98	84	116	124	103	25	14	79
103	U.S.S.R.	22	10	65	17	125	70	53	121	118	122	107	96	20	39	89	35	16	3	7
104	U.A.R.	56	20	91	48	35	114	32	5	1	58	96	68	126	45	128	75	105	50	9
105	UNITED KINGDOM	15	102	95	22	30	61	23	79	29	93	33	76	42	50	121	45	70	51	74
106	UNITED STATES	6	128	1	2	115	95	31	84	27	124	43	26	82	72	2	1	87	19	86
107	UPPER VOLTA	96	48	20	98	71	124	49	103	89	81	18	51	95	64	50	58	104	112	85
108	URUGUAY	23	101	69	83	120	76	26	44	69	42	49	61	56	91	34	85	118	79	88
109	VENEZUELA	51	84	128	63	127	86	126	40	116	106	124	72	11	118	65	31	34	90	22
110	NORTH VIETNAM	59	15	7	66	20	89	89	63	111	103	113	20	117	53	120	51	57	92	119
111	SOUTH VIETNAM	68	22	81	85	23	68	111	74	112	125	106	9	105	40	53	97	100	76	126
112	YEMEN	80	62	10	84	105	7	24	20	5	51	15	70	101	16	12	63	69	5	84
113	YUGOSLAVIA	45	11	80	18	45	41	16	127	85	110	87	63	8	79	56	83	23	71	3
114	BOTSWANA	100	94	5	73	104	14	22	83	31	88	120	38	49	128	83	112	47	99	87
115	GAMBIA	107	123	4	90	49	15	19	52	67	95	122	30	80	73	24	76	67	48	59
116	KUWAIT	33	56	70	114	99	102	118	47	19	59	123	123	124	18	57	106	17	35	68
117	LESOTHO	98	122	9	109	56	11	29	70	53	99	121	19	75	124	100	87	55	58	53
118	MALAWI	82	85	14	97	54	98	65	104	41	48	127	104	93	57	16	27	76	83	48
119	MALTA	30	93	34	122	3	42	73	57	17	30	116	82	91	98	6	128	54	70	26
120	TANZANIA	116	89	25	42	103	105	92	58	50	12	117	125	74	86	112	77	97	87	118
121	ZAMBIA	84	97	16	58	100	35	25	115	32	98	128	60	94	117	47	108	68	45	75
122	KENYA	121	104	39	39	64	55	101	69	77	54	126	124	78	113	108	114	32	73	77
123	W. SAMOA	67	70	42	128	37	101	82	89	101	97	84	15	97	74	46	61	64	80	96
124	SINGAPORE	35	110	50	121	11	92	70	117	28	63	125	46	76	9	119	18	106	30	89
125	GUYANA	76	87	87	117	109	59	128	80	71	78	67	33	13	31	72	19	27	41	97
126	MALAYSIA	79	124	86	67	29	38	78	129	115	123	93	24	106	35	81	33	35	23	105
127	MALDIVE ISLANDS	83	80	13	113	88	84	59	19	73	108	47	42	98	47	51	100	44	22	113
128	BARBADOS	49	109	74	129	8	28	66	72	15	41	90	28	9	81	63	17	1	62	46
129	TAIWAN	64	33	129	56	2	107	6	94	2	129	129	118	100	30	13	81	116	33	93

1. Adelman and Morris: Sociocultural concomitants of the indus-
    trialization-urbanization process; Evolution of participant political
    institutions; Character of leadership; Degree of social and political
    stability.
2. Alker: East-West; North-South.
3. Alker: Self-determination; Cold war; UN supranationalism; Mos-
    lem question.
4. Alker and Russett: East-West; North-South.
5. Alker and Russett: East-West; Relations of UN members with
    Spain; Support for UN information programs; Opposition to UN-
    sponsored Palestine partition; Trusteeship for S.W. Africa; Geno-
    cide convention; Secretariat taxes; Third General Assembly in
    Europe.
6. Alker and Russett: East-West; North-South.
7. Alker and Russett: Self-determination; UN supranationalism; Pal-
    estine and Tunisia; Indians and race; Cold war; Liberal interven-
    tionists; Factor 7; Factor 8; Factor 9.
8. Alker and Russett: East-West; North-South.
9. Alker and Russett: Cold war; Self-determination; Small power
    supranationalism; South African problems; Capital punishment
    compromise; Western humanitarianism; Mediation in Near East.
10. Alker and Russett: East-West; North-South.
11. Alker and Russett: Self-determination; UN supranationalism;
    Cold war; Palestine and Western Iranian mediation attempts;
    Composition of Palestine conciliation committee; Voting on mar-
    riage rules in the third committee; French concerns for nuclear
    testing and the Algerian problem; Colonial doubts about Rwanda
    and Burundi; Anti-Common Market.
12. Banks and Gregg: Polyarchic; Elitist; Centrist; Personalist; Tradi-
    tional.
13. Cattell: Size; Cultural pressure; Enlightened affluence; Conser-
    vative patriarchal solidarity; Emancipated urban rationalism;
    Thoughtful industriousness; Vigorous, self-willed order; Bourgeois
    philistinism; Progressiveness; Fastidiousness; Buddhism-Mongo-
    lism; Cultural integration and morale.
14. Cattell and Gorsuch: Vigorous development v. Lack of exploita-
    tion of resources; Underdeveloped backwardness v. Complex

social systems; Morale and morality; Patriarchal solidarity v. Stimulating atmosphere; Size; Democracy v. Totalitarianism; Occidental v. Buddhist-Mongolism; Narrowness v. Enlightened affluence; Cultural pressure v. Direct expression of drives; Anomie v. Channeled emotionality; Medical development; Emancipated urban rationalism v. Unsophisticated stability; Bourgeois Philistinism v. Reckless Bohemianism; Factor 14.

15. Cattell, Breul, and Hartman: Enlightened affluence; Vigorous order; Cultural pressure and complexity; Size; Emancipated rationalism; Classical patriarchalism; Oriental pattern; Metropolitan laxity; Bourgeois Philistinism; Mechanic's culture; Morality.

16. Gregg and Banks: Access; Differentiation; Consensus; Sectionalism; Legitimation; Interest; Leadership.

17. Rummel: Revolution; Subversion; Turmoil.

18. Rummel: War; Diplomatic; Belligerent.

19. Rummel; Turmoil; Revolutionary; Subversive.

20. Rummel: Aggressive; Instability; Error; Diplomatic; Subversion; Belligerent.

21. Rummel: Economic development; Size; Political system; Density; Catholic culture; Foreign conflict; Internal conflict.

22. Russett: Economic development; Communism; Size; Catholic culture; Intensive agriculture.

23. Russett: Factor 1; Factor 2; Factor 3; Factor 4.

24. Russett: Western community; Brazzaville Africans; Afro-Asians; Communist bloc; Conservative Arabs; Iberia.

25. Sawyer: Size; Wealth; Politics.

26. Tanter: Diplomatic; War; Belligerency.

27. Tanter: Turmoil; Internal war.

28. Tanter: Turmoil; Diplomatic; Internal war; War; Factor 5; Factor 6; Factor 7.

29. Vincent: Economic-Political; Politics; U.S. Relations; Belligerency; Dispersion; Population pressure.

30. Vincent: Underdeveloped; Democracy; U.S. relations; Smallness; Diffusion; Executive leadership; Turmoil; Religious and linguistic heterogeneity; Peaceful; Militarism; Mobilization; Internal peace; Voting intensity; Catholicism; Communist China economic relations; Racial homogeneity; Personalissimo; U.N. supranationalism; Friendly diplomatic activity.

LOCATION OF STATES ON THE 19 FACTOR DIMENSIONS
USING DIFFERENT FACTOR SCORE FORMULAS

Explanation

Table 29 gives the scores of states on the 19 rotated factor dimensions computed by the incomplete formula $F = ZA$ where loadings in the range of $-.49$ to $.49$ are set to zero. As indicated earlier, the application of this formula enhances the relationship of the heavily loading variables to the factor scores, but the disadvantage lies in loss of the orthogonal character of the factor score distributions. The effect is *similar* to rotating to an oblique solution. The primary value of such scores is that they are defined only by the heaviest loading variables and, thus, the focus of attention is narrowed.

Table 30 gives the scores of states on the 19 rotated factor dimensions computed by the incomplete formula $F = ZA$ where loadings in the range of $-.49$ to $.49$ are set to zero and all other loadings are set to ± 1.0. Like the previous incomplete formula, this formula also tends to produce scores which depart from orthogonality. The major difference between the two incomplete formulas is that the first weights the variables to the extent of their loadings in computing the scores, while the second gives all heaviest loading variables equal weight. Thus, for example, the first incomplete formula gives more weight to Newspaper Circulation and Radios per 1,000 Population, in defining the scores on the first dimension, than to the Percentage Literate in the Population Age 15 and Over, or Per Capita Gross National Product. The second formula gives all such variables equal weight by substituting unity for the loading. It should be apparent, then, if only one variable loads heavily on a dimension, the original scores "become" the "factor scores" in the application of this formula.

Tables 31 and 32 give the intercorrelations of the factor scores computed by the incomplete methods. It will be recalled by reference to Table 24 that, in using the complete solution, all dimensions were basically orthogonal to all other dimensions, even though the factor scores included mean estimates for a number of states on a number of variables. This stands in contrast to the intercorrelations in respect to the incomplete solutions. Thus in the case of the first incomplete solution, Factor Dimensions 1 and 2, concerning "Underdeveloped" and "Democracy," correlate $-.48$. This means that the "Underdeveloped" scores have a modest degree of predictive power in respect to the "Democracy" scores in the sense that developed states, as defined by the heaviest loading variables, tend to be nonauthoritarian and under-developed states authoritarian. Also, a modest degree of relationship exists between the first dimension and the third and fourth dimensions. In respect to the third dimension, the "developed" states tend to have "closer U.S. relations," and in respect to the fourth dimension "under-developed" states tend to be "smaller." The remaining dimensions can be similarly interpreted. Because of the high degree of correlation between the first incomplete solution and the second (which sets all nonzero loadings to ± 1.0) the general pattern of intercorrelations for the second are similar to the first.

Tables 33 and 34 show the predictive power of the raw factor scores, computed by the incomplete methods, to the original variables. The pre-dictive power of the complete solution in respect to the original variables has already been given in Table 25. As expected, the incomplete solu-tions tend to enhance relationship of the factor scores to the heaviest loading variables. For example, considering the first incomplete solu-tion, Inhabitants per Physician loaded .91 on the first factor dimension, but the correlation of the original variables with the factor scores com-puted by this incomplete solution is .92. Similarly, an enhancing effect can be noted in connection with Percentage of Labor Force Employed in Agriculture (loading = .64, correlation = .70), All Christians as a Percentage of Total Population (loading = .62, correlation = .65). The degree to which the heaviest loading variables are enhanced, of course, depends in part upon their loadings and in part upon their relationship to the other variables.

Tables 35 and 36 give the predictive power of the rank factor scores, of the incomplete methods, to the original variables for those research-ers who only wish to make rank assumptions. The predictive power of the rank factor scores in the case of the complete solution has already been given in Table 27.

Tables 37, 38, and 39 give the factor scores where regression esti-mates on the original variables have been made, in place of the mean estimates, for the researcher who prefers a regression estimate solu-tion. The regression estimates were made by locating that variable with the highest correlation with the variable on which the subject had missing information, from the entire group of variables, and then esti-mating his missing scores from his score on that variable.

Tables 40, 41, and 42 give the correlations of the factor scores to the estimate deck, Tables 43, 44, and 45 give the intercorrelations among the factor scores, and Table 46 lists the estimation deck.

Bibliography

Adelman, Irma, and Cynthia Taft Morris. 1967. *Society, Politics, and Economic Development: A Quantitative Approach.* Baltimore: Johns Hopkins Press.

Ahmavaara, Y. 1954. "The Mathematical Theory of Factorial Invariance Under Selection." *Psychometrika* 19: 27–38.

Alker, Hayward R., Jr. 1964. "Dimensions of Conflict in the General Assembly." *American Political Science Review* 58: 642–57.

Alker, Hayward R., Jr., and Bruce M. Russett. 1965. *World Politics in the General Assembly.* New Haven: Yale University Press.

Anderson, T. W. 1958. *An Introduction to Multivariate Statistical Analysis.* New York: John Wiley and Sons.

Baggaley, Andrew R. 1964. *Intermediate Correlational Methods.* New York: John Wiley and Sons.

Baggaley, Andrew R., and Raymond B. Cattell. 1954–56. "A Comparison of Exact and Approximate Linear Function Estimates of Oblique Factor Scores." *British Journal of Statistical Psychology* 7–9: 83–86.

Banks, Arthur S., and Phillip M. Gregg. 1965. "Grouping Political Systems: Q-Factor Analysis of *A Cross Polity Survey.*" *American Behavioral Scientist* 9: 3–6.

Bartlett, M. S. 1941. "The Statistical Significance of Canonical Correlations." *Biometrika* 32: 29–38.

————. 1950. "Tests of Significance in Factor Analysis." *British Journal of Psychology, Statistical Section* 3: 77–85.

————. 1951. "A Further Note on Tests of Significance in Factor Analysis." *British Journal of Psychology, Statistical Section* 4: 1–2.

Berry, Brian. 1960. "An Inductive Approach to the Regionalization of Economic Development." In *Essays on Geographic and Economic Development,* edited by Norton Ginsberg. Chicago: University of Chicago Press.

Burt, C. L. 1937. "Correlations Between Persons." *British Journal of Psychology* 28: 56–96.

————. 1941. *The Factors of the Mind: An Introduction to Factor-analysis in Psychology.* New York: Macmillan.

————. 1950. "The Factorial Analysis of Qualitative Data." *British Journal of Psychology, Statistical Section* 3: 166–85.

————. 1952. "Tests of Significance in Factor Analysis." *British Journal of Psychology, Statistical Section* 5: 109–85.

Carroll, J. B. 1953. "An Analytical Solution for Approximating Simple Structure in Factor Analysis." *Psychometrika* 18: 23–38.

————. 1957. "Biquartimin Criterion for Rotation to Oblique Simple Structure in Factor Analysis." *Science* 126: 1114–15.

————. 1958. "Oblimin Rotation Solution in Factor Analysis." *Computing Program for the IBM 704.* Mimeographed.

Cattell, Raymond B. 1952. *Factor Analysis: An Introduction and Manual for the Psychologist and Social Scientist.* New York: Harper and Row.

————. 1959. "The Dimensions of Culture Patterns by Factorization of National Characters." *Journal of Abnormal and Social Psychology* 44: 443–69.

————. 1962. "The Basis of Recognition and Interpretation of Factors." *Educational and Psychological Measurement* 22: 667–95.

————. 1965. "Factor Analysis: An Introduction to Essentials." *Biometrics* 21: 190–215.

————. 1966a. "The Meaning and Strategic Use of Factor Analysis." In *Handbook of Multivariate Experimental Psychology,* edited by Raymond B. Cattell, pp. 174–243. Chicago: Rand McNally & Co.

————. 1966b. "The Scree Test for the Number of Factors." *Multivariate Behavioral Research* 1: 245–76.

Cattell, Raymond B., and R. L. Gorsuch. 1965. "The Definition and Measurement of National Morale and Morality." *Journal of Social Psychology* 67: 77–96.

Cattell, Raymond B.; H. Breul; and H. Parker Hartman. 1952. "An Attempt at More Refined Definition of the Cultural Dimensions of Syntality in Modern Nations." *American Sociological Review* 17: 408–21.

Cooley, W. W., and P. R. Lohnes. 1962. *Multivariate Procedures for the Behavioral Sciences.* New York: John Wiley and Sons.

Clyde, Dean J.; Elliot M. Cramer; and Richard J. Sharin. 1966. *Multivariate Statistical Programs.* Coral Gables: University of Miami Press.

Fruchter, B. 1954. *Introduction to Factor Analysis.* New York: Van Nostrand.

Fruchter, B., and E. Jennings. 1962. "Factor Analysis." In *Computer Applications in the Behavioral Sciences,* edited by H. Borko, pp. 238–65. Englewood Cliffs, N.J.: Prentice-Hall.

Glass, Gene V., and Thomas O. Maguire. 1966. "Abuses of Factor Scores." *American Educational Research Journal* 3: 297–304.

Gregg, Phillip M., and Arthur S. Banks. 1965. "Dimensions of Political Systems: Factor Analysis of *A Cross Polity Survey.*" *American Political Science Review* 59: 602–14.

Guertin, Wilson H. 1968. "Comparison of Three Methods of Handling Missing Observations." *Psychological Reports* 22: 896.

Gutman, L. 1954. "Some Necessary Conditions for Common-Factor Analysis." *Psychometrika* 19: 149–61.

————. 1955. "The Determinacy of Factor Score Matrices with Implications for Five Other Basic Problems of Common-Factor Theory." *British Journal of Statistical Psychology* 8: 65–81.

————. 1956. " 'Best Possible' Systematic Estimates of Communalities." *Psychometrika* 21: 273–85.

————. 1957. "A Necessary and Sufficient Formula for Matrix Factoring." *Psychometrika* 22: 79–81.

————. 1958. "What Lies Ahead for Factor Analysis?" *Educational and Psychological Measurement* 18: 497–515.

Hadley, G. 1964. *Linear Algebra.* Reading, Mass.: Addison-Wesley Publishing Co.

Harman, Harry H. 1967. *Modern Factor Analysis.* Chicago: University of Chicago Press.

Harris, C. W. 1963. "Canonical Factor Models for the Description of Change." In

Problems in Measuring Change, edited by C. W. Harris, pp. 138–55. Madison: University of Wisconsin Press.

————. 1964. "Four Models for Factor Analysis." Paper presented at the American Psychological Association Convention, September 8, 1964.

Horn, John L. 1965a. "An Empirical Comparison of Methods for Estimating Factor Scores." *Educational and Psychological Measurement* 25: 313–23.

————. 1965b. *Factor Analysis of Data Matrices.* New York: Holt, Rinehart and Winston.

————. 1965c. "A Rationale and Test for the Number of Factors in Factor Analysis." *Psychometrika* 30: 179–85.

Horn, John L., and Wilbur C. Miller. 1966. "Evidence on Problems in Estimating Common Factor Scores." *Educational and Psychological Measurement* 26: 617–22.

Horst, Paul. 1961a. "Generalized Canonical Correlations and Their Applications to Experimental Data." Seattle: University of Washington. Mimeographed.

————. 1961b. "Relations Among m Sets of Measures." *Psychometrika* 26: 129–49.

Hotelling, Harold. 1933. "Analysis of a Complex of Statistical Variables into Principal Components." *Journal of Educational Psychology* 24: 417–41, 498–520.

————. 1935. "The Most Predictable Criterion." *Journal of Educational Psychology* 26: 139–42.

————. 1936. "Relations Between Two Sets of Variates." *Biometrika* 28: 321–77.

————. 1957. "The Relations of the Newer Multivariate Statistical Methods to Factor Analysis." *British Journal of Statistical Psychology* 10: 69–79.

Kaiser, Henry F. 1958. "The Varimax Criterion for Analytic Rotation in Factor Analysis." *Psychometrika* 23: 187–200.

————. 1959. "Computer Program for Varimax Rotation in Factor Analysis." *Educational and Psychological Measurement* 19: 413–20.

————. 1960. "The Application of Electronic Computers to Factor Analysis." *Educational and Psychological Measurement* 20: 141–51.

————. 1962. "Formulas for Component Scores." *Psychometrika* 27: 83–87.

Kendall, M. G. 1957. *A Course in Multivariate Analysis.* London: Charles Griffin and Co.

Lawley, D. N. 1943. "The Application of the Maximum Likelihood Method to Factor Analysis." *British Journal of Psychology* 33: 172–75.

Levin, J. 1963. *Three-Mode Factor Analysis.* Urbana: Department of Psychology, University of Illinois.

————. 1965. "Three-Mode Factor Analysis." *Psychological Bulletin* 64: 442–52.

Meredith, W. 1964a. "Notes on Factorial Invariance." *Psychometrika* 29: 177–86.

————. 1964b. "Rotation to Achieve Factorial Invariance." *Psychometrika* 29: 187–206.

Neuhaus, J. O., and C. Wrigley. 1954. "The Quartimax Method: An Analytical Approach to Orthogonal Simple Structure." *British Journal of Statistical Psychology* 7: 81–91.

Rao, C. R. 1955. "Estimation and Tests of Significance in Factor Analysis." *Psychometrica* 20: 93–111.

Royce, J. R. 1958. "The Development of Factor Analysis." *Journal of General Psychology* 58: 139–64.

Rummel, Rudolph J. 1963a. "Dimensions of Conflict Behavior Within and Between Nations." In *General Systems,* edited by Ludwig von Bertalanffy and Anatol Rapoport. Yearbook of the Society for the Advancement of General Systems Theory. Ann Arbor: University of Michigan Press.

————. 1963b. "Testing Some Possible Predictors of Conflict Behavior Within and Between Nations." Paper 1 in Peace Research Society, Chicago Conference, pp. 79–111.

————. 1965. "A Field Theory of Social Action with Application to Conflict Within Nations." In *General Systems*, edited by Ludwig von Bertalanffy and Anatol Rapoport. Yearbook of the Society for General Systems Research 10: 183–96.

————. 1966a. "The Dimensionality of Nations Project." In *Comparing Nations*, edited by R. L. Merrit and S. Rokkan, pp. 102–29. New Haven: Yale University Press.

————. 1966b. "Dimensions of Conflict Behavior Within Nations, 1946–59." *Journal of Conflict Resolution* 10: 65–73.

————. 1967a. "Dimensionality of Nations Project: Research Report No. 8." Paper prepared for presentation before the Symposium on Political Science Research, Institute for Defense Analysis. Washington: June 1967.

————. 1967b. "Dimensions of Dyadic War, 1820–1952." *Journal of Conflict Resolution* 11: 176–83.

————. 1967c. "Some Attribute and Behavioral Patterns of Nations." *Journal of Peace Research* 2: 196–206.

————. 1967d. "Understanding Factor Analysis." *Journal of Conflict Resolution* 11: 444–80.

————. 1968. "The Relationship Between National Attributes and Foreign Conflict Behavior." In *Quantitative International Politics: Insights and Evidence*, edited by J. David Singer, pp. 187–214. New York: The Free Press.

————. 1969. "Indicators of Cross-National and International Patterns." *American Political Science Review* 63: 127–47.

————. 1970. *Applied Factor Analysis*. Evanston: Northwestern University Press.

Russett, Bruce M. 1965. "Delineating International Regions." Carnegie-IDRC Joint Study Group on Measurement Problems, Paper No. G60. Indiana University, February 1965.

————. 1966. "Discovering Voting Groups in the United Nations." *American Political Science Review* 60: 327–39.

————. 1967. *International Regions and the International System*. Chicago: Rand McNally.

————. 1968. "Delineating International Regions." In *Quantitative International Politics: Insights and Evidence*, edited by J. David Singer, pp. 317–52. New York: The Free Press.

Sawyer, Jack. 1967. "Dimensions of Nations: Size, Wealth, and Politics." *American Journal of Sociology* 73: 145–72.

Solomon, H. 1960. "A Survey of Mathematical Models in Factor Analysis." In *Mathematical Thinking in the Measurement of Behavior*, edited by H. Solomon, pp. 269–314. Glencoe, Ill.: Free Press.

Spiegel, Murray R. 1961. *Theory and Problems of Statistics*. New York: Schaum Publishing Co.

Tanter, Raymond. 1966. "Dimensions of Conflict Behavior Within and Between Nations, 1958–60." *Journal of Conflict Resolution* 10: 41–64.

Thomson, Godfrey H. 1951. *The Factorial Analysis of Human Ability*. Boston: Houghton Mifflin.

Thompson, G. 1947. "The Maximum Correlation of Two Weighted Batteries." *British Journal of Psychology, Statistical Section*. Part I: 27–34.

Thurstone, L. L. 1947. *Multiple Factor Analysis: A Development and Expansion of the Vectors of Mind*. Chicago: University of Chicago Press.

Triandis, H. C. 1964. "Exploratory Factor Analyses of the Behavioral Component of Social Attitudes." *Journal of Abnormal and Social Psychology* 68: 420–30.

Tucker, L. R. 1963. "Implications of Factor Analysis of Three-Way Matrices for Measurement of Change." In *Problems in Measuring Change,* edited by C. W. Harris, pp. 122–37. Madison: University of Wisconsin Press.

Vincent, Jack E. 1968. "National Attributes as Predictors of Delegate Attitudes at the United Nations." *American Political Science Review* 62: 916–31.

UNIVERSITY OF FLORIDA MONOGRAPHS

Social Sciences

1. *The Whigs of Florida, 1845–1854,* by Herbert J. Doherty, Jr.

2. *Austrian Catholics and the Social Question, 1918–1933,* by Alfred Diamant

3. *The Siege of St. Augustine in 1702,* by Charles W. Arnade

4. *New Light on Early and Medieval Japanese Historiography,* by John A. Harrison

5. *The Swiss Press and Foreign Affairs in World War II,* by Frederick H. Hartmann

6. *The American Militia: Decade of Decision, 1789–1800,* by John K. Mahon

7. *The Foundation of Jacques Maritain's Political Philosophy,* by Hwa Yol Jung

8. *Latin American Population Studies,* by T. Lynn Smith

9. *Jacksonian Democracy on the Florida Frontier,* by Arthur W. Thompson

10. *Holman Versus Hughes: Extension of Australian Commonwealth Powers,* by Conrad Joyner

11. *Welfare Economics and Subsidy Programs,* by Milton Z. Kafoglis

12. *Tribune of the Slavophiles: Konstantin Aksakov,* by Edward Chmielewski

13. *City Managers in Politics: An Analysis of Manager Tenure and Ter-*mination, by Gladys M. Kammerer, Charles D. Farris, John M. DeGrove, and Alfred B. Clubok

14. *Recent Southern Economic Development as Revealed by the Changing Structure of Employment,* by Edgar S. Dunn, Jr.

15. *Sea Power and Chilean Independence,* by Donald E. Worcester

16. *The Sherman Antitrust Act and Foreign Trade,* by Andre Simmons

17. *The Origins of Hamilton's Fiscal Policies,* by Donald F. Swanson

18. *Criminal Asylum in Anglo-Saxon Law,* by Charles H. Riggs, Jr.

19. *Colonia Barón Hirsch, A Jewish Agricultural Colony in Argentina,* by Morton D. Winsberg

20. *Time Deposits in Present-Day Commercial Banking,* by Lawrence L. Crum

21. *The Eastern Greenland Case in Historical Perspective,* by Oscar Svarlien

22. **Jacksonian Democracy and the Historians,** by Alfred A. Cave

23. *The Rise of the American Chemistry Profession, 1850–1900,* by Edward H. Beardsley

24. *Aymara Communities and the Bolivian Agrarian Reform,* by William E. Carter

25. *Conservatives in the Progressive Era: The Taft Republicans of 1912,* by Norman M. Wilensky